Wave Propagation in Solid and Porous Half-Space Media

Hamid R. Hamidzadeh • Liming Dai
Reza N. Jazar

Wave Propagation in Solid and Porous Half-Space Media

Springer

Hamid R. Hamidzadeh
Tennessee State University
Nashville, TN, USA

Liming Dai
University of Regina
Regina, SK, Canada

Reza N. Jazar
RMIT University
Bundoora, VIC, Australia

ISBN 978-1-4939-5200-7 ISBN 978-1-4614-9269-6 (eBook)
DOI 10.1007/978-1-4614-9269-6
Springer New York Heidelberg Dordrecht London

Printed on acid-free paper

Springer is part of Springer Science+Business Media (www.springer.com)

Dedicated to our wives Azar, Xinming, and *Mojgan*

Trying is not doing; doing is trying.

Preface

The main scope of this book is to present the established analytical and experimental techniques to address the dynamic responses of elastic as well as porous half-space media when they are subjected to dynamic loads and the related topics in a concise and suitable manner. The book introduces the reader to the dynamic response of the surface of an elastic half-space excited by concentrated vertical or tangential force. Based on the presented analyses, it also addresses the dynamic response of a rigid massless footing of arbitrary shape resting on the surface of an elastic half-space medium for three modes of vertical, horizontal, and rocking vibrations. The book also presents solutions to the three pure modes of vibration for massive rectangular foundations by employing the impedance matching technique and provides design charts for these modes of vibrations. The solution for these modes is extended to develop a solution to the dynamics of simultaneous horizontal and rocking motions of a rectangular foundations resting on the surface of elastic half-space medium. Moreover, the book presents the required theoretical background needed for analysis of interaction of two rectangular foundations founded on the surface of an elastic half-space. In addition to the theoretical topics, the book describes a finite model to simulate an elastic half-space and introduces experimental techniques to verify the presented solution. Furthermore, experimental methods are presented to determine the two important elastic properties of shear modulus and Poisson's ratio for the medium. In order to verify the present theoretical results some experiments, procedures, and results are also provided.

This book presents the required theoretical background needed to develop mathematical models and their solutions for the above topics. Furthermore, it offers the engineering information and quantitative data needed for design analysis and applications of the presented analytical procedures for different disciplines such as: mechanical, civil, and bioengineering. The book in its entirety constitutes as an extensive guidance for its reader. It also provides a systematic solution for the dynamic analysis of elastic, porous, and layered half-space media. It also extends the provided analytical solution to address a variety of practical problems in engineering and to determine the essential elastic properties of the medium. The book is intended to lay the foundation for understanding mathematical modeling, vibration analysis, and the design of engineering systems which can be modeled by a half-space medium in a complete and succinct manner. Throughout the book, an attempt has been made to provide a conceptual framework that includes exposure to the required background in mathematics and the fundamentals of the theory of elasticity. The knowledge of the presented topics will enable the reader to pursue further advances in the field.

Level of the Book

The primary audience of this book is the graduate students in mechanical engineering, engineering mechanics, civil engineering, bioengineering, ocean engineering, mathematics, and science disciplines. In particular, it is geared toward the students interested in enhancing their knowledge by taking the second graduate course in the areas of vibration of continuous systems, application of wave propagation, and soil dynamics. The presented topics have been prepared to serve as an aid to engineering designers. It can also be utilized as a guide for professional engineers in research and industry who are seeking to expand their expertise and are expected to extend their knowledge for setting design specifications and ensuring their fulfillment.

Organization of the Book

The book is presented in ten chapters: introduction, fundamentals of elasticity, vibration analysis for single-layer cylinders, modal analysis for single-layer cylinders, vibration of multilayer thick cylinders, constrained-layer damping for cylindrical structures, and vibration of thick cylindrical panels. Furthermore, it offers helpful and significant tabulated results, which can be used as design guidelines for these structures.

To make effective use of the presented topics, the following procedure is suggested. The realization of the topics may require a review of certain theoretical concepts and methods which can be achieved through references in Chaps. 1 through 5. To become acquainted with the state of the art in this particular field and learn about the historical background on this topic, the reader should begin with Chap. 1, which lists an extensive number of key references with brief discussions on their methodology, required assumptions, and their achievements. Chapter 2 reviews the succinct fundamental theoretical background and concepts needed from the theory of elasto-dynamics, which will enable the reader to follow the derivation of the required governing equations and their solutions in Chaps. 3 and 4. Chapter 5 is intended to present numerical results for the non-dimensional frequency responses of rigid rectangular foundations resting on an elastic half-space for three modes of vertical, horizontal, and rocking vibrations, as well as coupled horizontal and rocking vibrations. Chapter 6 presents a finite size experimental model for a semi-infinite elastic half-space model and the experimental procedures for verifying the theoretical results. It also presents available techniques for determining the dynamic properties of the medium needed for analytical analysis. Chapters 7 and 8 provide analytical method to determine dynamic response of a rigid foundation subjected to a distance blast and to identify position of a vertical exciting force on the surface of an elastic half-space medium using sensor fusion, respectively. Chapter 9 presents an overview of techniques established for analyzing Surface Vibration of a multilayered elastic medium due to harmonic concentrated force. Chapter 10 will cover the three-dimensional wave propagation in porous media.

Method of Presentation

The scope of each chapter is clearly outlined and the governing equations are derived with an adequate explanation of the procedures. The covered topics are logically and completely presented without unnecessary overemphasis. The topics are presented in a book form rather than in the style of a handbook. Tables, charts, equations, and references are used in abundance. Proofs and derivations are often emphasized and the physical model and final results are accompanied with illustrations and interpretations. Certain specific information that is required in carrying out the design analysis in detail has been stressed.

Prerequisites

The book is written for graduate students, so the assumption is that the readers are familiar with the fundamentals of differential equations, as well as a basic knowledge of linear algebra, Fourier transform, and numerical methods. The presented topics are aimed to establish a conceptual framework that enables the reader to pursue further advances in the field. Although the governing equations will be derived with adequate explanations of the procedures, it is assumed that the readers have a working knowledge of theory of elasticity, fluid–structure interaction, and vibration engineering.

Unit System

Through the chapters, for the sake of generality, computed results and the required parameters are provided in non-dimensional forms. Nevertheless, the system of units adopted for case studies is, unless otherwise stated, the British Gravitational system of units (BG). The units of degree (deg) or radian (rad) are utilized for variables representing angular quantities.

Acknowledgements

We wish to express our thanks to our colleagues who have assisted in the development of this book. We are indebted to our students, friends, and colleagues for their constructive comments and suggestions on the draft manuscript of this book.

Nashville, TN, USA — Hamid R. Hamidzadeh
Regina, SK, Canada — Liming Dai
Bundoora, VIC, Australia — Reza N. Jazar

Contents

List of Figures

Chapter 1
Introduction

Abstract This chapter will discuss some of the issues of dynamics of soils and foundations from a practical point of view. Since this topic is quite broad, a brief description of methodology will be outlined, while details will be given for a few procedures that have proven to be effective and accurate. One of the main objectives of this chapter is to survey different available techniques for solving the dynamic response of foundations when subjected to harmonic loadings. Special attention is directed to the dynamic response of the surface of the medium due to concentrated dynamic loads, response of foundations, coupled vibrations of foundations, interaction between two foundations, experimental aspects of soils and foundations, and laboratory simulations.

Keywords Surface forces • Concentrated forces • Contact stress distribution • Mixed boundary value problems • Elastic medium

The possible occurrence of extreme dynamic excitation, either natural or man-made, has a major influence on the design of buildings and machine foundations. A primary concern in designing foundations is the knowledge of how they are expected to respond when subjected to dynamic loadings. The validity of the mathematical analysis depends entirely on how well the mathematical model simulates the behavior of the real foundation. Over the past decades our ability to analyze mathematical models for dynamics of foundations has been improved by the use of different analytical and numerical techniques. In most of these analyses it is common to assume that the footing is rigid and the medium is a homogeneous elastic half-space. Extensive efforts have been confined in the development of procedures and computer simulations to tackle some practical problems that arise in this field, while other important problems have been neglected. It should be noted that interaction between foundations for noncircular footings was not treated in a satisfactory manner and significant deficiencies remain in most of the previous analyses.

H.R. Hamidzadeh et al., *Wave Propagation in Solid and Porous Half-Space Media*,
DOI 10.1007/978-1-4614-9269-6_1, © Springer Science+Business Media New York 2014

This chapter will discuss some of the issues of dynamics of soils and foundations from a practical point of view. Since this topic is quite broad, a brief description of methodology will be outlined, while details will be given for a few procedures that have proven to be effective and accurate. One of the main objectives of this chapter is to survey different available techniques for solving the dynamic response of foundations when subjected to harmonic loadings. Special attention is directed to the dynamic response of the surface of the medium due to concentrated dynamic loads, response of foundations, coupled vibrations of foundations, interaction between two foundations, experimental aspects of soils and foundations, and laboratory simulations.

Before addressing the abovementioned problems in soil dynamics, it is essential to define the soil medium. To the authors' knowledge, diverse dynamic analyses for the soil–foundation interaction have been conducted using the following modeling approaches for the soil medium: subgrade reaction model, Winkler foundations, elastic half-space medium, which can be elastic isotropic or viscoelastic, Gibson soil model, layered medium, porous medium, and medium with particulate materials.

Among these models the subgrade reaction model can only be used for lumped systems and is developed using experimental results on different soils. Detailed information about this model are provided by Terzaghi (1955) and Barkan (1962). Winkler's foundation (1987) is a weak model which cannot account for the geometrical damping of the half-space medium. The half-space model is the most realistic model which can be extended to the viscoelastic one by considering complex elastic moduli for the medium. This model is adopted throughout Chaps. 2–6. The half-space medium can also be considered to be porous. Gibson's model (1967) allows for variation of the shear modulus within the depth of the elastic half-space. This model has rarely been used for real design analysis. In the layered soil medium model, the medium is divided into thin layers and each thin layer considered to be an elastic or viscoelastic medium with specific mechanical properties. See Kausel and Roesset (1981). The porous half-space medium represents volumetrically interacting solid–fluid aggregates, which can be modeled using continuum porous media theories by allowing for both solid-matrix deformation and fluid flow (Morand and Ohayon 1955). To study the mechanical behavior of sand and other granular soil medium, Tavarez and Plesha (2007) used the discrete element method to account for the discontinuous characteristic of some geomaterials. It should be noted that this book also addresses the porous and the layered stratum in different chapters.

1.1 Surface Response Due to Concentrated Forces

In the field of propagation of disturbances on the surface of an elastic half-space, the first mathematical attempt was made by Lamb (1904). He gave integral representations for the vertical and radial displacements of the surface of an elastic half-space due to a concentrated vertical harmonic force. Evaluation of these integrals involves considerable mathematical difficulties, due to the evaluation of a Cauchy principal integral and certain infinite integrals with oscillatory integrands. Nakano (1930) considered the same problem for a normal and tangential force distribution on the surface. Barkan (1962) presented a series solution for the evaluation of integrals for the vertical displacement caused by a vertical force on the surface, which was given by Shekhter (1948). Pekeris (1955a, b) gave a greatly improved solution to this problem when the surface motion is produced by a vertical point load varying with time, like the Heaviside function. Elorduy et al. (1967) developed a solution by applying Duhamel's integral to obtain the harmonic response of the surface of an elastic half-space due to a vertical harmonic point force. Heller and Weiss (1967) studied the far field ground motion due to an energy source on the surface of the ground.

Among the investigators who considered the three-dimensional problem for a tangential point force, Chao (1960) presented an integral solution to this problem for an applied force varying with time, like the Heaviside unit function. Papadopulus (1963) and Aggarwal and Ablow (1967) have presented solutions, in integral expressions, to a class of three-dimensional pulse propagation in an elastic half-space. Johnson (1974) used Green's functions for solving Lamb's problem, and Apsel (1979) employed Green's functions to formulate the procedure for layered media. Kausel and Roesset (1975) reported an explicit solution for dynamic response of layered media. Davies and Banerjee (1983) used Green's functions to determine responses of the medium due to forces which were harmonic in time with a constant amplitudes. The solution was derived from the general analysis for impulsive sources. Kobayashi and Nishimura (1980) utilized the Fourier transform to develop a solution for this problem and expressed the results in terms of the full-space Green's functions, which include infinite integrals of exponential and Bessel's function products. Banerjee and Mamoon (1990) provided a solution for a periodic point force in the interior of a three-dimensional, isotropic elastic half-space by employing the methods of synthesis and superposition. The solution was obtained in the Laplace transform as well as the frequency domain.

Hamidzadeh (1978) presented mathematical procedures for determination of the dynamic response of surface of an elastic half-space subjected to harmonic loadings and provided numerical results for displacement of any point on the surface in terms of properties of the medium and of the exciting force. The solution was analytically formulated by employing double Fourier transforms and was presented by integral expressions. Hamidzadeh (1987) and Hamidzadeh and Chandler (1991) provided dimensionless response for an elastic half-space and compared their results with other available approximate results. Verruijt (2010) provided mathematical procedures for analyzing the vertical displacement of the surface of an elastic half-space due to the surface point load, and moving vertical loads. Meral and Royston (2009) studied shear and surface wave motion in and on a viscoelastic material representative of biological tissue. They considered the surface wave motion on a half-space caused by a finite rigid circular disk located on the surface and oscillating normal to it and determined the compression, shear, and surface wave motion in a half-space generated by a subsurface finite dipole. In their study, they assumed fractional order Voigt model of viscoelasticity in their theoretical analysis. They concluded that their theoretical results had a better agreement with their measured results over a limited frequency range.

1.2 Dynamic Response of Foundations

Advances in the development of solutions for soil–foundation interaction problems are categorized in the following sections based on the formulation procedures.

1.2.1 Assumed Contact Stress Distributions

The first attempt to solve the vertical vibration of a massive circular base on the surface of an elastic medium was made by Reissner (1936). He adopted Lamb's (1904) approach and developed a solution by assuming a uniform stress distribution on the surface of the medium. He established an estimated solution for determining the vertical steady state response of circular footings. He also calculated the displacement of the center of the base and introduced the amplitude of vibration in terms of non-dimensional parameters. It has been proven that his results over-estimated the amplitude due to his consideration of the displacement at the center of the base. Reissner and Sagoci (1944) presented a static solution for the torsional oscillation of a disc on the medium. Miller and Pursey (1954, 1955) considered the vertical response of a circular base due to a force uniformly distributed on the contact surface. Quinlan (1953) and Sung (1953) independently extended Riessner's (1936, 1944) approach to solve the problem of vertical vibration of circular and infinitely long rectangular footings. In their analyses they considered three different harmonic stress distributions: uniform, parabolic, and stress produced by a rigid base under static conditions. They showed that the vibration characteristics of semi-infinite media effectively vary with the type of stress distributions and elastic properties of the medium.

Arnold et al. (1955) considered four vibrational modes (vertical, horizontal, torsional, and rocking) for a circular base on the surface of elastic media. By assuming harmonic static stress distributions for all modes, they evaluated the dynamic responses using an averaging technique. They also verified this work with experimental results using a finite model for the infinite medium. Bycroft (1956,1959) followed the same approach for four modes of vibration by determining complex functions to represent the in-phase and out-of-phase components of displacement of a rigid massless circular plate. Bycroft (1977) later carried out some tests to verify his previous theoretical work. Thomson and Kobori (1963) and Kobori et al. (1966a, b, 1968, 1970, 1971) considered the dynamic response of a rectangular base. They provided computational results for components of the complex displacement functions by assuming a uniform stress distribution on the contact area of the base and medium. Their analysis was for an elastic half-space, viscoelastic half-space, and layered viscoelastic media. Verruijt (2010) addressed the problems of the line and strip loads on the surface of an elastic half-space and presented analytical solutions to these problems. The problem of wave propagation in soil medium caused by the vertical vibration of circular footing on the surface, assuming constant uniform vertical stress distribution, was represented by Verruijt (2010). He also provided the analysis to determine a simple lumped parameter system for this problem. It should be noted that the assumption of uniform stress distribution can only be valid for a rigid footing on a very soft soil.

1.2.2 Mixed Boundary Value Problems

Harding and Sneddon (1945) and Sneddon (1959) gave a static solution for a rigid circular punch pressed into an elastic half-space. By the use of the Hankel transform the appropriate mixed boundary value problem was reduced to a pair of dual integral equations representing the stress distribution and uniform displacement under the rigid punch.

Several investigations have been conducted to extend the static solution to the corresponding dynamic problems. Awojobi and Grootenhuis (1965) and Awojobi (1964, 1966, 1969, 1972a, b) used the Hankel transform to present the complete dynamic problem by dual integral equations. They gave an analytical solution for the vertical and torsional oscillations of a circular body, and the vertical and rocking modes of an infinitely long strip foundation. The evaluation of stress distributions and uniform displacements was based on the extension of the Titchmarsh's (1937) dual integral equation and a method of successive approximations. Robertson (1966,1967) followed the same procedure and reduced the dual integral equations into Fredholm integral equations and gave series solutions to these equations for the vertical and torsional response of a rigid circular disc. Gladwell (1968a, b, 1969) developed a solution for the mixed boundary value problems for circular bases resting on an elastic half-space or elastic strata. He solved the integral equations and presented the displacements of four different modes of vibration in series forms. Karasudhi et al. (1968) treated the vertical, horizontal, and rocking oscillations of a rigid strip footing, on an elastic half-space, by reducing the dual integral equations into the Fredholm integral equations. Housner and Castellani (1969) conducted an analytical solution based on the work done by the total dynamic force and determined the weighted average vertical displacement for a cylindrical body. To determine the free field displacements of an elastic medium, for four different modes of vibration, for a cylindrical body, Richardson (1969) followed Bycroft's (1956) method and provided a solution to this problem. Luco and Westmann (1971) solved the mixed boundary value problems for four modes of vibration by considering a massless circular base. Their procedure reduced the resulting dual integral equations to the Fredholm integral equations. They calculated the complex displacement functions for a wide range of frequency factors. In a separate publication (1972) they followed the same procedure for the determination of the response of a rigid strip footing for the three modes of vertical, horizontal, and rocking vibrations. The vibration of a circular base was also treated by Veletsos and Wei (1971) for horizontal and rocking vibrations. Bycroft (1977) extended his earlier work to present approximate results for the complex displacement functions at higher frequency factors. Veletsos and Verbic (1974) introduced the vibration of a viscoelastic foundation. Clemmet (1974) included hysteretic damping in the Richardson (1969) solution. Luco (1976) provided a solution for a rigid circular foundation on a viscoelastic half-space medium.

These investigations were based on circular or infinitely long strip foundations. Few investigators have paid attention to the dynamic responses of a rectangular foundation, due to the difficulty of the asymmetric problem. Elorduy et al. (1967) introduced a numerical technique based on the uniform displacements for a number of points on the contact surface of the rectangular footings. In their analysis they employed an approximate solution for the surface motion of a medium due to a vertical point force. They gave complex displacement functions for vertical and rocking modes with different ratios of length to width of rectangular footings. By extending the Bycroft's (1977) idea of an equivalent circular base for a rectangular foundation, Tabiowo (1973) and Awojobi and Tabiowo (1976) gave a solution to this problem. They also introduced another solution by superimposing the solution of two orthogonal infinitely long strips and gave the displacement at the intersection of these strips for different frequencies. Wong and Luco (1976,1977,1978) solved this problem for the three modes of vertical, horizontal, and rocking vibrations. They used the approach reported by Kobori and Suzuki (1970) to provide an approximate solution to a footing which was divided into a number of square subregions. They assumed that the stress distribution for each subregion is uniform with unknown magnitude, while all the subregions experienced uniform displacements. Their solution considered the coupling effect for viscoelastic medium and the complex stiffness coefficients were tabulated (1971) for different loss factors.

Hamidzadeh (1978) and Hamidzadeh and Grootenhuis (1981) presented an improved version of Elorduy's method to obtain the dynamic responses for three modes of vertical, horizontal, and rocking vibration for rectangular foundations. In their analysis a uniform displacement for each node of rectangular subregions was assumed. They utilized the impedance matching technique to formulate the dimensionless response for foundations. As reported by Hamidzadeh and Grootenhuis, (1981) vertical, horizontal, and rocking displacements of a rigid massless base resting on an elastic half-space can be expressed conveniently in terms of two nondimensional displacement functions, one in-phase with the motion and the other in quadrature.

An approximate method for computation of compliance functions of rigid plates resting on elastic or viscoelastic half-space excited in all directions was reported by Rucker (1982). The proposed method provides compliances for vertical, horizontal, rocking, and torsional motion for rectangular foundations. Another approximate solution for harmonic response of an arbitrary shape foundation on an elastic half-space was reported by Chow (1987). In his analysis the contact area was discretized into square subregions and the influence of the square subregion was approximated by that of an equivalent circular base. He then compared his results with those of Wong and Luco (1977) and Hamidzadeh and Grootenhuis (1981). The dynamic stiffness of a rigid rectangular foundation on the half-space was considered by Triantafyllidis (1986), who provided solutions for the mixed-boundary value problem. The problem was formulated in terms of coupled Fredholm integral equations of the first kind. The displacement boundary value conditions were satisfied using the Bubnov–Galerkin method. The solution yielded the influence functions and the stiffness functions characterizing the dynamic interaction between the foundation and half-space. The presented analytical method considered the coupling between normal and shear stress distributions acting on the contact area between the footing and the half-space. The problem of wave propagation in soil medium caused by the vertical vibration of circular footing on the surface, assuming constant uniform vertical stress distribution, was represented by Verruijt (2010). He also provided the analysis to determine a simple lumped parameter system for this problem. It should be noted that the assumption of uniform stress distribution can only be valid for a rigid footing on a very soft soil. Le Houedec (2001) presented a review article on the transmission of vertical vibrations of a rigid footing over the surface of the ground by considering two different models of elastic half-space and an elastic layer medium. The vibration response of a shear wall on a rigid semicircular footing on a single layer of soil on a rigid base was studied by Liang et al. (2013).

1.2.3 Lumped Parameter Models

Based on the theoretical work for the response of a rigid massless circular plate, Hsieh (1962) was able to present equivalent mass-spring-dashpot models for all four modes of vibration. The calculated dynamic parameters for each mode were frequency dependent. Following the Hsieh's approach, Lysmer (1965) provided frequency independent values for equivalent spring and damping constants. The idea of developing an equivalent mass-spring-dashpot model for a rigid mass on the surface of a half-space has attracted many investigators such as Lysmer and Richart (1966), Weissmann (1966), Whitman and Richart (1967), Hall (1967), Veletsos and Wei (1971), Roesset et al. (1973), Oner and Janbu (1975), Hall et al. (1975), and Veletsos (1975). Approximate expressions for the dynamic stiffness coefficients in the frequency domain are well established and are summarized in a text by Wolf (1975). Lumped parameter models to represent the soil structure interaction for embedded foundations were developed by Wolf (1985). Dobry and Gazetas (1986a, b) presented a method to determine the effective dynamic stiffness and damping coefficient of a rigid footing by considering variations of foundation shape, soil type, and length ratio of the base. The analytical methods employed were those of Wong and Luco (1977) and they verified some of their computations by several in-situ experiments. A different approach based on the subgrade reaction method is also reported by many investigators such as Terzaghi (1943,1955), Barkan (1962), and Girard (1968). Maravas et al. (2008) presented an analytical solution for a single-degree-of-freedom oscillators founded on footings and piles on compliant ground. Truong (2010) considered the effects of material damping, geometrical damping, and dynamic soil mass and stiffness for horizontal and vertical vibrations of rigid footings resting on elastic half-space medium. Among the recent investigators for development of the lumped parameter method, Safak (2006) used the z-transform to determine impedance functions for soil–foundation interaction. Andersen (2010) presented an extensive assessment of lumped parameter models for rigid foundations and Wang et al. (2013) used the complex Chebyshev polynomials to develop the lumped parameter model for surface circular foundations, embedded square foundations, and pile group foundations. The parameters in the models were determined by curve-fitting to the established analytical or measurement results.

1.2.4 Computational Methods

The finite element method (FEM) has been applied to discretized foundations; the most crucial problem for discretization of foundations by FEM is transmitting waves through artificial finite boundaries. This is considered to be a drawback which is due to the difficulty encountered in proper modeling of an unbounded soil medium and satisfying the wave radiation condition. The problem of a rigid circular base on an elastic half-space has been considered numerically by many investigators such as Duns and Butterfield (1967), Lysmer and Kuhlemeyer (1971) and Seed et al. (1975). However, these solutions have not been generalized to cover all modes of vibration due to numerical limitations. Day and Frazier (1979) suggested the use of artificial boundaries far away from the region of interest to avoid the undesirable wave reflections. Bettess and Zienkiewicz (1977) recommended the use of infinite elements and Roesset and Ettouney (1977) and Kausel and Tassoulas (1981) proposed transmitting or non-reflecting boundaries to circumvent the problem. Chuhan and Chongbin (1987) presented an approximate solution for a viscoelastic medium and strip foundation by considering two-dimensional wave equations in conjunction with the use of Galerkin weighted residual approximations. A frequency-dependent compatible infinite element was presented, then by coupling the infinite elements with ordinary finite elements the system was used to simulate the propagation of waves.

The boundary element method (BEM) has been used as an effective numerical technique for solving elastodynamic problems. In this method, boundary integral representation provides an exact formulation and the only approximations are those due to the numerical implementation of the integral equations. The technique is suitable for infinite and semi-infinite domain due to employment of Green's functions which satisfy the radiation condition at far field. The first application of this technique on soil–structure interaction was performed by Dominguez (1978) and Dominguez and Roesset (1978) in frequency domain. Karabalis and Beskos

(1984) determined the frequency and time domain solutions for dynamic stiffness of a rectangular foundation resting on an elastic half-space medium. Spyrakos and Beskos (1986a, b) used the time domain BEM to consider dynamic response of two-dimensional rigid and flexible foundations.

Numerical solutions for a layered medium are reported by Luco and Apsel (1983) and Chapel and Tsakalidis (1985). In these solutions formulations for Green's functions were made using the Hankel transform for each layer. Kausel (1981) presented an explicit closed-form solution for the Green's functions corresponding to dynamic loads acting on layered strata. These functions embody all the essential mechanical properties of the medium. The solution is based on a discretization of the medium in the direction of layering, which results in a formula yielding algebraic expressions for the displacement of all layers. Determination of response for a layered medium to dynamic loads, prescribed at some location in the soil, can be achieved by using the stiffness matrix method presented by Kausel and Roesset (1975). In their solution the external loads, applied at the layer interfaces, are related to the displacements at these locations through stiffness matrices which are functions of both frequency and wave number. The proposed method can treat simultaneous solutions for multiple loadings. Israil and Ahmad (1989) investigated the dynamic response of rigid strip foundations on different media of a viscoelastic half-space, viscoelastic strata on a half-space, and viscoelastic strata on a rigid bed. An advanced boundary element algorithm was developed by incorporating isoparametric quadratic elements. The effect of Poisson's ratio and material damping, layer depth, embedment, and type of contact at the foundation–medium interface was studied. Jeong et al. (2013) provided a systematic method for determining the impedance functions for arbitrarily shaped foundations resting on (or embedded in) heterogeneous soil medium by employing the finite element method to consider the arbitrary heterogeneity of soil and foundation types and geometries. In their analyses, the semi-infinite boundaries were treated by utilizing perfectly matched layers. Rayhani and Naggar (2008) numerically and experimentally simulated the response of rectangular ten-story buildings on uniform and layered soft soils subjected to several earthquakes. In their analysis they assumed a nonlinear elastic–plastic model for the soil. Liang et al. (2013) used the indirect-boundary-element method and non-singular Green's functions of distributed loads to consider the in-plane, dynamic soil–structure interaction for incident-plane P and SV waves for a shear wall on a rigid foundation that was embedded in a soil layer over bedrock. Yang and Hung (2009) presented an extensive study of wave propagation in an elastic half-space medium induced by a moving train. In their analysis they considered the finite/infinite element method.

1.3 Coupled Vibrations of Foundations

In the analyses for the pure rocking or the horizontal mode, it is usual to assume that the medium is rigidly stiff for either shear or compression deformations. However, it is known from characteristics of the soil that it can resist elastically for both applied compression and shear. Therefore, the problem of horizontal or rocking motion for a massive footing on the surface of elastic half-space media should be considered as a coupled motion. Wong and Luco (1976,1978), Rucker (1982), and Triantafyllidis (1986) have presented solutions for coupled vibrations of rectangular foundations on an elastic and viscoelastic half-space.

Another approximate solution is based on simultaneous horizontal and rocking motions. Hall (1967) is among the early investigators who used the solution for pure sliding and rocking of circular bases. He followed the Hsieh (1962) approach in deriving the equations for simultaneous motion and presented numerical results for certain cases. Richart and Whitman (1967) compared the experimental results obtained by Fry (1963) with a similar prediction that Hall introduced for a circular base. They found that their comparisons were satisfactory. Karasudhi et al. (1968) and Luco and Westmann (1972) provided an approximate solution to the problem of the coupled motion of an infinitely long rigid strip. Veletsos and Wei (1971) introduced an approximate solution to evaluate the stiffness and damping coefficients for a massless rigid circular footing. They also compared these coefficients with their corresponding value for simultaneous motion. Ratay (1971) considered simultaneous motions when the circular base is excited by a harmonic horizontal force. He studied the variation of the frequency response curve due to variation of involved parameters. Beredugo and Novak (1972) studied the simultaneous motions of a circular base embedded in the surface of a homogeneous elastic medium.

Using the finite lumped parameter model for circular embedded foundation, Krizek and Gupta (1972) gave a solution to this problem. Clemmet (1974) improved the horizontal translation model and considered the rocking displacement of the base due to a horizontal force. Wolf (2002) and many others studied the practical problem of soil–structure interaction by allowing simultaneous motions for circular foundations. A number of publications on the simultaneous motions of rectangular foundations are available. These are based on the subgrade reaction method discussed by Barkan (1962) and Girard (1968). Wolf (1975,1985) investigated the problem by allowing simultaneous motions for the circular foundation. It should be noted that for all modes except torsional, Wolf's analysis was done by applying relaxed boundary conditions, this allowed one of the components of the surface tractions under the foundation to be zero. In his analysis Wolf considered the coupled horizontal and rocking motions of a rectangular foundation based on the Wong and Luco (1976) theoretical results. Hamidzadeh and Minor (1993) utilized the procedure reported in reference (1981) and developed a method for simultaneous horizontal and rocking motions of square foundations on an elastic half-space. They compared their results with those of coupled responses determined by Wong and Luco (1976,1978). The comparison indicates satisfactory agreement for low dimensionless frequencies. Azarhoosh and Amiri conducted a parametric study on the elastic response of soil–structure systems having shallow foundations subjected to a pulse-type motion and near-fault records and presented a procedure for assessing near-fault and soil–structure interaction effects on design of structures. Tileylioglu et al. (2008) used an earthquake recordings for a structure model to evaluate inertial soil–structure interaction effects. The test structure was rested on a reinforced concrete foundation with no embedment. Parametric system identification was conducted using seismic monitoring utilizing uni-axial accelerometers on the roof and foundation as well as vertical sensors at the corners of the foundation. Then, by determining the frequencies of the fundamental vibration mode for a fixed- and flexible-base condition, they were able to predict the stiffness, damping, and inertia due to the effect of soil–structure interaction. Chopra (2007) and Chin (2008) presented the dynamic response of a foundation under general multi-directional forced excitation. Anastasopoulos and Kontoroupi (2014) presented an approximate method for analyzing the rocking response of a single degree of freedom system resting on compliant soil by considering inelastic property of soil, as well as foundation uplifting. They concluded that their proposed simplified approach should be considered as a substitute for more sophisticated analytical methods.

1.4 Interactions Between Foundations

To the authors' knowledge little attention has been directed to the dynamics of foundations on the surface of a homogeneous elastic half-space. The available solutions are limited to circular bases, infinitely long strips, or for foundations which are very far apart. Iljitchov (1967) introduced this problem and gave a poor estimation of the effect of vertical vibration of one foundation on the other. This problem was considered in detail by Richardson (1969), Richardson et al. (1971), and Warburton et al. (1971,1972). They studied the dynamic responses of two circular bases and provided numerical results for active and passive bases. Their solution was based on averaging techniques for the displacements of both footings. Lee and Wesley (1973) presented a solution to the dynamic responses of a group of flexible structures on the surface of an elastic half-space medium. MacCalden and Matthiesen (1973) presented theoretical and experimental results for far field bases. Utilizing the method reported by Richardson et al. (1971), Clemmet (1974) improved the solution for horizontal and rocking motion of a circular base and introduced hysteretic damping for the media. He verified his results for the vibration of passive and active bases with the experimental results of Tabiowo (1973). Snyder et al. (1975) employed a two-dimensional FEM to study this problem for circular and infinitely long rigid strip bases. Hamidzadeh (1978) investigated the interactions of two foundations resting on the surface of a homogeneous elastic half-space. In his analysis the two rectangular foundations were separated by a certain distance. The mathematical model was developed by considering displacement components for each base. The first component was due to the base vibration and the second one caused by the induced displacement due to the reaction force at the other base. The analysis allowed displacement and rotation in every direction. The results of analysis were provided for two different cases of active–active and active–passive foundations. Puangnak et al. (2012) studied the development of analytical model to consider the effect of structure–soil–structure interaction and in particular they considered special two cases of interacting surface footings within a single structure and footing–soil–footing interaction. It should be noted that the effect of neighboring structures on the response of a soil–foundation–superstructure system has not yet been adequately addressed by investigators. This is largely due to lack of understanding and analytical tools.

1.5 Experimental Studies

Many different laboratory and in-situ experiments have been carried out in this field. These experimental works can be divided into two main groups: First, the measurement of dynamic properties for the medium and second, the measurement of frequency response for foundations.

For measurement of shear modulus various techniques have been developed by investigators. Among researchers Awojobi and Tabiowo (1976) employed refraction and reflection surveys. Jones (1958,1959) measured the velocity of Rayleigh waves to evaluate the shear modulus. Maxwell and Fry (1967) measured the velocity of shear and compression waves to determine the shear modulus and Poisson's ratio. Stokoe and Richart (1974) and Beeston and McEvilly (1977) measured the velocity of shear waves using cross-hole tests. The calculated response characteristics of vertical vibration of a rigid circular footing were used by Jones (1958), Dawance and Guillot (1963), and Grootenhuis and Awojobi (1965) to determine shear modulus and Poisson's ratio. In addition to the above in-situ experiments, many laboratory tests have also been conducted. Hardin and Drnevich (1972a, b) and Cunny and Fry (1973) recommended resonant column tests. Lawrence (1965) used a pulse technique to measure shear modulus. Theirs and Seed (1968) and Kovacs et al. (1971) used a cyclic simple shear test for low frequency cyclic loading. They measured the shear modulus and the damping ratio of the soil for very small strains by recording the free response of the vertical vibration of a circular footing.

In development of experimental methods for determination of frequency response for foundations, few experiments have been performed to determine the frequency response of a footing on the surface of an elastic half-space. This is due to the difficulties involved in creating a suitable environment for the field test and establishing a finite model for an infinite elastic medium. Jones (1958), Kanai and Yoshizawa (1961), Bycroft (1956), Dawance and Guillot (1963), and Awojobi (1976) presented some experimental frequency responses for a circular footing in the field. Eastwood (1953), Arnold et al. (1955), Chae (1960), and Tabiowo (1973) established a finite model for a half-space and did some tests on the dynamic response for a circular bases. Eastwood (1953) and Tabiowo (1973) gave a number of experimental results for the response of a rectangular base, but Eastwood did not determine the dynamic elastic constants. Kanai and Yoshizawa (1961) tested an actual building for the rocking mode.

Hamidzadeh (1978,2010) and Hamidzadeh and Grootenhuis (1981) reported experimental results using a laboratory model which simulated an elastic half-space medium subjected to dynamic loading. The model was used in an attempt to experimentally determine the two important elastic properties of shear modulus and Poisson's ratio for the medium. The finite size model was also used to conduct experiments to verify the validity of established theories. In the case of two circular footings on an elastic medium, MacCalden (1973) gave results for the vibrations of active and passive circular bases using in-situ tests.

Extensive surveys of experimental studies for the interaction between soil and structure are reported by Luco et al. (1987,1988) and Wong et al. (2007). All the reported investigations were performed on the Millikan Library Building which has been the subject of a large number of forced vibration tests. These experimental results showed that forced vibration tests can be used to obtain estimates of the foundation impedance functions. The transmission of vibrations on the surface of the ground in the far field due to harmonic load acting over a strip was investigated by Peplow et al. (1999). They considered a vibration attenuation device which could modify the modal wave propagation regime of the ground by introducing an artificial stiffened layer within the elastic layered half-space medium. Ashlock (2011) conducted in-situ experimental work to study simultaneous vertical and coupled lateral-rocking vibrations of surface footing on soil medium subjected to dynamic loading. He also evaluated the validity of the homogeneous half-space theory for modeling the multi-modal response of the soil–foundation system. Pak et al. (2012) conducted experimental studies to determine the dynamics of square footings resting on a sand stratum by means of centrifuge modeling. In their investigation, they considered the horizontal, vertical, and rocking response characteristics of different foundations and determined bearing pressure conditions, the response time histories, and frequency response functions of the system. Experimental studies of dynamics of soil–foundation interaction have been reported extensively by Pak et al. (2008, 2011) and Pak and Ashlock (2011).

1.6 Layered Elastic Medium

The early investigations on dynamic of layered stratum subjected to surface load provided integral formulations with no numerical evaluation (see Harkrider 1964; Apsel 1979). An alternate procedure for analysis of layered medium is based on the stiffness matrix approach that was developed by Kausel and Roesset (1981). In their procedure, the external loads applied at the layer interfaces were related to the displacements at these locations. The above solution is limited to the cases that for the applied load closed-form solution for the stress distributions and the displacements at interface are possible. Israil and Ahmad (1989) considered the dynamic response of rigid strip foundation on viscoelastic strata on a half-space and on a rigid bed by employing BEM. In their analysis the effect of material damping, layered depth, and the type of contact between the foundation and the medium was studied. Numerical solutions for layered medium are also reported by Luco and Aspel (1983). Swaddiwudhipong et al. (1991) studied the steady-state dynamic responses of a massless foundation on layered media by considering the finite layer method and presented the horizontal, vertical, and rocking vibration responses for these modes. Hsu and Schoenberg (1990) utilized ultrasonic experiments to investigate the elastic behavior of systems of closely spaced elastic plates with roughened surfaces and they measured the compressional and shear velocities in directions normal and parallel to the plates. Quasicompressional and quasishear velocities were measured at a variety of oblique angles. Le Houedec (2001) presented a review article on the transmission of vertical vibrations of a rigid footing over the surface of the ground by considering two different models of elastic half-space and an elastic layer medium. Impedance functions for a rigid massless circular footing resting on a soil layer on a rigid base subjected to vertical harmonic excitation were determined by considering the one-dimensional wave propagation in cones. Moreover, the effects of mass ratio, Poisson's ratio, depth of the layer, and hysteretic material damping ratio on the response of the foundation were presented by Pradhan (2004).

Utilizing these stiffness matrix method and transforming stresses and displacements across each layer, an approximate solution for the dynamic response of the surface of multi-layered elastic mediums subjected to harmonic concentrated surface loading is provided in the following chapters.

Chapter 2
Governing Equations

Abstract In this chapter fundamental governing equations for propagation of a harmonic disturbance on the surface of an elastic half-space is presented. The elastic media is assumed to be isotropic, continuous and infinite. Although this subject has been treated in various approaches in a number of references, it may be appropriate to develop a general procedure for solution of a number of essential soil dynamic problems based on the first principles of elasticity. The main scope of this chapter is to present a general systematic solution for addressing the problems considered in the next two chapters. This general solution is presented in two-dimensional Fourier domain in terms of elastic dilatation, and two elastic rotation components, which are used for decoupling of the involved partial differential equations. This chapter is also presenting the required fundamental of elasto-dynamics to achieve the set goal for the chapter.

Keywords Stress–strain relation • Elastic rotations • Boundary stresses • Boundary displacements

In this chapter fundamental governing equations for propagation of a harmonic disturbance on the surface of an elastic half-space is presented. The elastic media is assumed to be isotropic, continuous and infinite. Although this subject has been treated in various approaches in a number of references such as Lamb (1904), Timoshenko and Goodier (1951), Sneddon (1951), Ewing and Jardetzky (1957), and many others, it may be appropriate to develop a general procedure for solution of a number of essential soil dynamic problems based on the first principles of elasticity. The main scope of this chapter is to present a general systematic solution for addressing the problems considered in the next two chapters. This general solution is presented in two-dimensional Fourier domain in terms of elastic dilatation, and two elastic rotation components, which are used for decoupling of the involved partial differential equations. This chapter is also presenting the required fundamental of elasto-dynamics to achieve the set goal for the chapter.

H.R. Hamidzadeh et al., *Wave Propagation in Solid and Porous Half-Space Media*,
DOI 10.1007/978-1-4614-9269-6_2,

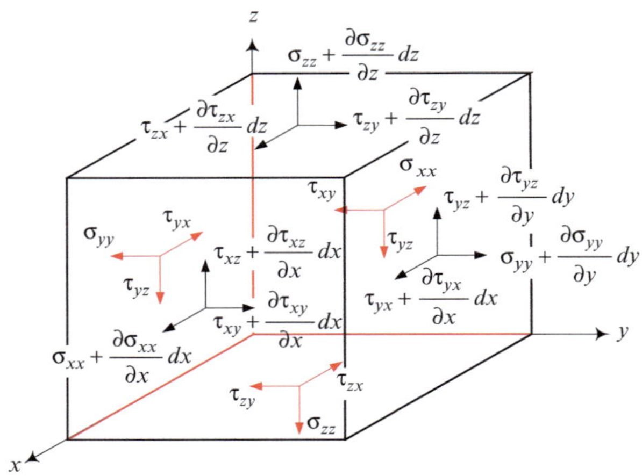

Fig. 2.1 System of stresses on an infinitesimal element in the elastic medium

2.1 Derivation of Equations of Motion

The equations of motion for an elastic half-space medium can be obtained by deriving the equilibrium equations for an infinitesimal element in terms of the applied stresses. This element is assumed to be an elastic body with an applied orthogonal stresses as shown in Fig. 2.1. The equations of motion for the element can be expressed by writing Newton's second law in three directions (see Shames and Cozzarelli 1991). These equations are:

$$\frac{\partial \sigma_{xx}}{\partial x} + \frac{\partial \tau_{yx}}{\partial y} + \frac{\partial \tau_{zx}}{\partial z} + X = \rho \frac{\partial^2 u}{\partial t^2} \tag{2.1}$$

$$\frac{\partial \sigma_{yy}}{\partial y} + \frac{\partial \tau_{zy}}{\partial z} + \frac{\partial \tau_{xy}}{\partial x} + Y = \rho \frac{\partial^2 v}{\partial t^2} \tag{2.2}$$

$$\frac{\partial \sigma_{zz}}{\partial z} + \frac{\partial \tau_{xz}}{\partial x} + \frac{\partial \tau_{yz}}{\partial y} + Z = \rho \frac{\partial^2 w}{\partial t^2} \tag{2.3}$$

where ρ is the mass density of the medium; σ_{ij} and τ_{ij} are direct and shear stresses on the plane perpendicular to j direction and along the direction of i; X, Y, and Z are internal forces per unit volume in the x, y, and z directions; u, v, and w are displacements of a point in the medium in the x, y, and z directions.

2.2 Stress–Strain Relation

For a homogeneous isotropic material, i.e. a material having the same properties in all directions, Hooke's Law can be written by the following equations:

$$\varepsilon_{xx} = \frac{1}{E}\left[\sigma_{xx} - \upsilon\left(\sigma_{yy} + \sigma_{zz}\right)\right] \tag{2.4}$$

$$\varepsilon_{yy} = \frac{1}{E}\left[\sigma_{yy} - \upsilon\left(\sigma_{zz} + \sigma_{xx}\right)\right] \tag{2.5}$$

$$\varepsilon_{zz} = \frac{1}{E}\left[\sigma_{zz} - \upsilon\left(\sigma_{xx} + \sigma_{yy}\right)\right] \tag{2.6}$$

where E and υ are elastic modulus and Poisson ratio of the medium. Adding Eqs. (2.4)–(2.6) yields:

$$E\left(\varepsilon_{xx} + \varepsilon_{yy} + \varepsilon_{zz}\right) = (1 - 2\upsilon)\left(\sigma_{xx} + \sigma_{yy} + \sigma_{zz}\right) \tag{2.7}$$

After substituting $\sigma_{yy} + \sigma_{zz}$ from Eq. (2.4) into Eq. (2.7) and simplifying, this yields:

$$\sigma_{xx} = \frac{E\upsilon}{(1+\upsilon)(1-2\upsilon)}\varepsilon + \frac{E}{(1+\upsilon)}\varepsilon_{xx} \tag{2.8}$$

$$\varepsilon = \varepsilon_{xx} + \varepsilon_{yy} + \varepsilon_{zz} \tag{2.9}$$

Introducing Lame constants (λ and G) the above equation can be written as:

$$\sigma_{xx} = \lambda\varepsilon + 2G\varepsilon_{xx} \tag{2.10}$$

Following the same procedure, σ_{yy} and σ_{zz} become

$$\sigma_{yy} = \lambda\varepsilon + 2G\varepsilon_{yy} \tag{2.11}$$

$$\sigma_{zz} = \lambda\varepsilon + 2G\varepsilon_{zz} \tag{2.12}$$

The familiar equations relating shear stresses to strains are:

$$\tau_{xy} = \tau_{yx} = G\gamma_{xy} \tag{2.13}$$

$$\tau_{yz} = \tau_{zy} = G\gamma_{yz} \tag{2.14}$$

$$\tau_{zx} = \tau_{xz} = G\gamma_{zx} \tag{2.15}$$

2.3 Strains in Terms of Displacements

Since strains generally vary from point to point, the mathematical definitions of strain must relate to an infinitesimal element. Consider an element in the (x, y) plane (Fig. 2.2). During straining, point A experiences displacements u and v in the x and y directions. Displacements of other points are also shown in Fig. 2.2.

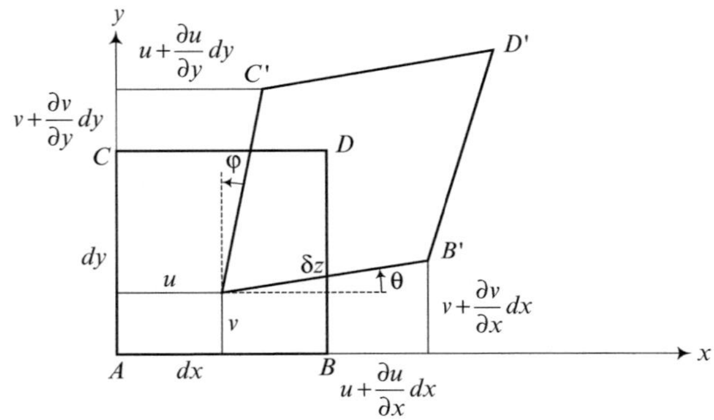

Fig. 2.2 Elastic deformation of the cubic element in the (x, y) plane

On this basis, the mathematical expressions for the linear strains as described by Ford and Alexander (1963) and many others are

$$\varepsilon_{xx} = \frac{\partial u}{\partial x} \tag{2.16}$$

$$\varepsilon_{yy} = \frac{\partial v}{\partial y} \tag{2.17}$$

$$\varepsilon_{zz} = \frac{\partial w}{\partial z} \tag{2.18}$$

These three strains are direct strains, but shear strains due to small deformation can be presented in terms of rotation. On the other hand, angles of rotation of any side of the element are very small, therefore, by definition:

$$\gamma_{xy} = \theta + \varphi = \frac{\left(v + \frac{\partial v}{\partial x}dx\right) - v}{\frac{\partial u}{\partial x} + dx} + \frac{\left(u + \frac{\partial u}{\partial y}dy\right) - u}{\frac{\partial v}{\partial y} + dy} \tag{2.19}$$

Because $dx >> \frac{\partial u}{\partial x}$ and $dy >> \frac{\partial v}{\partial y}$, Eq. (2.19) becomes

$$\gamma_{xy} = \frac{\partial v}{\partial x} + \frac{\partial u}{\partial y} \tag{2.20}$$

Consequently, other shear strains are given by

$$\gamma_{yz} = \frac{\partial w}{\partial y} + \frac{\partial v}{\partial z} \tag{2.21}$$

$$\gamma_{zx} = \frac{\partial w}{\partial x} + \frac{\partial u}{\partial z} \tag{2.22}$$

2.4 Elastic Rotations in Terms of Displacements

Based on Fig. 2.2 the elastic rotation about the z-axis is defined to be:

$$\omega_z = \frac{1}{2}(\theta - \varphi) \tag{2.23}$$

Considering that both angles of θ and φ are very small then this component of elastic rotation can be expressed as

$$\omega_z = \frac{1}{2}\left(\frac{\partial v}{\partial x} - \frac{\partial u}{\partial y}\right) \tag{2.24}$$

Similarly, other rotations are

$$\omega_x = \frac{1}{2}\left(\frac{\partial w}{\partial y} - \frac{\partial v}{\partial z}\right) \tag{2.25}$$

$$\omega_y = \frac{1}{2}\left(\frac{\partial u}{\partial z} - \frac{\partial w}{\partial x}\right) \tag{2.26}$$

2.5 Equations of Motion

Now, having substituted Eqs. (2.10)–(2.12) and (2.16)–(2.18) into Eqs. (2.1)–(2.3), the equations of motion for the element become

$$\frac{\partial}{\partial x}(\lambda\varepsilon + 2G\varepsilon_{xx}) + \frac{\partial}{\partial y}G\gamma_{yx} + \frac{\partial}{\partial z}G\gamma_{zx} + X = \rho\frac{\partial^2 u}{\partial t^2} \tag{2.27}$$

$$\frac{\partial}{\partial y}(\lambda\varepsilon + 2G\varepsilon_{yy}) + \frac{\partial}{\partial z}G\gamma_{yz} + \frac{\partial}{\partial x}G\gamma_{xy} + Y = \rho\frac{\partial^2 v}{\partial t^2} \tag{2.28}$$

$$\frac{\partial}{\partial z}(\lambda\varepsilon + 2G\varepsilon_{zz}) + \frac{\partial}{\partial x}G\gamma_{xz} + \frac{\partial}{\partial y}G\gamma_{yz} + Z = \rho\frac{\partial^2 w}{\partial t^2} \tag{2.29}$$

Substituting for strains in Eqs. (2.27)–(2.29), in terms of volumetric strain and displacements, the equations of motion become:

$$(\lambda + G)\frac{\partial \varepsilon}{\partial x} + G\nabla^2 u + X = \rho\frac{\partial^2 u}{\partial t^2} \tag{2.30}$$

$$(\lambda + G)\frac{\partial \varepsilon}{\partial y} + G\nabla^2 v + Y = \rho\frac{\partial^2 v}{\partial t^2} \tag{2.31}$$

$$(\lambda + G)\frac{\partial \varepsilon}{\partial z} + G\nabla^2 w + Z = \rho\frac{\partial^2 w}{\partial t^2} \tag{2.32}$$

where ∇^2 is the Laplacian operator.

If the body force is neglected and variations of the displacements are harmonic with circular frequency of Ω, then Eqs. (2.30)–(2.32) can be written in the following form:

$$(\lambda + G)\frac{\partial \varepsilon}{\partial x} + G\nabla^2 u = -\rho\Omega^2 u \tag{2.33}$$

$$(\lambda + G)\frac{\partial \varepsilon}{\partial y} + G\nabla^2 v = -\rho\Omega^2 v \tag{2.34}$$

$$(\lambda + G)\frac{\partial \varepsilon}{\partial z} + G\nabla^2 w = -\rho\Omega^2 w \tag{2.35}$$

The above equations can be represented in terms of dilatation and rotations by differentiating Eqs. (2.33)–(2.35) with respect to x, y, and z, then adding them together:

$$(\lambda + G)\nabla^2\varepsilon + \left(G\nabla^2 + \rho\Omega^2\right)\left(\frac{\partial u}{\partial x} + \frac{\partial v}{\partial y} + \frac{\partial w}{\partial z}\right) = 0 \tag{2.36}$$

Taking into account the definition of volumetric strain, the above equation becomes

$$\left(\nabla^2 + 2G\right)\nabla^2\varepsilon + \rho\Omega^2\varepsilon = 0 \tag{2.37}$$

or

$$\left(\nabla^2 + \Omega_1^2\right)\varepsilon = 0 \tag{2.38}$$

where

$$\Omega_1^2 = \frac{\Omega^2}{C_1^2} \tag{2.39}$$

$$C_1^2 = \frac{\lambda + 2G}{\rho} \tag{2.40}$$

C_1 is the velocity of waves of dilatation in the media. Subtracting the derivative of (2.34) with respect to x from the derivative of (2.33) with respect to y gives

$$(\lambda + G)\frac{\partial^2}{\partial x \partial y}\varepsilon + G\nabla^2\frac{\partial u}{\partial y} - (\lambda + G)\frac{\partial^2}{\partial x \partial y}\varepsilon - G\nabla^2\frac{\partial u}{\partial x} = -\rho\Omega^2\left(\frac{\partial u}{\partial y} - \frac{\partial v}{\partial x}\right) \tag{2.41}$$

After simplifying the above equation yields:

$$\left(\nabla^2 + \Omega_2^2\right)\omega_z = 0 \tag{2.42}$$

where

$$\Omega_2^2 = \frac{\Omega^2}{C_2^2} \tag{2.43}$$

$$C_2^2 = \frac{G}{\rho} \tag{2.44}$$

C_2 is velocity of waves of distortion.

Finally, subtracting the derivative of Eq. (2.34) with respect to z from the derivative of Eq. (2.35) with respect to y results:

$$\left(\nabla^2 + \Omega_2^2\right)\omega_x = 0 \tag{2.45}$$

2.6 Displacements in Terms of Dilatation and Rotation Components

Since the general solutions of equations of motion are given in terms of dilatation and rotation components, in order to determine displacements, they must be given in terms of dilatation and rotation components. Using Eqs. (2.24)–(2.26), the Laplacian operator of the displacement in the x-direction in terms of dilatation and rotation components is

$$\begin{aligned}
\nabla^2 u &= \frac{\partial^2 u}{\partial x^2} + \frac{\partial^2 v}{\partial y^2} + \frac{\partial^2 w}{\partial z^2} \\
&= \frac{\partial \varepsilon_{xx}}{\partial x} + \frac{\partial}{\partial y}\left(\frac{\partial v}{\partial x} - 2\omega_z\right) + \frac{\partial}{\partial z}\left(\frac{\partial w}{\partial x} + 2\omega_y\right) \\
&= \frac{\partial \varepsilon_{xx}}{\partial x} + \frac{\partial \varepsilon_{yy}}{\partial x} + \frac{\partial \varepsilon_{zz}}{\partial x} + 2\frac{\partial \omega_y}{\partial z} - 2\frac{\partial \omega_z}{\partial y} \\
&= \frac{\partial \varepsilon}{\partial x} + 2\frac{\partial \omega_y}{\partial z} - 2\frac{\partial \omega_z}{\partial y}
\end{aligned} \tag{2.46}$$

Similarly, the corresponding expression for the Laplacian operator of v and w are

$$\nabla^2 v = \frac{\partial \varepsilon}{\partial y} + 2\frac{\partial \omega_z}{\partial x} - 2\frac{\partial \omega_x}{\partial z} \tag{2.47}$$

$$\nabla^2 w = \frac{\partial \varepsilon}{\partial z} + 2\frac{\partial \omega_x}{\partial y} - 2\frac{\partial \omega_y}{\partial x} \tag{2.48}$$

By substituting these equations into Eqs. (2.33)–(2.35), the displacements will be found in terms of dilatation and components of rotations:

$$(\lambda + G)\frac{\partial \varepsilon}{\partial x} + G\left(\frac{\partial \varepsilon}{\partial x} + 2\frac{\partial \omega_y}{\partial z} - 2\frac{\partial \omega_z}{\partial y}\right) = -\rho\Omega^2 u \tag{2.49}$$

$$(\lambda + G)\frac{\partial \varepsilon}{\partial y} + G\left(\frac{\partial \varepsilon}{\partial y} + 2\frac{\partial \omega_z}{\partial x} - 2\frac{\partial \omega_x}{\partial z}\right) = -\rho\Omega^2 v \tag{2.50}$$

$$(\lambda + G)\frac{\partial \varepsilon}{\partial z} + G\left(\frac{\partial \varepsilon}{\partial z} + 2\frac{\partial \omega_x}{\partial y} - 2\frac{\partial \omega_y}{\partial x}\right) = -\rho\Omega^2 w \tag{2.51}$$

After some rearrangement and reduction, the above equations become:

$$u = -\frac{1}{\Omega_1^2}\frac{\partial \varepsilon}{\partial x} + \frac{2}{\Omega_2^2}\frac{\partial \omega_z}{\partial y} - \frac{2}{\Omega_2^2}\frac{\partial \omega_y}{\partial z} \tag{2.52}$$

$$v = -\frac{1}{\Omega_1^2}\frac{\partial \varepsilon}{\partial y} + \frac{2}{\Omega_2^2}\frac{\partial \omega_x}{\partial z} - \frac{2}{\Omega_2^2}\frac{\partial \omega_z}{\partial x} \tag{2.53}$$

$$w = -\frac{1}{\Omega_1^2}\frac{\partial \varepsilon}{\partial z} + \frac{2}{\Omega_2^2}\frac{\partial \omega_y}{\partial x} - \frac{2}{\Omega_2^2}\frac{\partial \omega_x}{\partial y} \tag{2.54}$$

2.7 Stresses in Terms of Dilatation and Rotation Components

In order to get the stresses in terms of the dilatation and rotation components, Eq. (2.11) can be written as follows:

$$\sigma_{yy} = \lambda\varepsilon + 2G\varepsilon_{yy} = (\lambda + 2G)\,\varepsilon_{yy} + \lambda\varepsilon_{xx} + \lambda\varepsilon_{zz} \tag{2.55}$$

and, in terms of displacement components, it is

$$\sigma_{yy} = (\lambda + 2G)\frac{\partial v}{\partial y} + \lambda\frac{\partial w}{\partial z} + \lambda\frac{\partial u}{\partial x} \tag{2.56}$$

Having substituted derivatives of Eqs. (2.52)–(2.54) in the above relation, it yields:

$$\begin{aligned}\sigma_{yy} = &-\left(\frac{\lambda + 2G}{\Omega_1^2}\frac{\partial^2 \varepsilon}{\partial y^2}\right) + \frac{2(\lambda + 2G)}{\Omega_2^2}\left(\frac{\partial^2 \omega_x}{\partial y \partial z} - \frac{\partial^2 \omega_z}{\partial y \partial x}\right)\\ &-\frac{\lambda}{\Omega_1^2}\frac{\partial^2 \varepsilon}{\partial x^2} + \frac{2\lambda}{\Omega_2^2}\left(\frac{\partial^2 \omega_z}{\partial x \partial y} - \frac{\partial^2 \omega_y}{\partial x \partial z}\right)\\ &-\frac{\lambda}{\Omega_1^2}\frac{\partial^2 \varepsilon}{\partial z^2} + \frac{2\lambda}{\Omega_2^2}\left(\frac{\partial^2 \omega_y}{\partial x \partial z} - \frac{\partial^2 \omega_x}{\partial y \partial z}\right)\end{aligned} \tag{2.57}$$

simplification, the above equation yields

$$\sigma_{yy} = -\left[\frac{\lambda + 2G}{\Omega_1^2}\frac{\partial^2\varepsilon}{\partial y^2} - \frac{4G}{\Omega_2^2}\left(\frac{\partial^2\omega_x}{\partial y\partial z} - \frac{\partial^2\omega_z}{\partial x\partial y}\right) + \frac{\lambda}{\Omega_1^2}\left(\frac{\partial^2\varepsilon}{\partial x^2} + \frac{\partial^2\varepsilon}{\partial z^2}\right)\right] \tag{2.58}$$

Similarly, the other direct stresses are

$$\sigma_{xx} = -\left[\frac{\lambda + 2G}{\Omega_1^2}\frac{\partial^2\varepsilon}{\partial x^2} - \frac{4G}{\Omega_2^2}\left(\frac{\partial^2\omega_z}{\partial x\partial y} - \frac{\partial^2\omega_y}{\partial x\partial z}\right) + \frac{\lambda}{\Omega_1^2}\left(\frac{\partial^2\varepsilon}{\partial y^2} + \frac{\partial^2\varepsilon}{\partial z^2}\right)\right] \tag{2.59}$$

and

$$\sigma_{zz} = -\left[\frac{\lambda + 2G}{\Omega_1^2}\frac{\partial^2\varepsilon}{\partial z^2} - \frac{4G}{\Omega_2^2}\left(\frac{\partial^2\omega_y}{\partial x\partial z} - \frac{\partial^2\omega_x}{\partial y\partial z}\right) + \frac{\lambda}{\Omega_1^2}\left(\frac{\partial^2\varepsilon}{\partial y^2} + \frac{\partial^2\varepsilon}{\partial x^2}\right)\right] \tag{2.60}$$

Shear stresses in the dilatation strain and the elastic rotation components can be obtained as presented in the following. For instance, the shear stress, τ_{xy}, in terms of displacements, is:

$$\tau_{xy} = G\left(\frac{\partial u}{\partial y} + \frac{\partial v}{\partial x}\right) \tag{2.61}$$

Substituting for u and v from Eqs. (2.52)–(2.54), this shear stress be presented by the following equation

$$\tau_{xy} = G\left[-\frac{2}{\Omega_1^2}\frac{\partial^2\varepsilon}{\partial x\partial y} + \frac{2}{\Omega_2^2}\left(\frac{\partial^2\omega_z}{\partial y^2} + \frac{\partial^2\omega_x}{\partial x\partial z} - \frac{\partial^2\omega_z}{\partial x^2} - \frac{\partial^2\omega_y}{\partial y\partial z}\right)\right] \tag{2.62}$$

Differentiating Eqs. (2.24)–(2.26) with respect to x, y, and z,

$$\frac{\partial\omega_x}{\partial x} = \frac{1}{2}\left(\frac{\partial^2 w}{\partial y\partial x} - \frac{\partial^2 v}{\partial z\partial x}\right) \tag{2.63}$$

$$\frac{\partial\omega_y}{\partial y} = \frac{1}{2}\left(\frac{\partial^2 u}{\partial z\partial y} - \frac{\partial^2 w}{\partial x\partial y}\right) \tag{2.64}$$

$$\frac{\partial\omega_z}{\partial z} = \frac{1}{2}\left(\frac{\partial^2 v}{\partial x\partial z} - \frac{\partial^2 u}{\partial y\partial z}\right) \tag{2.65}$$

And adding them all it yields:

$$\frac{\partial\omega_x}{\partial x} + \frac{\partial\omega_y}{\partial y} + \frac{\partial\omega_z}{\partial z} = 0 \tag{2.66}$$

The above equations demonstrate that the ω_y depends on the other two elastic rotations ω_x and ω_z. After substituting $\partial\omega_y/\partial y$ from the above equation into Eq. (2.62) and simplifying, the shear stress τ_{xy} is obtained:

$$\tau_{xy} = 2G\left[-\frac{1}{\Omega_1^2}\frac{\partial^2\varepsilon}{\partial x\partial y} + \frac{1}{\Omega_2^2}\left(\frac{\partial^2}{\partial y^2} - \frac{\partial^2}{\partial x^2} - \frac{\partial^2}{\partial z^2}\right)\omega_z + \frac{2}{\Omega_2^2}\frac{\partial^2\omega_x}{\partial x\partial z}\right] \tag{2.67}$$

Following the same procedure, the shear stress τ_{yz} will be presented by the following equation:

$$\tau_{yz} = 2G\left[-\frac{1}{\Omega_1^2}\frac{\partial^2\varepsilon}{\partial y\partial z} + \frac{1}{\Omega_2^2}\left(\frac{\partial^2}{\partial z^2} - \frac{\partial^2}{\partial x^2} - \frac{\partial^2}{\partial y^2}\right)\omega_x + \frac{2}{\Omega_2^2}\frac{\partial^2\omega_z}{\partial x\partial z}\right] \tag{2.68}$$

2.8 Fourier Transformation of Equations of Motion, Boundary Stresses, and Displacements

In order to reduce the three partial differential equations of (2.38), (2.42), and (2.45) into three ordinary differential equations in terms of y, complex Fourier transformation was used. The boundary conditions must then be treated in the same way, so that, instead of having relations in partial derivatives with respect to x, y, and z, relations are obtained in terms of the derivatives with respect to y only.

By applying the two-dimensional Fourier transforms (see Appendix A) on the equations of motion, the reduced equations become

$$\frac{\partial^2}{\partial y^2} - \left(p^2 + q^2 - \Omega_1^2\right)\bar{\varepsilon} = 0 \tag{2.69}$$

$$\frac{\partial^2}{\partial y^2} - \left(p^2 + q^2 - \Omega_2^2\right)\bar{\omega}_x = 0 \tag{2.70}$$

$$\frac{\partial^2}{\partial y^2} - \left(p^2 + q^2 - \Omega_2^2\right)\bar{\omega}_z = 0 \tag{2.71}$$

where $\bar{\varepsilon}$, $\bar{\omega}_x$, and $\bar{\omega}_z$ are the corresponding double complex Fourier transformations of ε, ω_x, and ω_z.

The transformed stresses at the surface of the half-space ((x, z) plane) in terms of the transform of the dilatation and elastic rotation components are

$$\bar{\sigma}_{yy} = -\left[\frac{\lambda + 2G}{\Omega_1^2}\frac{\partial^2\bar{\varepsilon}}{\partial y^2} + \frac{4iGq}{\Omega_2^2}\frac{\partial\bar{\omega}_x}{\partial y} - \frac{4iGq}{\Omega_2^2}\frac{\partial\bar{\omega}_z}{\partial y} - \frac{\lambda\bar{\varepsilon}}{\Omega_1^2}\left(p^2 + q^2\right)\right] \tag{2.72}$$

For the shear stresses,

$$\bar{\tau}_{xy} = 2G\left[\frac{ip}{\Omega_1^2}\frac{d\bar{\varepsilon}}{dy} + \frac{1}{\Omega_2^2}\left(\frac{d^2}{d^2} + p^2 - q^2\right)\bar{\omega}_z - \frac{2pq}{\Omega_2^2}\bar{\omega}_x\right] \tag{2.73}$$

and

$$\bar{\tau}_{yz} = 2G\left[\frac{iq}{\Omega_1^2}\frac{d\bar{\varepsilon}}{dy} - \frac{1}{\Omega_2^2}\left(\frac{d^2}{d^2} - p^2 + q^2\right)\bar{\omega}_x + \frac{2pq}{\Omega_2^2}\bar{\omega}_z\right] \tag{2.74}$$

The double complex Fourier transform of the displacements will be transforms of Eqs. (2.52)–(2.54)

$$\bar{u} = \frac{i}{\Omega_1^2}p\bar{\varepsilon} + \frac{2}{\Omega_2^2}\frac{\partial\bar{\omega}_z}{\partial y} + \frac{2i}{\Omega_2^2}q\bar{\omega}_y \tag{2.75}$$

$$\bar{v} = \frac{2}{\Omega_2^2}\left(ip\bar{\omega}_z - iq\bar{\omega}_x\right) - \frac{1}{\Omega_1^2}\frac{\partial\bar{\varepsilon}}{\partial y} \tag{2.76}$$

$$\bar{w} = \frac{i}{\Omega_1^2}q\bar{\varepsilon} - \frac{2}{\Omega_2^2}\frac{\partial\bar{\omega}_x}{\partial y} - \frac{2i}{\Omega_2^2}p\bar{\omega}_y \tag{2.77}$$

$\bar{\omega}_y$ can be obtained by transforming equation (2.26).

$$\bar{\omega}_y = -\frac{1}{2}iq\,\bar{u} + \frac{1}{2}ip\,\bar{w} \tag{2.78}$$

Substituting $\bar{\omega}_y$ from the above equation into Eqs. (2.75) and (2.77), and solving for $\bar{u}$ and $\bar{w}$ results in:

$$\bar{u} = \frac{i}{\Omega_1^2}p\bar{\varepsilon} + \frac{2pq/\Omega_2^2}{\Omega_2^2 - q^2 - p^2}\frac{\partial\bar{\omega}_x}{\partial y} + \frac{2\left(\Omega_2^2 - p^2\right)/\Omega_2^2}{\Omega_2^2 - q^2 - p^2}\frac{\partial\bar{\omega}_z}{\partial y} \tag{2.79}$$

$$\bar{v} = -\frac{1}{\Omega_1^2}\frac{d\bar{\varepsilon}}{dy} - \frac{2iq}{\Omega_2^2}\bar{\omega}_x + \frac{2ip}{\Omega_2^2}\bar{\omega}_z \tag{2.80}$$

$$\bar{w} = \frac{i}{\Omega_1^2}q\bar{\varepsilon} - \frac{2pq/\Omega_2^2}{\Omega_2^2 - q^2 - p^2}\frac{\partial\bar{\omega}_z}{\partial y} - \frac{2\left(\Omega_2^2 - q^2\right)/\Omega_2^2}{\Omega_2^2 - q^2 - p^2}\frac{\partial\bar{\omega}_x}{\partial y} \tag{2.81}$$

2.9 General Solution of Transformed Equations of Motion

The general solution of Eqs. (2.69)–(2.71) as a function of y is

$$\bar{\varepsilon} = A_\varepsilon \exp\left(-\gamma_1 y\right) \tag{2.82}$$

$$\bar{\omega}_x = A_x \exp\left(-\gamma_2 y\right) \tag{2.83}$$

$$\bar{\omega}_z = A_z \exp\left(-\gamma_3 y\right) \tag{2.84}$$

where values of γ_1, γ_2, and γ_3 must be negative to satisfy the boundary conditions requirements for $\bar{\varepsilon}$, $\bar{\omega}_x$, and $\bar{\omega}_z$ as y approaches to infinity. Substituting the above in Eqs. (2.69)–(2.71) results in:

$$\gamma_1 = p^2 + q^2 - \Omega_1^2 \tag{2.85}$$

$$\gamma_2 = p^2 + q^2 - \Omega_2^2 \tag{2.86}$$

$$\gamma_3 = p^2 + q^2 - \Omega_2^2 \tag{2.87}$$

From the last two equations, it is obvious that

$$\gamma_2 = \gamma_3 \tag{2.88}$$

Values of A_ε, A_x, and A_z are functions of p and q and depend on the boundary conditions of the problem. This means that they are dependent on the complex double Fourier transform of the stresses, which act as external excitations on the surface of the half-space medium. In order to evaluate these arbitrary functions, Eqs. (2.72)–(2.74) must be satisfied by the boundary conditions that are expressed by three stress components σ_{yy}, τ_{xy}, and τ_{yz}, which are applied on the surface of elastic half-space (x, z) where y is zero. Substituting $y = 0$ in Eqs. (2.82)–(2.84) and placing them into Eqs. (2.72)–(2.74) the double complex Fourier transform of the stresses on the surface of half-space will be presented by the following equations:

$$\frac{-2q^2 - 2p^2 + \Omega_2^2}{\Omega_1^2} A_\varepsilon + \frac{4i\gamma_2 q}{\Omega_2^2} A_x - \frac{4i\gamma_2 p}{\Omega_2^2} A_z = \frac{\bar{\sigma}_{yy}(p, q)}{G} \tag{2.89}$$

$$\frac{-2i\gamma_1 p}{\Omega_1^2} A_\varepsilon - \frac{4pq}{\Omega_2^2} A_x + \frac{2\left(2p^2 - \Omega_2^2\right)}{\Omega_2^2} A_z = \frac{\bar{\tau}_{xy}(p, q)}{G} \tag{2.90}$$

$$\frac{-2i\gamma_1 q}{\Omega_1^2} A_\varepsilon + \frac{2\left(-2q^2 + \Omega_2^2\right)}{\Omega_2^2} A_x + \frac{4pq}{\Omega_2^2} A_z = \frac{\bar{\tau}_{yz}(p, q)}{G} \tag{2.91}$$

The solutions to the above set of equations are given by the values of A_ε, A_x, and A_z. These values can be expressed as

$$A_\varepsilon = \frac{D_\varepsilon}{D} \tag{2.92}$$

$$A_x = \frac{D_x}{D} \tag{2.93}$$

$$A_z = \frac{D_z}{D} \tag{2.94}$$

where

$$D = \begin{bmatrix} \dfrac{-2q^2 - 2p^2 + \Omega_2^2}{\Omega_1^2} & \dfrac{4i\gamma_2 q}{\Omega_2^2} & \dfrac{-4i\gamma_2 p}{\Omega_2^2} \\ \dfrac{-2i\gamma_1 p}{\Omega_1^2} & \dfrac{-4pq}{\Omega_2^2} & \dfrac{2\left(2p^2 - \Omega_2^2\right)}{\Omega_2^2} \\ \dfrac{-2i\gamma_1 q}{\Omega_1^2} & \dfrac{2\left(\Omega_2^2 - 2q^2\right)}{\Omega_2^2} & \dfrac{4pq}{\Omega_2^2} \end{bmatrix} \tag{2.95}$$

or

$$D = \frac{4}{\Omega_1^2 \Omega_2^2} \Phi(p, q) \tag{2.96}$$

and

$$\Phi(p, q) = \left[2\left(p^2 + q^2\right) - \Omega_2^2\right]^2 - 4\gamma_1\gamma_2\left(p^2 + q^2\right) \tag{2.97}$$

This is the well-known function associated with Rayleigh surface waves, also

$$D_\varepsilon = \begin{vmatrix} \bar{\sigma}_{yy}/G & 4i\gamma_2 q & -4i\gamma_2 p \\ \bar{\tau}_{xy}/G & -4pq & 2\left(2p^2 - \Omega_2^2\right) \\ \bar{\tau}_{yz}/G & 2\left(\Omega_2^2 - 2q^2\right) & 4pq \end{vmatrix} \frac{1}{\Omega_2^4} \tag{2.98}$$

$$D_x = \begin{vmatrix} -2q^2 - 2p^2 + \Omega_2^2 & \bar{\sigma}_{yy}/G & -4i\gamma_2 p \\ -2i\gamma_1 p & \bar{\tau}_{xy}/G & 2\left(2p^2 - \Omega_2^2\right) \\ -2i\gamma_1 q & \bar{\tau}_{yz}/G & 4pq \end{vmatrix} \frac{1}{\Omega_1^2 \Omega_2^2} \tag{2.99}$$

$$D_z = \begin{vmatrix} -2q^2 - 2p^2 + \Omega_2^2 & 4i\gamma_2 q & \bar{\sigma}_{yy}/G \\ -2i\gamma_1 p & -4pq & \bar{\tau}_{xy}/G \\ -2i\gamma_1 q & 2\left(\Omega_2^2 - 2q^2\right) & \bar{\tau}_{yz}/G \end{vmatrix} \frac{1}{\Omega_1^2 \Omega_2^2} \tag{2.100}$$

Equations (2.82) through (2.84) along with Eqs. (2.89)–(2.91) present the general solution for the elastic dilatation $\bar{\varepsilon}$ and elastic rotations $\bar{\omega}_x$ and $\bar{\omega}_z$ for the surface of the medium due to applied stresses on the surface in the Fourier domain. These general equations will be utilized to determine the displacements on the surface of an elastic half-space for the two specific surface stress distributions caused by a vertical/horizontal harmonic point force that will be addressed in Chaps. 3 and 4, respectively.

Chapter 3
Surface Response of an Elastic Half-Space Due to a Vertical Harmonic Point Force

Abstract This chapter will review the surface response of an elastic space to a vertical harmonic point force. The response of an elastic half-space subjected to a vertical point load has been studied by many authors. Among them the original formulation of the problem is due to Lamb (Philos Trans R Soc 203(A):1–42, 1904) who provided integral expressions for the harmonic displacement of the surface of an elastic half-space. Barkan (Dynamics of bases and foundations. McGraw-Hill, New York, 1962) gave an approximate solution for vertical displacement. For the case of surface motions produced by a point load varying with time like the Heaviside function, a great simplification is achieved by Pekeris (Proc Natl Acad Sci USA 41:629–639, 469–480, 1955) for Poisson ratio of 0.25 and Awojobi and Sobayo (1977) through the use of the Bateman Pekeris integral theorem. Elorduy et al. (Proceedings of the international symposium on wave propagation and dynamic properties of Earth materials, Albuquerque, University of New Mexico, pp 105–123, 1967) used Pekeris's results and by applying numerical Duhamel's integral he estimated the vertical response due to a harmonic point force. In this chapter an analytical solution for the frequency response of any point on the surface of the elastic half-space generated by a harmonic vertical point force is discussed. Several approximate solutions to this problem are obtained by the integral transform method, where inversion of the transform is quite involved. To achieve numerical results for this case, an effective numerical technique has been developed for calculation of the integrals represented in inversion of the transformed relations.

Keywords Harmonic point force • Elastic surface response • Waves on half elastic space • Displacements evaluation

The response of an elastic half-space subjected to a vertical point load has been studied by many authors. Among them the original formulation of the problem is due to Lamb (1904) who provided integral expressions for the harmonic displacement of the surface of an elastic half-space. Barkan (1962) gave an approximate

H.R. Hamidzadeh et al., *Wave Propagation in Solid and Porous Half-Space Media*,
DOI 10.1007/978-1-4614-9269-6_3,

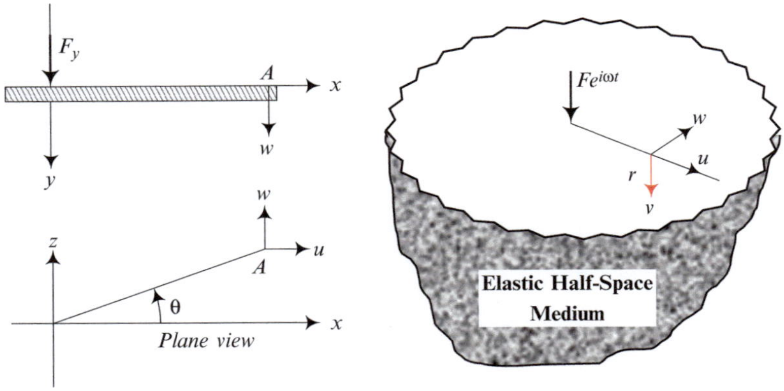

Fig. 3.1 Diagram for displacements of a point "A" on the surface of an elastic half-space due to a vertical point force

solution for vertical displacement. For the case of surface motions produced by a point load varying with time, like the Heaviside function, a great simplification is achieved by Pekeris (1955a, b) for Poisson ratio of 0.25 and Awojobi (1977) through the use of the Bateman Pekeris integral theorem. Elorduy et al. (1967) used Pekeris's results and by applying numerical Duhamel's integral he estimated the vertical response due to a harmonic point force. In this chapter an analytical solution for the frequency response of any point on the surface of the elastic half-space generated by a harmonic vertical point force is discussed in detail (see Fig. 3.1). Several approximate solutions to this problem are obtained by the integral transform method, where inversion of the transform is quite involved. To achieve numerical results for this case, an effective numerical technique has been developed for calculation of the integrals represented in inversion of the transformed relations. These techniques are discussed in Appendices B and C.

3.1 Boundary Conditions of the Problem

The elastic medium is defined as infinitely deep and occupying an infinitely wide horizontal plane, which forms its only boundary, $y = 0$. It is assumed to be homogeneous and isotropic. This definition ensures that the propagated waves from a source die out and are not reflected, thus eliminating any possible interference of waves.

Boundary stresses on the surface of the half-space are

$$\tau_{xy} = 0 \qquad 0 \leq |x| < \infty \wedge 0 \leq |z| < \infty \tag{3.1}$$

$$\tau_{zy} = 0 \qquad 0 \leq |x| < \infty \wedge 0 \leq |z| < \infty \tag{3.2}$$

The amplitude of the normal harmonic applied stress caused by the point force can be expressed by Dirac's delta function with magnitude of σ, as shown below.

$$\sigma_{yy} = \sigma\,\delta\,(x, z)\,e^{i\Omega t} \tag{3.3}$$

This means that σ_{yy} will vanish everywhere except the point at which the force is applied. The mathematical presentation of the point force in terms of the above stress can be written by the following expression:

$$F_y = \int_{-\infty}^{\infty}\int_{-\infty}^{\infty} \sigma\,\delta\,(x, z)\,e^{i\Omega t}\,dx\,dz = \sigma \tag{3.4}$$

The above relation indicates that the amplitude of the applied point force is equal to the value of σ.

Double complex Fourier transformations of these stresses with respect to x, then z are

$$\bar{\sigma}\,(p, q) = \frac{1}{2\pi}\int_{-\infty}^{\infty}\int_{-\infty}^{\infty} \sigma\,\delta\,(x, z)\,e^{ipx}\,e^{iqz}\,dx\,dz = \frac{\sigma}{2\pi} = \frac{F_y}{2\pi} \tag{3.5}$$

$$\bar{\tau}_{xy} = 0 \tag{3.6}$$

$$\bar{\tau}_{zy} = 0 \tag{3.7}$$

3.2 Integral Representations of Displacements

Having substituted for the transforms of the stresses from Eqs. (3.5)–(3.7) into Eqs. (2.98)–(2.100), values of D_ε, D_x, and D_z can be determined by

$$D_\varepsilon = \frac{1}{\Omega_2^2}\bar{\sigma}\,(p, q)\,\left(2p^2 + 2q^2 - \Omega_2^2\right) \tag{3.8}$$

$$D_x = \frac{-1}{\Omega_1^2}\bar{\sigma}\,(p, q)\,i\gamma_1 q \tag{3.9}$$

$$D_z = \frac{1}{\Omega_1^2}\bar{\sigma}\,(p, q)\,i\gamma_1 p \tag{3.10}$$

Substituting for these values in Eqs. (2.92)–(2.94), arbitrary values of Eqs. (2.82)–(2.84) become

$$A_\varepsilon = \frac{-\Omega_1^2\bar{\sigma}\,(p, q)\,\left(2p^2 + 2q^2 - \Omega_2^2\right)}{G\Phi\,(p, q)} \tag{3.11}$$

$$A_x = \frac{i\Omega_2^2\bar{\sigma}(p,q)\,q\gamma_1}{G\Phi(p,q)} \tag{3.12}$$

$$A_z = \frac{-i\Omega_2^2\bar{\sigma}(p,q)\,p\gamma_1}{G\Phi(p,q)} \tag{3.13}$$

To obtain the solution for the displacements in the Fourier transformation domain, substitute for A_ε, A_x, and A_z from the last relations into (2.82)–(2.84) and then the results will be put into (2.79)–(2.81). After simplification, the displacements in the Fourier domain for $y = 0$ become

$$\bar{u}(p,o,q) = ip\frac{\bar{\sigma}(p,q)}{G\Phi(p,q)}\left(-2\gamma_2^2 + 2\gamma_1\gamma_2 - \Omega_2^2\right) \tag{3.14}$$

$$\bar{v}(p,o,q) = \frac{\bar{\sigma}(p,q)}{G\Phi(p,q)}\gamma_1\Omega_2^2 \tag{3.15}$$

$$\bar{w}(p,o,q) = -iq\frac{\bar{\sigma}(p,q)}{G\Phi(p,q)}\left(-2\gamma_2^2 + 2\gamma_1\gamma_2 - \Omega_2^2\right) \tag{3.16}$$

Considering that the

$$\bar{\sigma}(p,q) = \frac{F_y}{2\pi} \tag{3.17}$$

the inverse transform of these three equations will give the integral representation of the displacements (see Appendix A).

$$u(x,z) = \frac{1}{2\pi}\int_{-\infty}^{\infty}\int_{-\infty}^{\infty}\frac{F_y ip\left(-2\gamma_2^2 + 2\gamma_1\gamma_2 - \Omega_2^2\right)}{2\pi G\Phi(p,q)}e^{i(px+qz)}dpdq \tag{3.18}$$

$$v(x,z) = \frac{1}{2\pi}\int_{-\infty}^{\infty}\int_{-\infty}^{\infty}\frac{F_y\gamma_1\Omega_2^2}{2\pi G\Phi(p,q)}e^{i(px+qz)}dpdq \tag{3.19}$$

$$w(x,z) = \frac{1}{2\pi}\int_{-\infty}^{\infty}\int_{-\infty}^{\infty}\frac{F_y iq\left(-2\gamma_2^2 + 2\gamma_1\gamma_2 - \Omega_2^2\right)}{2\pi G\Phi(p,q)}e^{i(px+qz)}dpdq \tag{3.20}$$

By changing coordinates systems (x,z) and (p,q) into polar coordinates (r,θ) and (δ,ϕ) using the following four relations

$$p = \delta\sin\phi \tag{3.21}$$

$$q = \delta\cos\phi \tag{3.22}$$

and

$$x = r\cos\theta \tag{3.23}$$

$$z = r\sin\theta \tag{3.24}$$

the displacement in terms of the new variables becomes

$$u(r,\theta) = \frac{iF_y}{2G\pi^2}\int_0^\infty \frac{-2\left(\delta^2-\Omega_2^2\right)+\left(\delta^2-\Omega_1^2\right)\sqrt{\delta^2-\Omega_1^2}-\Omega_2^2}{\left(2\delta^2-\Omega_2^2\right)^2-4\delta^2\sqrt{\left(\delta^2-\Omega_1^2\right)\left(\delta^2-\Omega_2^2\right)}}\delta^2 \times \int_0^{2\pi} \sin\phi\, e^{-i\delta r\sin(\theta+\phi)}\, d\phi\, d\delta \tag{3.25}$$

$$v(r,\theta) = \frac{iF_y\Omega_2^2}{2G\pi^2}\int_0^\infty \frac{\sqrt{\delta^2-\Omega_1^2}}{\left(2\delta^2-\Omega_2^2\right)^2-4\delta^2\sqrt{\left(\delta^2-\Omega_1^2\right)\left(\delta^2-\Omega_2^2\right)}}\delta \times \int_0^{2\pi} e^{-i\delta r\sin(\theta+\phi)}\, d\phi\, d\delta \tag{3.26}$$

$$w(r,\theta) = \frac{iF_y}{2G\pi^2}\int_0^\infty \frac{-2\left(\delta^2-\Omega_2^2\right)+2\sqrt{\left(\delta^2-\Omega_1^2\right)\left(\delta^2-\Omega_1^2\right)}-\Omega_2^2}{\left(2\delta^2-\Omega_2^2\right)^2-4\delta^2\sqrt{\left(\delta^2-\Omega_1^2\right)\left(\delta^2-\Omega_2^2\right)}}\delta^2 \times \int_0^{2\pi} \cos\phi\, e^{-i\delta r\sin(\theta+\phi)}\, d\phi\, d\delta \tag{3.27}$$

Results of the three integrals which appear at the end of the displacement equations can be expressed in terms of Bessel functions as

$$S_1 = \int_0^{2\pi} \sin\phi\, e^{-i\delta r\sin(\theta+\phi)}\, d\phi = -2\pi i\; J_1(\delta r)\;\cos\theta \tag{3.28}$$

$$S_2 = \int_0^{2\pi} e^{-i\delta r\sin(\theta+\phi)}\, d\phi = 2\pi\; J_o(\delta r) \tag{3.29}$$

$$S_3 = \int_0^{2\pi} \cos\phi\, e^{-i\delta r\sin(\theta+\phi)}\, d\phi = -2\pi i\; J_1(\delta r)\;\sin\theta \tag{3.30}$$

By introducing the following four variables

$$\varepsilon = \delta r \tag{3.31}$$

$$k = r\Omega_2 \tag{3.32}$$

$$h = r\Omega_1 \tag{3.33}$$

$$\gamma = \frac{h}{k} = \frac{\Omega_1}{\Omega_2} = \sqrt{\frac{\lambda+2G}{G}} = \sqrt{\frac{1-2\upsilon}{2(1-\upsilon)}} \tag{3.34}$$

and substituting for S_1, S_2, and S_3, the displacement equations can be presented in terms of a single integral as

$$u(r,\theta) = \frac{F_y \cos\theta}{2\pi G r} \int_0^{\infty} \frac{\varepsilon^2 \left[-2\varepsilon^2 + k^2 + 2\sqrt{\varepsilon^2 - h^2}\right]}{\left(2\varepsilon^2 - k^2\right)^2 - 4\varepsilon^2 \sqrt{\left(\varepsilon^2 - k^2\right)\left(\varepsilon^2 - h^2\right)}} J_1(\varepsilon)\, d\varepsilon \quad (3.35)$$

$$v(r,\theta) = \frac{F_y k^2}{2\pi G r} \int_0^{\infty} \frac{\varepsilon \sqrt{\varepsilon^2 - h^2}}{\left(2\varepsilon^2 - k^2\right)^2 - 4\varepsilon^2 \sqrt{\left(\varepsilon^2 - k^2\right)\left(\varepsilon^2 - h^2\right)}} J_0(\varepsilon)\, d\varepsilon \quad (3.36)$$

$$w(r,\theta) = \frac{F_y \sin\theta}{2\pi G r} \int_0^{\infty} \frac{\varepsilon^2 \left[-2\varepsilon^2 + k^2 + 2\sqrt{\left(\varepsilon^2 - k^2\right)\left(\varepsilon^2 - h^2\right)}\right]}{\left(2\varepsilon^2 - k^2\right)^2 - 4\varepsilon^2 \sqrt{\left(\varepsilon^2 - k^2\right)\left(\varepsilon^2 - h^2\right)}} J_1(\varepsilon)\, d\varepsilon \quad (3.37)$$

The denominator common to these three integrals is known as the Rayleigh function.

$$F(\varepsilon) = \left(2\varepsilon^2 - k^2\right)^2 - 4\varepsilon^2 \sqrt{\left(\varepsilon^2 - k^2\right)\left(\varepsilon^2 - h^2\right)} \quad (3.38)$$

3.3 Real Root of Rayleigh's Function

The real root of the Rayleigh function $F(\varepsilon)$ plays a major role in the analysis of Rayleigh wave. This root can be calculated easier by multiplying the Rayleigh function by the following $f(\varepsilon)$, which is

$$f(\varepsilon) = \left(2\varepsilon^2 - k^2\right)^2 + 4\varepsilon^2 \sqrt{\left(\varepsilon^2 - k^2\right)\left(\varepsilon^2 - h^2\right)} \quad (3.39)$$

Multiplying $f(\varepsilon)$ by $F(\varepsilon)$, it yields:

$$F(\varepsilon)\, f(\varepsilon) = \left(2\varepsilon^2 - k^2\right)^4 + 16\varepsilon^4 \left(\varepsilon^2 - k^2\right)\left(\varepsilon^2 - h^2\right) = 0 \quad (3.40)$$

Introducing

$$\gamma = \frac{h}{k} \quad (3.41)$$

and simplifying, Eq. (3.40) becomes:

$$F(\varepsilon)\, f(\varepsilon) = k^8 \left[1 - 8\frac{\varepsilon^2}{k^2} + \left(24 - 16\gamma^2\right)\frac{\varepsilon^4}{k^4} - 16\left(1 - \gamma^2\right)\frac{\varepsilon^6}{k^6}\right] = 0 \quad (3.42)$$

The above equation is a cubic equation in $(\varepsilon/k)^2$ and since $k^2 > h^2$, it will obviously have a real root between one and infinity. Any other root should lie in the

$(0, \gamma^2)$ interval. Any root in the first range will make $\sqrt{(\varepsilon^2 - k^2)}$ and $\sqrt{(\varepsilon^2 - h^2)}$ real and positive, therefore $f(\varepsilon)$ cannot be zero. Roots in the second range of $(0, \gamma^2)$ will make $\sqrt{(\varepsilon^2 - k^2)}$ and $\sqrt{(\varepsilon^2 - h^2)}$ positive and imaginary, which cannot be roots of $F(\varepsilon)$. Thus, the only roots of the Rayleigh function can be

$$\varepsilon = \pm\kappa \qquad |\kappa| > k \tag{3.43}$$

This value is dependent on the value of γ, which itself is a function of Poisson's ratio υ.

$$\gamma = \frac{h}{k} = \frac{\Omega_1}{\Omega_2} = \sqrt{\frac{1 - 2\upsilon}{2(1 - \upsilon)}} \tag{3.44}$$

3.4 System of Free Waves

These types of waves were first studied by Lord Rayleigh (1885) and are also known as Rayleigh waves. Their significance is that it will propagate on the surface of an elastic half-space and is considerable at a distance far away from the applied force. On the other hand, their influence decreases rapidly with depth. It should be noted that Rayleigh waves will diverge over the free surface and that they are a general solution of this boundary value problem.

A system of waves with the characteristics discussed before can be obtained by substituting zeros for both stresses on the surface and Rayleigh function, in Eqs. (3.37)–(3.35). A is the limit of the following ratio as applied force and $F(\kappa)$ both approach to zeros.

$$\frac{F_y}{F(\kappa)} = A \tag{3.45}$$

where A is an arbitrary value. Then, free displacement wave will be given by substituting zero of the Rayleigh function into Eqs. (3.37)–(3.35).

$$u_R(\varepsilon) = \frac{A\cos\theta}{Gr}\kappa^2\left[-2\kappa^2 + k^2 + 2\sqrt{(\kappa^2 - k^2)(\kappa^2 - h^2)}\right]J_1(\kappa) \tag{3.46}$$

$$v_R(\varepsilon) = \frac{Ak^2}{Gr}\kappa\sqrt{(\kappa^2 - h^2)}J_0(\kappa) \tag{3.47}$$

$$w_R(\varepsilon) = \frac{A\sin\theta}{Gr}\kappa^2\left[-2\kappa^2 + k^2 + 2\sqrt{(\kappa^2 - k^2)(\kappa^2 - h^2)}\right]J_1(\kappa) \tag{3.48}$$

3.5 Evaluation of Displacements

In order to evaluate these integrals, follow the Lamb (1904) approach and introduce the following relations for the Bessel function integral representation.

$$J_0(\varepsilon) = \frac{-1}{\pi}\int_0^{\infty}[\exp(i\varepsilon\cosh\zeta) - \exp(-i\varepsilon\cosh\zeta)]\,d\zeta \tag{3.49}$$

$$J_1(\varepsilon) = \frac{-1}{\pi}\int_0^{\infty}[\exp(i\varepsilon\cosh\zeta) + \exp(-i\varepsilon\cosh\zeta)]\,d\zeta \tag{3.50}$$

Substituting these two Bessel functions into Eqs. (3.37)–(3.35), they then become

$$u(r,\theta) = \frac{-F_y}{2\pi^2 Gr}\int_0^{\infty}\int_{-\infty}^{\infty}\frac{\varepsilon^2\left[2\varepsilon^2 - k^2 - 2\sqrt{(\kappa^2-k^2)(\kappa^2-h^2)}\right]}{F(\varepsilon)} \times e^{i\varepsilon\cosh\zeta}\,d\varepsilon\,\cosh\zeta\,d\zeta \tag{3.51}$$

$$v(r,\theta) = \frac{-iF_y}{2\pi^2 Gr}\int_0^{\infty}\int_{-\infty}^{\infty}\frac{\varepsilon k^2\sqrt{\varepsilon^2-h^2}}{F(\varepsilon)}e^{i\varepsilon\cosh\zeta}\,d\varepsilon\,d\zeta \tag{3.52}$$

$$w(r,\theta) = u(r,\theta)\tan\theta \tag{3.53}$$

Two infinite integrals with respect to β will be obtained by performing the operation $-i\partial/\partial x$ upon J_1 and J_2 integrals and then replacing x by $\cosh u$. These two integrals of J_1 and J_2 have been solved by Lamb (1904) by means of contour integration. An appropriate method of integration and its contour are given in Appendix B. Consequently, the displacement functions u, v, and w can be presented by the following equations:

$$u(r,\theta) = \frac{-F_y\cos\theta}{2\pi^2 Gr}\int_0^{\infty}\left[2\pi\kappa H_1\sin(\kappa\cosh\zeta)\,d\zeta + 4k^2\int_h^k\frac{\sigma^2(2\sigma^2-k^2)\alpha\beta}{F(\sigma)f(\sigma)}e^{-i\kappa\sigma\cosh\zeta}\,d\sigma\right]\cosh\zeta\,d\zeta \tag{3.54}$$

$$v(r,\theta) = \frac{-iF_y\cos\theta}{2\pi^2 Gr}\int_0^{\infty}\left[2\pi\kappa H_2\sin(\kappa\cosh\zeta)\,d\zeta - 2k^2\int_h^{\infty}\frac{\sigma\alpha\beta}{F(\sigma)f(\sigma)}e^{-i\sigma\cosh\zeta}\,d\sigma - 2k^2\int_h^k\frac{\sigma(2\sigma^2-k^2)\alpha}{F(\sigma)f(\sigma)}e^{-i\sigma\cosh\zeta}\,d\sigma\right]d\zeta \tag{3.55}$$

The integrals in the above equation can be represented by the following integrals.

$$Q_1 = \int_0^{\infty} 2\pi\kappa H_1 \sin(\kappa \cosh\zeta) \cosh\zeta \, d\zeta = \pi^2 \kappa H_1 Y_1(\kappa) \tag{3.56}$$

$$Q_2 = \int_0^{\infty} 2\pi\kappa H_2 \sin(\kappa \cosh\zeta) \, d\zeta = \pi^2 \kappa H_2 J_0(\kappa) \tag{3.57}$$

$$Q_3 = \int_0^{\infty} \exp(-i\sigma \cosh\zeta) \, d\zeta = -\frac{\pi}{2} i H_0^{(2)}(\sigma) \tag{3.58}$$

$$Q_4 = \int_0^{\infty} \exp(-i\sigma \cosh\zeta) \cosh\zeta \, d\zeta = -\frac{\pi}{2} i H_1^{(2)}(\sigma) \tag{3.59}$$

The displacements of the surface of an elastic half-space finally can be given by substituting the last integrations into Eqs. (3.54)–(3.54).

$$\begin{aligned} u(r,\theta) = \frac{F_y \cos\theta}{2\pi G r} & \left[\pi\kappa H_1 Y_1(\kappa) \right. \\ & \left. +2k^2 \int_h^k \frac{\sigma^2 (2\sigma^2 - k^2) \sqrt{(\kappa^2 - k^2)(\kappa^2 - h^2)}}{F(\sigma) f(\sigma)} H_1^{(2)}(\sigma) \, d\sigma \right] \end{aligned} \tag{3.60}$$

$$\begin{aligned} v(r,\theta) = \frac{F_y \cos\theta}{2\pi G r} & \left[i\pi\kappa H_2 J_0(\kappa) + k^2 \mathbb{C}P \int \int_k^{\infty} \frac{\sigma\sqrt{(\kappa^2 - h^2)}}{F(\sigma)} H_0^{(2)}(\sigma) \, d\sigma \right. \\ & \left. +k^2 \int_h^k \frac{\sigma (2\sigma^2 - k^2) \sqrt{(\kappa^2 - k^2)}}{F(\sigma) f(\sigma)} H_0^{(2)}(\sigma) \, d\sigma \right] \end{aligned} \tag{3.61}$$

$$\begin{aligned} w(r,\theta) = \frac{F_y \sin\theta}{2\pi G r} & \left[\pi\kappa H_1 Y_1(\kappa) + \right. \\ & \left. +2k^2 \int_h^k \frac{\sigma^2 (2\sigma^2 - k^2) \sqrt{(\kappa^2 - k^2)(\kappa^2 - h^2)}}{F(\sigma) f(\sigma)} H_1^{(2)}(\sigma) \, d\sigma \right] \end{aligned} \tag{3.62}$$

It should be noted that $\left(\mathbb{C}P \int\right)$ is used to indicate the Cauchy principal value of integration, $Y_1(\kappa)$ is the Bessel function of the second kind, and $H_0^{(2)}(\sigma)$ and $H_1^{(2)}(\sigma)$ are Bessel functions of the third kind, which are introduced in Eqs. (B.9) and (B.14) in the Appendix B.

In order to complete the solution, all the standing waves should be eliminated because the final solution should only consist of waves propagated outwards from the exciting point. In the above equations, only two terms of $\frac{F_y \kappa}{2Gr} H_2 J_0(\kappa)$

and $\frac{iF_y\kappa}{2Gr}H_1 J_1(\kappa)$ can be standing waves. Therefore, the necessary portion of the Rayleigh waves (a system of free waves) must be superimposed on the solution. By equating these standing waves to the Rayleigh displacements, the arbitrary coefficient of free waves becomes

$$A = \frac{F_y}{2F'(\kappa)} \tag{3.63}$$

Adding the system of free waves yields the final displacements:

$$\begin{aligned} u(r,\theta) = \frac{F_y \cos\theta}{2\pi G r} \Big[& i\pi\kappa H_1 H_1^{(2)}(\kappa) \\ & + 2k^2 \int_h^k \frac{\sigma^2 (2\sigma^2 - k^2) \sqrt{(\kappa^2 - k^2)(\kappa^2 - h^2)}}{F(\sigma) f(\sigma)} H_1^{(2)}(\sigma)\, d\sigma \Big] \end{aligned} \tag{3.64}$$

$$\begin{aligned} v(r,\theta) = \frac{F_y}{2\pi G r} \Bigg[& k^2 P \int_k^\infty \frac{\sigma \sqrt{(\kappa^2 - h^2)}}{F(\sigma)} H_0^{(2)}(\sigma)\, d\sigma \\ & + k^2 \int_h^k \frac{\sigma (2\sigma^2 - k^2)^2 \sqrt{(\kappa^2 - k^2)}}{F(\sigma) f(\sigma)} H_0^{(2)}(\sigma)\, d\sigma \Bigg] \end{aligned} \tag{3.65}$$

$$\begin{aligned} w(r,\theta) = \frac{F_y \sin\theta}{2\pi G r} \Big[& i\pi\kappa H_1 H_1^{(2)}(\kappa) \\ & + 2k^2 \int_h^k \frac{\sigma^2 (2\sigma^2 - k^2) \sqrt{(\kappa^2 - k^2)(\kappa^2 - h^2)}}{F(\sigma) f(\sigma)} H_1^{(2)}(\sigma)\, d\sigma \Big] \end{aligned} \tag{3.66}$$

These equations are complete solutions for displacements on the surface of an elastic half-space caused by a vertical harmonic force acting on the surface of a medium. For simplicity of numerical integration, the following non-dimensional parameter may be introduced:

$$\sigma = k\eta \tag{3.67}$$

$$h = k\gamma \tag{3.68}$$

After substituting for σ, the displacement relation becomes

$$\begin{aligned} u(r,\theta) = \frac{F_y \cos\theta}{2\pi G r} \Big[& i\pi\kappa H_1 H_1^{(2)}(\kappa) \\ & + 2k \int_\gamma^1 \frac{\eta^2 (2\eta^2 - 1) \sqrt{(\eta^2 - \gamma^2)(\eta^2 - 1)}}{(2\eta^2 - 1)^4 - 16\eta^4 (\eta^2 - \gamma^2)(\eta^2 - 1)} H_1^{(2)}(k\eta)\, d\eta \Big] \end{aligned} \tag{3.69}$$

$$v(r,\theta) = \frac{F_y}{2\pi Gr}\left[kP\int_1^{\infty}\frac{\eta^2\sqrt{\eta^2-\gamma^2}\,H_0^{(2)}(k\eta)}{\left(2\eta^2-1\right)^2-4\eta^2\sqrt{(\eta^2-\gamma^2)(\eta^2-1)}}d\eta\right.$$
$$\left.+k\int_{\gamma}^{1}\frac{\eta\left(2\eta^2-1\right)^2\sqrt{\eta^2-\gamma^2}}{\left(2\eta^2-1\right)^4-16\eta^4(\eta^2-\gamma^2)(\eta^2-1)}H_0^{(2)}(k\eta)\,d\eta\right] \tag{3.70}$$

and finally

$$w(r,\theta) = \frac{F_y\sin\theta}{2\pi Gr}\left[i\pi\kappa H_1 H_1^{(2)}(\kappa)\right.$$
$$\left.+2k\int_{\gamma}^{1}\frac{\eta^2\left(2\eta^2-1\right)\sqrt{(\eta^2-\gamma^2)(\eta^2-1)}}{\left(2\eta^2-1\right)^4-16\eta^4(\eta^2-\gamma^2)(\eta^2-1)}H_1^{(2)}(k\eta)\,d\eta\right] \tag{3.71}$$

The above displacements of the surface can be represented by

$$u(r,\theta) = \frac{F_y\cos\theta}{Gr}(u_3+iu_4)\exp(i\Omega t) \tag{3.72}$$
$$v(r,\theta) = \frac{F_y}{Gr}(v_3+iv_4)\exp(i\Omega t) \tag{3.73}$$
$$w(r,\theta) = \frac{F_y\sin\theta}{Gr}(w_3+iw_4)\exp(i\Omega t) \tag{3.74}$$

where

$$u_3+iu_4 = w_3+iw_4 \tag{3.75}$$

and

$$u_3+iu_4 = \left[i\pi\kappa H_1 H_1^{(2)}(\kappa)\right.$$
$$\left.+2k\int_{\gamma}^{1}\frac{\eta^2\left(2\eta^2-1\right)\sqrt{(\eta^2-\gamma^2)(\eta^2-1)}}{\left(2\eta^2-1\right)^4-16\eta^4(\eta^2-\gamma^2)(\eta^2-1)}H_1^{(2)}(k\eta)\,d\eta\right] \tag{3.76}$$

$$v_3+iv_4 = \left[kP\int_1^{\infty}\frac{\eta^2\sqrt{\eta^2-\gamma^2}H_0^{(2)}(k\eta)}{\left(2\eta^2-1\right)^2-4\eta^2\sqrt{(\eta^2-\gamma^2)(\eta^2-1)}}d\eta\right.$$
$$\left.+k\int_{\gamma}^{1}\frac{\eta\left(2\eta^2-1\right)^2\sqrt{\eta^2-\gamma^2}}{\left(2\eta^2-1\right)^4-16\eta^4(\eta^2-\gamma^2)(\eta^2-1)}H_0^{(2)}(k\eta)\,d\eta\right] \tag{3.77}$$

3.6 Numerical Integration for Displacements

To determine the displacements functions presented in Eqs. (3.76) and (3.77), the following three integrals must be evaluated by means of numerical methods:

$$A_1 = \int_1^{\infty} \frac{\eta^2 \sqrt{\eta^2 - \gamma^2}}{(2\eta^2 - 1)^2 - 4\eta^2 \sqrt{(\eta^2 - \gamma^2)(\eta^2 - 1)}} H_0^{(2)}(k\eta)\, d\eta \tag{3.78}$$

$$A_2 = \int_\gamma^{1} \frac{\eta (2\eta^2 - 1)^2 \sqrt{\eta^2 - \gamma^2}}{(2\eta^2 - 1)^4 - 16\eta^4 (\eta^2 - \gamma^2)(\eta^2 - 1)} H_0^{(2)}(k\eta)\, d\eta \tag{3.79}$$

$$A_3 = \int_\gamma^{1} \frac{\eta^2 (2\eta^2 - 1) \sqrt{(\eta^2 - \gamma^2)(\eta^2 - 1)}}{(2\eta^2 - 1)^4 - 16\eta^4 (\eta^2 - \gamma^2)(\eta^2 - 1)} H_1^{(2)}(k\eta)\, d\eta \tag{3.80}$$

3.7 Evaluation of A_1

Since the denominator of the integral has a root in the range of the integration, it cannot be evaluated by using the usual numerical integration methods. To find this principal value, the Longman (1958) technique can be applied. In this method, the range of $[1\text{to}2\eta_r - 1]$ and $[2\eta_r - 1\text{to}\infty)$ the integral is divided into two ranges and also $H_0^{(2)}(k\eta)$ is replaced by its components of the first and second kinds of Bessel functions $J_0(k\eta)$ and $Y_0(k\eta)$ Then,

$$\begin{aligned} A_1 = {} & P \int_1^{2\eta_r - 1} LJ(\eta)\, d\eta + \int_{2\eta_r - 1}^{\infty} LJ(\eta)\, d\eta \\ & -iP \int_1^{2\eta_r - 1} LY(\eta)\, d\eta + \int_{2\eta_r - 1}^{\infty} LY(\eta)\, d\eta \end{aligned} \tag{3.81}$$

where

$$\eta_r = \frac{\kappa}{k} \tag{3.82}$$

$$LJ(\eta) = \frac{\eta^2 \sqrt{\eta^2 - \gamma^2}}{(2\eta^2 - 1)^2 - 4\eta^2 \sqrt{(\eta^2 - \gamma^2)(\eta^2 - 1)}} J_0(k\eta) \tag{3.83}$$

and

$$LY(\eta) = \frac{\eta^2 \sqrt{\eta^2 - \gamma^2}}{(2\eta^2 - 1)^2 - 4\eta^2 \sqrt{(\eta^2 - \gamma^2)(\eta^2 - 1)}} Y_0(k\eta) \tag{3.84}$$

The first and third integrals in (3.82) can be evaluated by applying the Gauss quadrature method and the Longman (1958) method for determining the Cauchy principal value of the integral (see Appendix C). The second and fourth infinite integrals can be obtained by using another Longman (1956,1957) technique which is used for the integration of oscillatory integrands with infinite upper limits.

3.8 Evaluation of A_2

Dividing this integral into its real and imaginary parts, it becomes

$$A_2 = \int_{\gamma}^{1} \frac{\eta\left(2\eta^2-1\right)^2\sqrt{\eta^2-\gamma^2}}{\left(2\eta^2-1\right)^4-16\eta^4\left(\eta^2-\gamma^2\right)\left(\eta^2-1\right)}\left[J_0\left(k\eta\right)+iY_0\left(k\eta\right)\right]\,d\eta \quad (3.85)$$

Krylov (1962) has shown that these integrals cannot be correctly evaluated by using the usual Gaussian quadrature method, because the remainder error of this method is proportional to the n-th derivative of the integrand at the boundaries. In this case, the derivative at the lower boundary is infinite. To deal with these integrals, Krylov (1962) has given an appropriate Gauss quadrature for these integrals. See Appendix C.

Transforming the integral into a suitable form for this method, A_2 becomes

$$A_2 = \int_{\gamma}^{1} \sqrt{\eta^2-\gamma^2}\phi_1\left(\eta\right)\,d\eta \quad (3.86)$$

$$\phi_1\left(\eta\right) = \frac{\eta\left(2\eta^2-1\right)^2\sqrt{\eta^2-\gamma^2}}{\left(2\eta^2-1\right)^4-16\eta^4\left(\eta^2-\gamma^2\right)\left(\eta^2-1\right)}\left[J_0\left(k\eta\right)+iY_0\left(k\eta\right)\right] \quad (3.87)$$

3.9 Evaluation of A_3

This integral has a similar condition to the last integral except it has infinite n-th derivatives in both boundaries. An appropriate quadrature table for this kind of integral is given by Krylov (1962) and Stroud and Secrest (1966). A suitable form of A_3 would be

$$A_2 = \int_{\gamma}^{1} \sqrt{\left(\eta^2-\gamma^2\right)\left(1-\eta^2\right)}\phi_2\left(\eta\right)\,d\eta \quad (3.88)$$

where

$$\phi_2(\eta) = \frac{\eta^2 \left(2\eta^2 - 1\right)^2 \sqrt{(\eta^2 + \gamma^2)(\eta^2 + 1)}}{(2\eta^2 - 1)^4 - 16\eta^4(\eta^2 - \gamma^2)(\eta^2 - 1)} H_1^{(2)}(k\eta) \tag{3.89}$$

Appendix C provides additional information for the numerical techniques used for integration of the presented integrals.

3.10 Results and Discussion

Calculated results of υ_3 and υ_4 are compared with Barkan's polynomial approximation (1962) for low frequency factors. The differences between the present results and Barkan's at various frequency factors are tabulated in Table 3.1. From these comparisons, it can be seen that the two calculated functions υ_3 and υ_4 are in very good agreement with Barkan's results for a frequency factor range of 0.0–3.4, as was expected. These results are also compared with the approximate values of Elorduy et al. (1967) for a Poisson ratio of 0.25. However, their results are satisfactory only within a frequency factor range of 0.0–1.0 and outside this range, the difference is appreciable. This difference could be caused by errors in the numerical integrations of Duhamel's integral using the approximate results of Pekeris (1955b).

The variations of frequency factors for the four different Poisson ratios of 0.0, 0.25, 0.31, and 0.5 for a wide range of frequency factor are given in Figs. 3.2, 3.3, 3.4, 3.5, 3.6, 3.7, 3.8, and 3.9 (Tables 3.2, 3.3, 3.4, and 3.5).

Table 3.1 Difference percentage for the displacement functions υ_3 and υ_4 at various Poisson's ratios and frequency factors between approximate results of Barkan (1962) and the present theory

	$\nu = 0$		$\nu = 0.25$		$\nu = 0.5$	
k	υ_3 (%)	υ_4 (%)	υ_3 (%)	υ_4 (%)	υ_3 (%)	υ_4 (%)
0.5	0.115	0.104	0.393	0.848	0.197	0.279
1.5	0.089	0.073	1.320	9.372	0.485	0.217
2.5	1.717	0.966	0.491	230.600	0.281	1.219
3.1	23.330	0.845	12.930	23.710	3.840	2.790
4.1	362.080	7.650	528.040	28.140	69.610	12.710

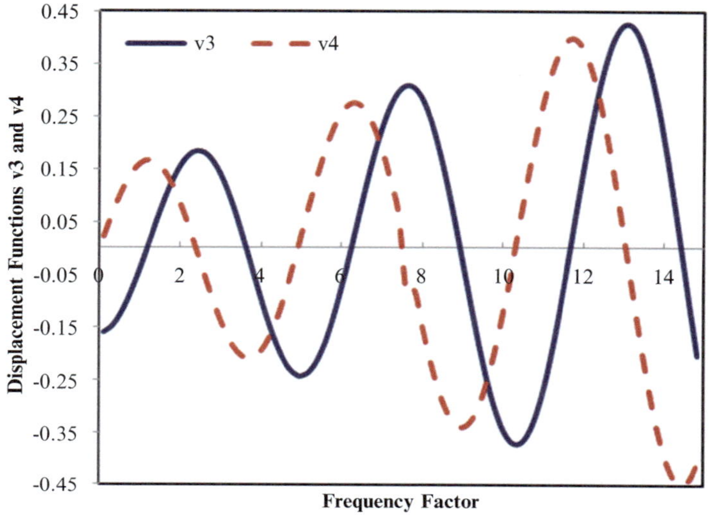

Fig. 3.2 Values of displacement functions v_3 and v_4 versus frequency factor for vertical harmonic point force on the surface of an elastic half-space with Poisson ratio of 0.0

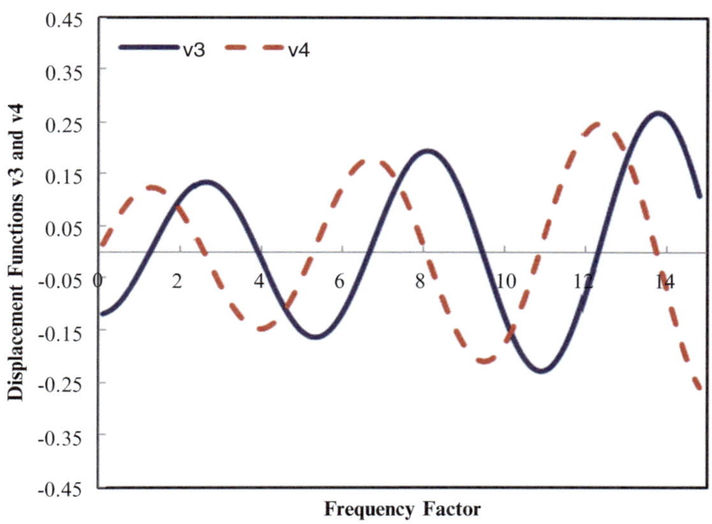

Fig. 3.3 Values of displacement functions v_3 and v_4 versus frequency factor for vertical harmonic point force on the surface of an elastic half-space with Poisson ratio of 0.25

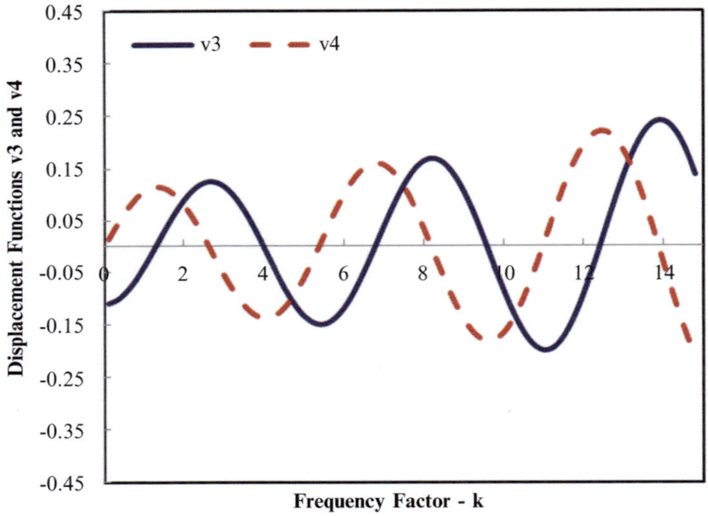

Fig. 3.4 Values of displacement functions v_3 and v_4 versus frequency factor for vertical harmonic point force on the surface of an elastic half-space with Poisson ratio of 0.31

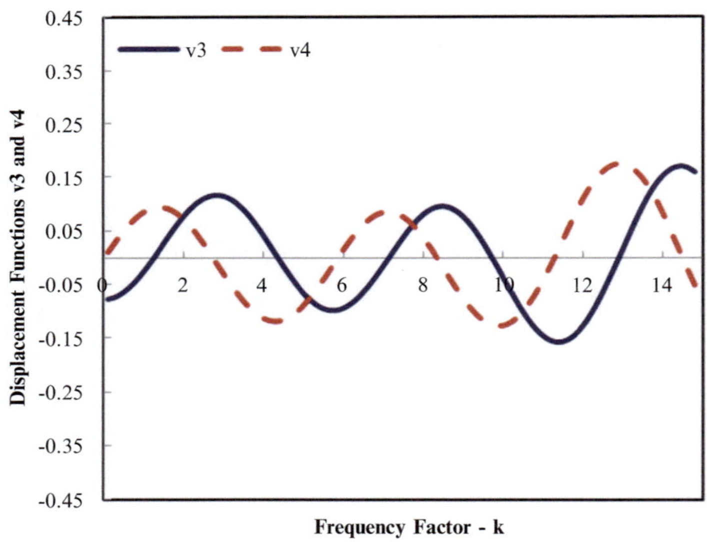

Fig. 3.5 Values of displacement functions v_3 and v_4 versus frequency factor for vertical harmonic point force on the surface of an elastic half-space with Poisson ratio of 0.50

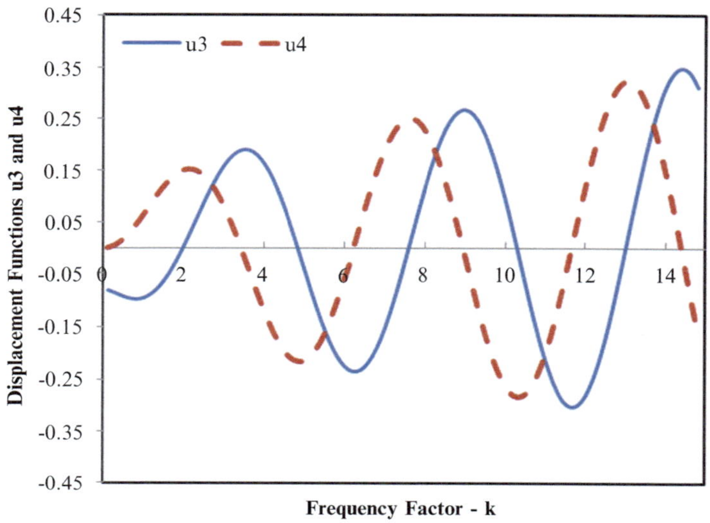

Fig. 3.6 Values of displacement functions u_3 and u_4 versus frequency factor for vertical harmonic point force on the surface of an elastic half-space with Poisson ratio of 0.0

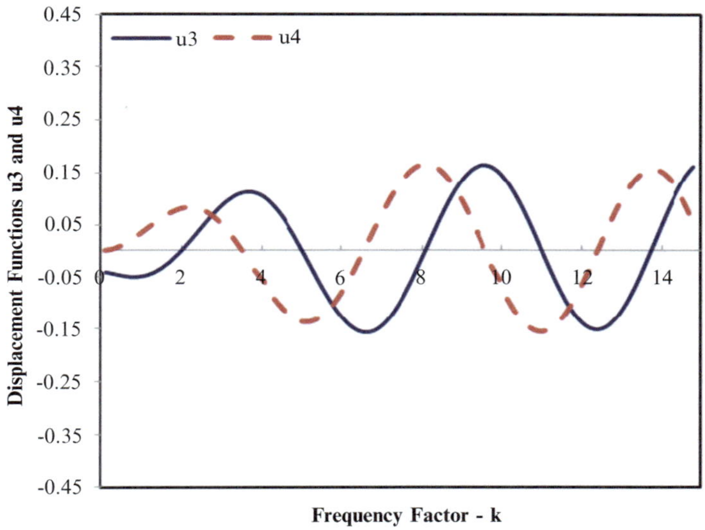

Fig. 3.7 Values of displacement functions u_3 and u_4 versus frequency factor for vertical harmonic point force on the surface of an elastic half-space with Poisson ratio of 0.25

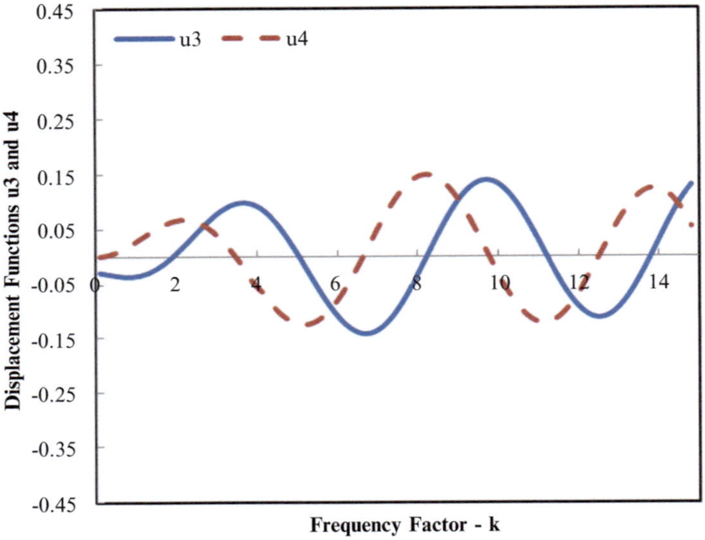

Fig. 3.8 Values of displacement functions u_3 and u_4 versus frequency factor for vertical harmonic point force on the surface of an elastic half-space with Poisson ratio of 0.31

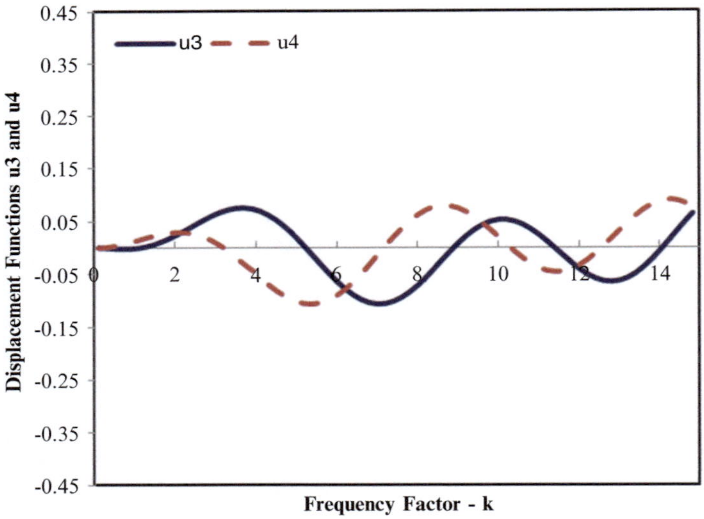

Fig. 3.9 Values of displacement functions u_3 and u_4 versus frequency factor for vertical harmonic point force on the surface of an elastic half-space with Poisson ratio of 0.50

Table 3.2 Values of displacement functions for different frequency factors for a vertical harmonic force

	Poisson ratio $\upsilon = 0.0$					
k	u_3	u_4	v_3	v_4	w_3	w_4
0.10	−0.080968	0.000780	−0.157725	−0.080968	0.021339	0.000780
0.30	−0.087042	0.006931	−0.146774	0.062606	−0.087042	0.006931
0.50	−0.093451	0.018763	−0.125486	0.099739	−0.093451	0.018763
0.70	−0.097476	0.035369	−0.095256	0.130264	−0.097476	0.035369
0.90	−0.097289	0.055460	−0.058014	0.152116	−0.097289	0.055460
1.10	−0.091673	0.077449	−0.016147	0.163771	−0.091673	0.077449
1.30	−0.079970	0.099555	0.027648	0.164339	−0.079970	0.099555
1.50	−0.062069	0.119919	0.070523	0.153624	−0.062069	0.119919
1.70	−0.038383	0.136729	0.109660	0.132141	−0.038383	0.136729
1.90	−0.009818	0.148336	0.142437	0.101093	−0.009818	0.148336
2.10	0.022292	0.153375	0.166601	0.062301	0.022292	0.153375
2.30	0.056269	0.150855	0.180400	0.018102	0.056269	0.150855
2.50	0.090195	0.140241	0.182704	−0.028791	0.090195	0.140241
2.70	0.122033	0.121498	0.173081	−0.075446	0.122033	0.121498
2.90	0.149755	0.095115	0.151826	−0.118892	0.149755	0.095115
3.10	0.171474	0.062087	0.119973	−0.156295	0.171474	0.062087
3.30	0.185570	0.023874	0.079217	−0.185135	0.185570	0.023874
3.50	0.190796	−0.017673	0.031844	−0.203361	0.190796	−0.017673
3.70	0.186373	−0.060413	−0.019402	−0.209526	0.186373	−0.060413
3.90	0.172052	−0.102033	−0.071479	−0.202883	0.172052	−0.102033
4.10	0.148149	−0.140181	−0.121223	−0.183443	0.148149	−0.140181
4.30	0.115547	−0.172605	−0.165531	−0.151989	0.115547	−0.172605
4.50	0.075663	−0.197288	−0.201549	−0.110047	0.075663	−0.197288
4.70	0.030381	−0.212573	−0.226848	−0.059807	0.030381	−0.212573
4.90	−0.018039	−0.217274	−0.239572	−0.004009	−0.018039	−0.217274
5.00	−0.042645	−0.215435	−0.240828	0.025003	−0.042645	−0.215435
5.20	−0.091009	−0.203260	−0.232763	0.083172	−0.091009	−0.203260
5.40	−0.136039	−0.180061	−0.210707	0.138684	−0.136039	−0.180061
5.60	−0.175192	−0.146744	−0.175510	0.188202	−0.175192	−0.146744
5.80	−0.206175	−0.104806	−0.128813	0.228647	−0.206175	−0.104806
6.00	−0.227079	−0.056264	−0.072966	0.257388	−0.227079	−0.056264
6.20	−0.236496	−0.003556	−0.010913	0.272402	−0.236496	−0.003556
6.40	−0.233612	0.050581	0.053969	0.272405	−0.233612	0.050581
6.60	−0.218268	0.103262	0.118052	0.256940	−0.218268	0.103262
6.80	−0.190985	0.151604	0.177659	0.226427	−0.190985	0.151604
7.00	−0.152951	0.192893	0.229276	0.182150	−0.152951	0.192893
7.20	−0.105973	0.224736	0.269749	0.126212	−0.105973	0.224736
7.40	−0.052387	0.245197	0.296480	0.061426	−0.052387	0.245197
7.60	0.005052	0.252921	0.307576	−0.088220	0.005052	0.252921
7.80	0.063321	0.247218	0.301981	−0.080757	0.063321	0.247218
8.00	0.119283	0.228115	0.279547	−0.150414	0.119283	0.228115

(continued)

Table 3.2 (continued)

	Poisson ratio $\upsilon = 0.0$					
k	u_3	u_4	v_3	v_4	w_3	w_4
8.20	0.169860	0.196372	0.241066	−0.213860	0.169860	0.196372
8.40	0.212204	0.153452	0.188241	−0.267415	0.212204	0.153452
8.60	0.243857	0.101461	0.123618	−0.307859	0.243857	0.101461
8.80	0.262892	0.043039	0.050452	−0.332628	0.262892	0.043039
9.00	0.268033	−0.018770	−0.027449	−0.339962	0.268033	−0.018770
9.20	0.258735	−0.080676	−0.105930	−0.329026	0.258735	−0.080676
9.40	0.235228	−0.139323	−0.180713	−0.299973	0.235228	−0.139323
9.60	0.198520	−0.191465	−0.247624	−0.253955	0.198520	−0.191465
9.80	0.150355	−0.234150	−0.302833	−0.193079	0.150355	−0.234150
10.00	0.093134	−0.264883	−0.343066	−0.120312	0.093134	−0.264883
10.20	0.029795	−0.281774	−0.365799	−0.039335	0.029795	−0.281774
10.40	−0.036338	−0.283649	−0.369408	0.045647	−0.036338	−0.283649
10.60	−0.101727	−0.270126	−0.353280	0.130123	−0.101727	−0.270126
10.80	−0.162811	−0.241656	−0.317861	0.209519	−0.162811	−0.241656
11.00	−0.216197	−0.199504	−0.264653	0.279443	−0.216197	−0.199504
11.20	−0.258848	−0.145703	−0.196155	0.335933	−0.258848	−0.145703
11.40	−0.288256	−0.082954	−0.115738	0.375673	−0.288256	−0.082954
11.60	−0.302581	−0.014487	−0.027487	0.396193	−0.302581	−0.014487
11.80	−0.300771	0.056097	0.064017	0.396011	−0.300771	0.056097
12.00	−0.282624	0.125021	0.153930	0.374737	−0.282624	0.125021
12.20	−0.248820	0.188530	0.237404	0.333108	−0.248820	0.188530
12.40	−0.200893	0.243095	0.309853	0.272970	−0.200893	0.243095
12.60	−0.141169	0.285609	0.367198	0.197198	−0.141169	0.285609
12.80	−0.072648	0.313560	0.406104	0.109556	−0.072648	0.313560
13.00	0.001144	0.325175	0.424164	0.014510	0.001144	0.325175
13.20	0.076335	0.319531	0.420048	−0.082999	0.076335	0.319531
13.40	0.148914	0.296616	0.393593	−0.177818	0.148914	0.296616
13.60	0.214940	0.257345	0.345828	−0.264851	0.214940	0.257345
13.80	0.270757	0.203527	0.278938	−0.339335	0.270757	0.203527
14.00	0.313194	0.137779	0.196164	−0.397101	0.313194	0.137779
14.20	0.339745	0.063402	0.101648	−0.434805	0.339745	0.063402
14.40	0.348710	−0.015786	0.000224	−0.450123	0.348710	−0.015786
14.60	0.339301	−0.095650	−0.102833	−0.441886	0.339301	−0.095650
14.80	0.311698	−0.171948	−0.202077	−0.410166	0.311698	−0.171948

Table 3.3 Values of displacement functions for different frequency factors for a vertical harmonic force

	Poisson ratio $\upsilon = 0.25$					
k	u_3	u_4	v_3	v_4	w_3	w_4
0.10	−0.040490	0.000390	−0.118365	0.014779	−0.040490	0.000390
0.30	−0.043583	0.003469	−0.111398	0.043482	−0.043583	0.003469
0.50	−0.046909	0.009409	−0.097642	0.069702	−0.046909	0.009409
0.70	−0.049116	0.017786	−0.077961	0.091925	−0.049116	0.017786
0.90	−0.049283	0.027997	−0.053457	0.108859	−0.049283	0.027997
1.10	−0.046779	0.039291	−0.025499	0.119502	−0.046779	0.039291
1.30	−0.041231	0.050818	0.004343	0.123196	−0.041231	0.050818
1.50	−0.032522	0.061678	0.034389	0.119664	−0.032522	0.061678
1.70	−0.020781	0.070973	0.062930	0.109028	−0.020781	0.070973
1.90	−0.006373	0.077863	0.088330	0.091805	−0.006373	0.077863
2.10	0.010125	0.081620	0.109110	0.068881	0.010125	0.081620
2.30	0.027954	0.081672	0.124030	0.041463	0.027954	0.081672
2.50	0.046222	0.077642	0.132161	0.011023	0.046222	0.077642
2.70	0.063945	0.069381	0.132935	−0.020787	0.063945	0.069381
2.90	0.080112	0.056976	0.126177	−0.052216	0.080112	0.056976
3.10	0.093734	0.040762	0.112127	−0.081513	0.093734	0.040762
3.30	0.103909	0.021309	0.091422	−0.107015	0.103909	0.021309
3.50	0.109872	−0.000605	0.065072	−0.127244	0.109872	−0.000605
3.70	0.111046	−0.24024	0.034408	−0.140983	0.111046	−0.024024
3.90	0.107078	−0.047861	0.001015	−0.147348	0.107078	−0.047861
4.10	0.097868	−0.070955	−0.033347	−0.145838	0.097868	−0.070955
4.30	0.083583	−0.092128	−0.066837	−0.136367	0.083583	−0.092128
4.50	0.064659	−0.110246	−0.097631	−0.119273	0.064659	−0.110246
4.70	0.041788	−0.124284	−0.124015	−0.095309	0.041788	−0.124282
4.90	0.015886	−0.133377	−0.144480	−0.065607	0.015886	−0.133377
5.00	0.002138	−0.135857	−0.152098	−0.049049	0.002138	−0.135857
5.20	−0.026191	−0.136395	−0.161504	−0.013554	−0.026191	−0.136395
5.40	−0.054547	−0.130850	−0.162622	0.023586	−0.054547	−0.130850
5.60	−0.081595	−0.119255	−0.155208	0.060446	−0.081595	−0.119255
5.80	−0.106021	−0.101941	−0.139466	0.095077	−0.106021	−0.101941
6.00	−0.126595	−0.079526	−0.116039	0.125613	−0.126595	−0.079526
6.20	−0.142240	−0.052896	−0.085983	0.150365	−0.142240	−0.052896
6.40	−0.152086	−0.023164	−0.050722	0.167913	−0.152086	−0.023164
6.60	−0.155519	0.008376	−0.011979	0.177187	−0.155519	0.008376
6.80	−0.152219	0.040311	0.028312	0.177526	−0.152219	0.040311
7.00	−0.142179	0.071177	0.068097	0.168719	−0.142179	0.071177
7.20	−0.125718	0.099526	0.105309	0.151023	−0.125718	0.099526
7.40	−0.103467	0.124001	0.137977	0.125159	−0.103467	0.124001
7.60	−0.076344	0.143403	0.164322	0.092284	−0.076344	0.143403
7.80	−0.045521	0.156752	0.182857	0.053935	−0.045521	0.156752
8.00	−0.012364	0.163339	0.192469	0.011961	−0.012364	0.163339

(continued)

Table 3.3 (continued)

	Poisson ratio $\upsilon = 0.25$					
k	u_3	u_4	v_3	v_4	w_3	w_4
8.20	0.021624	0.162767	0.192477	−0.031571	0.021624	0.162767
8.40	0.054881	0.154974	0.182679	−0.074472	0.054881	0.154974
8.60	0.085853	0.140243	0.163368	−0.114544	0.085853	0.140243
8.80	0.113076	0.119191	0.135327	−0.149692	0.113076	0.119191
9.00	0.135245	0.092748	0.099796	−0.178030	0.135245	0.092748
9.20	0.151285	0.062111	0.058419	−0.197981	0.151285	0.062111
9.40	0.160399	0.028694	0.013166	−0.208363	0.160399	0.028694
9.60	0.162116	−0.005941	−0.033758	−0.208455	0.162116	−0.005941
9.80	0.156318	−0.040155	−0.080026	−0.198040	0.156318	−0.040155
10.00	0.143249	−0.072315	−0.123297	−0.177425	0.143249	−0.072315
10.20	0.123506	−0.100873	−0.161339	−0.147437	0.123506	−0.100873
10.40	0.098017	−0.124440	−0.192138	−0.109386	0.098017	−0.124440
10.60	0.067999	−0.141859	−0.214007	−0.065011	0.067999	−0.141859
10.80	0.034899	−0.152262	−0.225673	−0.016399	0.034899	−0.152262
11.00	0.000329	−0.155119	−0.226350	0.034115	0.000329	−0.155119
11.20	−0.034015	−0.150266	−0.215781	0.084055	−0.034015	−0.150266
11.40	−0.066431	−0.137915	−0.194267	0.130933	−0.066431	−0.137915
11.60	−0.095299	−0.118653	−0.162656	0.172363	−0.095299	−0.118653
11.80	−0.119160	−0.093414	−0.122312	0.206189	−0.119160	−0.093414
12.00	−0.136788	−0.063435	−0.075053	0.230594	−0.136788	−0.063435
12.20	−0.147259	−0.030203	−0.023075	0.244189	−0.147259	−0.030203
12.40	−0.149992	0.004619	0.031159	0.246099	−0.149992	0.004619
12.60	−0.144790	0.039271	0.085030	0.236005	−0.144790	0.039271
12.80	−0.131849	0.071981	0.135891	0.214174	−0.131849	0.071981
13.00	−0.111756	0.101057	0.181199	0.181453	−0.111756	0.101057
13.20	−0.085461	0.124966	0.218639	0.139239	−0.085461	0.124966
13.40	−0.054241	0.142415	0.246243	0.089414	−0.054241	0.142415
13.60	−0.019633	0.152419	0.262489	0.034264	−0.019633	0.152419
13.80	0.016630	0.154354	0.266386	−0.023627	0.016630	0.154354
14.00	0.052712	0.147991	0.257530	−0.081500	0.052712	0.147991
14.20	0.086754	0.133512	0.236128	−0.136552	0.086754	0.133512
14.40	0.116972	0.111511	0.203002	−0.186075	0.116972	0.111511
14.60	0.141746	0.082967	0.159857	−0.227586	0.141746	0.082967
14.80	0.159697	0.049203	0.107716	−0.258953	0.159697	0.049203

Table 3.4 Values of displacement functions for different frequency factors for a vertical harmonic force

	Poisson ratio $\upsilon = 0.31$					
k	u_3	u_4	v_3	v_4	w_3	w_4
0.10	−0.000267	0.000156	−0.078713	0.010510	−0.000267	0.000156
0.30	−0.001408	0.001387	−0.074142	0.030864	−0.001408	0.001387
0.50	−0.002541	0.003755	−0.064898	0.049666	−0.002541	0.003755
0.70	−0.003114	0.007083	−0.051579	0.065945	−0.003114	0.007083
0.90	−0.002750	0.011113	−0.034856	0.078872	−0.002750	0.011113
1.10	−0.001189	0.015525	−0.015546	0.087776	−0.001189	0.015525
1.30	0.001725	0.019953	0.005410	0.092181	0.001725	0.019953
1.50	0.006040	0.024008	0.026987	0.091824	0.006040	0.024008
1.70	0.011703	0.027297	0.048127	0.086675	0.011703	0.027297
1.90	0.018557	0.029446	0.067789	0.076932	0.018557	0.029446
2.10	0.026355	0.030125	0.085000	0.063018	0.026355	0.030125
2.30	0.034765	0.029063	0.098899	0.045559	0.034765	0.029063
2.50	0.043393	0.026073	0.108784	0.025352	0.043393	0.026073
2.70	0.051800	0.021058	0.114141	0.003334	0.051800	0.021058
2.90	0.059520	0.014024	0.114671	−0.019471	0.059520	0.014024
3.10	0.066090	0.005086	0.110306	−0.041996	0.066090	0.005086
3.30	0.071072	−0.005537	0.101212	−0.063181	0.071072	−0.005537
3.50	0.074075	−0.017527	0.087784	−0.082028	0.074075	−0.017527
3.70	0.074777	−0.030478	0.070623	−0.097645	0.074777	−0.030478
3.90	0.072948	−0.043915	0.050514	−0.109293	0.072948	−0.043915
4.10	0.068459	−0.057308	0.028386	−0.116418	0.068459	−0.057308
4.30	0.061297	−0.070105	0.005265	−0.118684	0.061297	−0.070105
4.50	0.051569	−0.081746	−0.017767	−0.115984	0.051569	−0.081746
4.70	0.039499	−0.091701	−0.039633	−0.108449	0.039499	−0.091701
4.90	0.025424	−0.099486	−0.059304	−0.096447	0.025424	−0.099486
5.00	0.017769	−0.102432	−0.068022	−0.088943	0.017769	−0.102432
5.20	0.001537	−0.106222	−0.082712	−0.071389	0.001537	−0.106222
5.40	−0.015451	−0.107000	−0.093183	−0.051165	−0.015451	−0.107000
5.60	−0.032589	−0.104624	−0.098947	−0.029258	−0.032589	−0.104624
5.80	−0.049242	−0.099079	−0.099742	−0.006741	−0.049242	−0.099079
6.00	−0.064782	−0.090484	−0.095548	0.015279	−0.064782	−0.090484
6.20	−0.078608	−0.079085	−0.086585	0.035713	−0.078608	−0.079085
6.40	−0.090178	−0.065253	−0.073310	0.053542	−0.090178	−0.065253
6.60	−0.099034	−0.049464	−0.056393	0.067867	−0.099034	−0.049464
6.80	−0.104820	−0.032286	−0.036690	0.077956	−0.104820	−0.032286
7.00	−0.107306	−0.014351	−0.015199	0.083278	−0.107306	−0.014351
7.20	−0.106392	0.003669	0.006985	0.083530	−0.106392	0.003669
7.40	−0.102123	0.021094	0.028718	0.078656	−0.102123	0.021094
7.60	−0.094680	0.037260	0.048868	0.068854	−0.094680	0.037260
7.80	−0.084381	0.051557	0.066365	0.054563	−0.084381	0.051557
8.00	−0.071666	0.063446	0.080255	0.036452	−0.071666	0.063446

(continued)

Table 3.4 (continued)

	Poisson ratio $\upsilon = 0.31$					
k	u_3	u_4	v_3	v_4	w_3	w_4
8.20	−0.057077	0.072494	0.089747	0.015387	−0.057077	0.072494
8.40	−0.041237	0.078387	0.094253	−0.007609	−0.041237	0.078387
8.60	−0.024821	0.080946	0.093417	−0.031400	−0.024821	0.080946
8.80	−0.008532	0.080139	0.087140	−0.054790	−0.008532	0.080139
9.00	0.006940	0.076080	0.075584	−0.076584	0.006940	0.076080
9.20	0.020939	0.069027	0.059169	−0.095636	0.020939	0.069027
9.40	0.032879	0.059370	0.038558	−0.110912	0.032879	0.059370
9.60	0.042267	0.047614	0.014629	−0.121539	0.042267	0.047614
9.80	0.048731	0.034360	−0.011564	−0.126843	0.048731	0.034360
10.00	0.052031	0.020271	−0.038836	−0.126394	0.052031	0.020271
10.20	0.052074	0.006051	−0.065928	−0.120019	0.052074	0.006051
10.40	0.048920	−0.007594	−0.091559	−0.107823	0.048920	−0.007594
10.60	0.042780	−0.019983	−0.114493	−0.090184	0.042780	−0.019983
10.80	0.034002	−0.030498	−0.133589	−0.067740	0.034002	−0.030498
11.00	0.023066	−0.038607	−0.147862	−0.041366	0.023066	−0.038607
11.20	0.010550	−0.043890	−0.156528	−0.012134	0.010550	−0.043890
11.40	−0.002883	−0.046064	−0.159043	0.018732	−0.002883	−0.046064
11.60	−0.016527	−0.044991	−0.155134	0.049908	−0.016527	−0.044991
11.80	−0.029656	−0.040691	−0.144812	0.080032	−0.029656	−0.040691
12.00	−0.041563	−0.033339	−0.128379	0.107765	−0.041563	−0.033339
12.20	−0.051594	−0.023260	−0.106417	0.131852	−0.051594	−0.023260
12.40	−0.059173	−0.010916	−0.079765	0.151181	−0.059173	−0.010916
12.60	−0.063837	0.003117	−0.049486	0.164835	−0.063837	0.003117
12.80	−0.065253	0.018171	−0.016821	0.172134	−0.065253	0.018171
13.00	−0.063241	0.033515	0.016867	0.172675	−0.063241	0.033515
13.20	−0.057777	0.048395	0.050149	0.166349	−0.057777	0.048395
13.40	−0.049005	0.062061	0.081598	0.153348	−0.049005	0.062061
13.60	−0.037224	0.073805	0.109851	0.134165	−0.037224	0.073805
13.80	−0.022884	0.082994	0.133670	0.109572	−0.022884	0.082994
14.00	−0.006561	0.089103	0.152004	0.080590	−0.006561	0.089103
14.20	0.011061	0.091735	0.164033	0.048443	0.011061	0.091735
14.40	0.029227	0.090647	0.169210	0.014507	0.029227	0.090647
14.60	0.047137	0.085759	0.167291	−0.019750	0.047137	0.085759
14.80	0.063989	0.077164	0.158342	−0.052835	0.063989	0.077164

Table 3.5 Values of displacement functions for different frequency factors for a vertical harmonic force

	Poisson ratio $\upsilon = 0.50$					
k	u_3	u_4	v_3	v_4	w_3	w_4
0.10	−0.030812	0.000320	−0.108878	0.013491	−0.030812	0.000320
0.30	−0.033332	0.002843	−0.102636	0.039702	−0.033332	0.002843
0.50	−0.036027	0.007711	−0.090245	0.063713	−0.036027	0.007711
0.70	−0.037787	0.014576	−0.072489	0.084172	−0.037787	0.014576
0.90	−0.037856	0.022944	−0.050334	0.099928	−0.037856	0.022944
1.10	−0.035712	0.032198	−0.024983	0.110074	−0.035712	0.032198
1.30	−0.031047	0.041639	0.002184	0.114004	−0.031047	0.041639
1.50	−0.023758	0.050524	0.029683	0.111443	−0.023758	0.050524
1.70	−0.013947	0.058110	0.056000	0.102463	−0.013947	0.058110
1.90	−0.001908	0.063699	0.079674	0.087481	−0.001908	0.063699
2.10	0.011886	0.066683	0.099373	0.067241	0.011886	0.066683
2.30	0.026816	0.066577	0.113967	0.042775	0.026816	0.066577
2.50	0.042143	0.063057	0.122584	0.015349	0.042143	0.063057
2.70	0.057055	0.055981	0.124665	−0.013597	0.057055	0.055981
2.90	0.070706	0.045405	0.119990	−0.042525	0.070706	0.045405
3.10	0.082269	0.031586	0.108695	−0.069883	0.082269	0.031586
3.30	0.090977	0.014977	0.091268	−0.094179	0.090977	0.014977
3.50	0.096174	−0.003791	0.068529	−0.114065	0.096174	−0.003791
3.70	0.097349	−0.023937	0.041587	−0.128406	0.097349	−0.023937
3.90	0.094176	−0.044564	0.011787	−0.136341	0.094176	−0.044564
4.10	0.086531	−0.064705	−0.019355	−0.137337	0.086531	−0.064705
4.30	0.074513	−0.083369	−0.050232	−0.131210	0.074513	−0.083369
4.50	0.058443	−0.099591	−0.079227	−0.118148	0.058443	−0.099591
4.70	0.038855	−0.112483	−0.104796	−0.098702	0.038855	−0.112483
4.90	0.016481	−0.121282	−0.125548	−0.073761	0.016481	−0.121282
5.00	0.004524	−0.123953	−0.133744	−0.059592	0.004524	−0.123953
5.20	−0.020310	−0.125558	−0.145181	−0.028716	−0.020310	−0.125558
5.40	−0.045472	−0.121968	−0.149454	0.004253	−0.045472	−0.121968
5.60	−0.069841	−0.113139	−0.146215	0.037659	−0.069841	−0.113139
5.80	−0.092297	−0.099272	−0.135491	0.069796	−0.092297	−0.099272
6.00	−0.111776	−0.080810	−0.117698	0.098997	−0.111776	−0.080810
6.20	−0.127318	−0.058426	−0.093618	0.123717	−0.127318	−0.058426
6.40	−0.138124	−0.032995	−0.064365	0.142615	−0.138124	−0.032995
6.60	−0.143590	−0.005555	−0.031339	0.154624	−0.143590	−0.005555
6.80	−0.143347	0.022738	0.003848	0.159011	−0.143347	0.022738
7.00	−0.137282	0.050661	0.039446	0.155416	−0.137282	0.050661
7.20	−0.125545	0.076983	0.073656	0.143881	−0.125545	0.076983
7.40	−0.108553	0.100523	0.104717	0.124849	−0.108553	0.100523
7.60	−0.086972	0.120207	0.131000	0.099148	−0.086972	0.120207
7.80	−0.061690	0.135119	0.151088	0.067960	−0.061690	0.135119
8.00	−0.033778	0.144552	0.163852	0.032762	−0.033778	0.144552

(continued)

Table 3.5 (continued)

	Poisson ratio $\upsilon = 0.50$					
k	u_3	u_4	v_3	v_4	w_3	w_4
8.20	−0.004445	0.14804	0.168512	−0.004742	−0.004445	0.148040
8.40	0.025023	0.145384	0.164684	−0.042701	0.025023	0.145384
8.60	0.053314	0.136670	0.152400	−0.079210	0.053314	0.136670
8.80	0.079161	0.122263	0.132120	−0.112402	0.079161	0.122263
9.00	0.101397	0.102797	0.104709	−0.140545	0.101397	0.102797
9.20	0.119014	0.079147	0.071406	−0.162127	0.119014	0.079147
9.40	0.131213	0.052389	0.033764	−0.175935	0.131213	0.052389
9.60	0.137441	0.023755	−0.006418	−0.181123	0.137441	0.023755
9.80	0.137424	−0.005431	−0.047183	−0.177258	0.137424	−0.005431
10.00	0.131184	−0.033808	−0.086509	−0.164346	0.131184	−0.033808
10.20	0.119035	−0.060047	−0.122409	−0.142846	0.119035	−0.060047
10.40	0.101576	−0.082915	−0.153028	−0.113644	0.101576	−0.082915
10.60	0.079662	−0.101333	−0.176737	−0.078026	0.079662	−0.101333
10.80	0.054369	−0.114432	−0.192218	−0.037618	0.054369	−0.114432
11.00	0.026938	−0.121594	−0.198532	0.005690	0.026938	−0.121594
11.20	−0.001275	−0.122487	−0.195174	0.049831	−0.001275	−0.122487
11.40	−0.028873	−0.117082	−0.182100	0.092658	−0.028873	−0.117082
11.60	−0.054478	−0.105658	−0.159742	0.132047	−0.054478	−0.105658
11.80	−0.076802	−0.088793	−0.128991	0.166007	−0.076802	−0.088793
12.00	−0.094709	−0.067335	−0.091164	0.192770	−0.094709	−0.067335
12.20	−0.107271	−0.042367	−0.047944	0.210886	−0.107271	−0.042367
12.40	−0.113821	−0.015155	−0.001306	0.219299	−0.113821	−0.015155
12.60	−0.113980	0.012911	0.046574	0.217402	−0.113980	0.012911
12.80	−0.107685	0.040384	0.093423	0.205073	−0.107685	0.040384
13.00	−0.095196	0.065830	0.136976	0.182691	−0.095196	0.065830
13.20	−0.077083	0.087898	0.175086	0.151125	−0.077083	0.087898
13.40	−0.054204	0.105389	0.205832	0.111698	−0.054204	0.105389
13.60	−0.027670	0.117315	0.227612	0.066131	−0.027670	0.117315
13.80	0.001211	0.122948	0.239225	0.016468	0.001211	0.122948
14.00	0.030992	0.121865	0.239936	−0.035019	0.030992	0.121865
14.20	0.060156	0.113969	0.229516	−0.085934	0.060156	0.113969
14.40	0.087190	0.099497	0.208262	−0.133872	0.087190	0.099497
14.60	0.110658	0.079020	0.176988	−0.176529	0.110658	0.079020
14.80	0.129272	0.053414	0.136996	−0.211820	0.129272	0.053414

Chapter 4
Response of the Surface of an Elastic Half-Space Due to a Horizontal Harmonic Point Force

Abstract In this chapter, we review the response of an elastic half space to a horizontal harmonic point force. Due to complexities of the mathematical procedures for addressing the problem of dynamic response of surface of an elastic half-space subjected to concentrated shear force, the literature has been confined to the simpler problem of the axial symmetrical or two-dimensional case, where the concentrated line source acts on the surface. In the present chapter the three-dimensional problem is solved by following the procedure that was used in the last chapter and results are obtained by employing numerical techniques to retransform the displacements.

Keywords Harmonic point force • Elastic surface response • Rayleigh wave displacements • Waves on half elastic space • Displacements evaluation

Due to complexities of the mathematical procedures for addressing the problem of dynamic response of the surface of an elastic half-space subjected to concentrated shear force, the literature has been confined to the simpler problem of the axial symmetrical or two-dimensional case, where the concentrated line source acts on the surface. Among the investigators who considered the three-dimensional problem, Chao (1960) presented a solution for the displacement of an elastic half-space for a tangential applied force which varied with time as the Heaviside unit function. Papadopulos (1963) and Aggarwal and Ablow (1967) gave solutions to a class of three-dimensional pulse propagation in an elastic half-space but without any numerical results for a harmonic force.

In the present chapter, the three-dimensional problem is solved by following the procedure that was used in the last chapter, and results are obtained by employing numerical techniques to retransform the displacements. The system of coordinates and schematic diagram for the problem are shown in Fig. 4.1.

H.R. Hamidzadeh et al., *Wave Propagation in Solid and Porous Half-Space Media*,
DOI 10.1007/978-1-4614-9269-6_4,

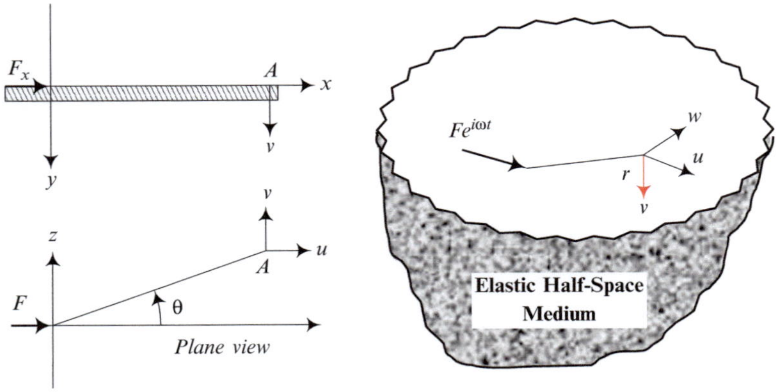

Fig. 4.1 Diagram for displacements of a point "A" on the surface of an elastic half-space due to a horizontal point force

4.1 Boundary Conditions for the Problem

The complete boundary conditions for this case are exactly similar to the last case, except for the stresses on surfaces, which are

$$\sigma_{yy} = 0 \quad |x| \geq 0 \ \hat{} \ |z| \geq 0 \tag{4.1a}$$

$$\tau_{yz} = 0 \quad |x| \geq 0 \ \hat{} \ |z| \geq 0 \tag{4.1b}$$

The point force is applied in the x direction, and the corresponding amplitude of stress due to this force is expressed by the Dirac delta function

$$\tau_{yx} = \tau\delta\,(x, z)\exp\,(i\,\Omega t) \tag{4.1c}$$

The physical conditions demand that the waves generated by this point force should die out as they travel through the half-space. If the amplitude of the harmonic tangential force is F_x, then the relation between this force and the corresponding shear stress is

$$F_x = \int_{-\infty}^{+\infty}\int_{-\infty}^{+\infty} \tau\delta\,(x, z)\,dxdz = \tau \tag{4.2}$$

The double complex Fourier transforms of the stresses on the surface with respect to x and z are

$$\overline{\sigma_{yy}} = 0 \tag{4.3a}$$

$$\overline{\tau_{yz}} = 0 \tag{4.3b}$$

$$\overline{\tau_{yx}} = \frac{1}{2\pi}\int_{-\infty}^{+\infty}\int_{-\infty}^{+\infty} F_x\delta\,(x,z)\exp(ipx)\exp(iqz)dzdx = \frac{F_x}{2\pi}\Bigg) \tag{4.3c}$$

4.2 Integral Representations of Displacements

$$D_\varepsilon = \frac{1}{2\Omega_2^2}\overline{\tau_{yx}}\left(4ipG_{\gamma 2}\right) \tag{4.4a}$$

$$D_x = \frac{Gpq}{\Omega_1^2\Omega_2^2}\overline{\tau_{yx}}\left(2\gamma_2^2+\Omega_2^2-2\gamma_1\gamma_2\right) \tag{4.4b}$$

$$D_z = \frac{-G}{2\Omega_1^2\Omega_2^2}\overline{\tau_{yx}}\left(\left(2\gamma_2^2+\Omega_2^2\right)\left(\Omega_2^2-2q^2\right)+4G\gamma_1\gamma_2q^2\right)\Big) \tag{4.4c}$$

Arbitrary values of Eq. 2.38 for this case can be evaluated by placing the transforms of the stresses into Eq. 2.46 and substituting the results into Eq. 2.42. These arbitrary values are

$$A_\varepsilon = \frac{2i\Omega_1^2\overline{\tau_{yx}}P\gamma_2}{G\Phi\,(p,q)} \tag{4.5a}$$

$$A_x = \frac{pq\left(2\gamma_2^2+\Omega_2^2-2\gamma_1\gamma_2\right)\overline{\tau_{yx}}}{G\Phi\,(p,q)} \tag{4.5b}$$

$$A_z = -\frac{\left(\left(2\gamma_2^2+\Omega_2^2\right)\left(\Omega_2^2-2q^2\right)+4\gamma_1\gamma_2q^2\right)\overline{\tau_{yx}}}{G\Phi\,(p,q)} \tag{4.5c}$$

The transforms of the displacements will be given by Eq. 2.38 after substituting for A_ε, A_x, and A_z from the above equations. By putting $y=0$ for surface displacements, these equations become

$$\overline{V} = \frac{-\overline{\tau_{yx}}}{G\Omega_2^2\Phi\,(p,q)}\{[ip\left(\left(2\gamma_2^2+\Omega_2^2\right)\left(\Omega_2^2-2q^2\right)+4\gamma_1\gamma_2q^2\right) \\ +2ipq^2\left(2\gamma_2^2+\Omega_2^2-2\gamma_1\gamma_2\right)]-2ip\gamma_1\gamma_2\Omega_2^2\} \tag{4.6a}$$

$$\overline{U} = \frac{-\overline{\tau_{yx}}}{2G\Phi\,(p,q)}\Bigg[4p^2\gamma_2-\frac{4p^2q^2\gamma_2\left(2\gamma_2^2+\Omega_2^2-2\gamma_1\gamma_2\right)}{\gamma_2^2\Omega_2^2} \\ +\frac{2\left(\Omega_2^2-p^2\right)\gamma_2}{\gamma_2^2\Omega_2^2}\left(\left(2\gamma_2^2+\Omega_2^2\right)\left(\Omega_2^2-2q^2\right)+4\gamma_1\gamma_2q^2\right)\Bigg] \tag{4.6b}$$

$$\overline{W} = -\frac{\overline{\tau_{yx}}}{2G\Phi(p,q)}\left[4pq\gamma_2 - \frac{2pq\gamma_2}{\gamma_2^2\Omega_2^2}\left(\left(2\gamma_2^2+\Omega_2^2\right)\left(\Omega_2^2-2q^2\right)+4\gamma_1\gamma_2 q^2\right)\right.$$
$$\left. + \frac{4\gamma_2 pq\left(\Omega_2^2-q^2\right)}{\gamma_2^2\Omega_2^2}\left(2\gamma_2^2+\Omega_2^2-2\gamma_1\gamma_2\right)\right] \quad (4.6c)$$

By inverting these equations into the *x*- and *z*-domain displacements of the free surface due to the tangential force on the surface of the elastic half-space media, they can be expressed by

$$V(x,z) = \frac{1}{2\pi}\int_{-\infty}^{+\infty}\int_{-\infty}^{+\infty}\frac{-F_x}{2\pi\Phi(p,q)G}ip\left(\Omega_2^2+\gamma_2^2-2\gamma_1\gamma_2\right)\exp(-iqz)\,dpdq \quad (4.7a)$$

$$U(x,z) = \frac{1}{2\pi}\int_{-\infty}^{+\infty}\int_{-\infty}^{+\infty} - \frac{F_x/2}{2\pi\Phi(p,q)G}\left[\frac{2\gamma_2^2\Omega_2^4 - 2\Omega_2^2 q^2\left(\Omega_2^2+4\gamma_2^2-4\gamma_1\gamma_2\right)}{\gamma_2\Omega_2^2}\right]$$
$$\exp(-ipx)\exp(-iqz)\,dpdq \quad (4.7b)$$

$$W(x,z) = \frac{1}{2\pi}\int_{-\infty}^{+\infty}\int_{-\infty}^{+\infty}\frac{-2pq\left(4\gamma_2^2+\Omega_2^2-4\gamma_1\gamma_2\right)F_x}{4\pi G\gamma_2\Phi(p,q)}$$
$$\exp(-ipx)\exp(-ipz)\,dqdq \quad (4.7c)$$

By introducing polar variables for *p*, *q*, *x*, and *z* as

$$p = \delta\sin\varphi \quad (4.8a)$$

$$q = \delta\cos\varphi \quad (4.8b)$$

and

$$x = r\cos\theta \quad (4.9a)$$

$$z = r\sin\theta \quad (4.9b)$$

the displacements then take the form

$$V(r,\theta) = \frac{-iF_x}{4\pi^2 G}\int_0^{\infty}\frac{\delta^2\left(2\delta^2-\Omega_2^2-2\sqrt{\left(\delta^2-\Omega_1^2\right)\left(\delta^2-\Omega_2^2\right)}\right.}{\Phi(\delta)}$$
$$\int_0^{2\pi}\sin\phi\exp\left[-i\delta r^*\sin(\theta+\phi)\right]d\phi d\delta \quad (4.10a)$$

$$U(r,\theta) = \frac{-F_x}{4\pi^2 G}\int_0^\infty \frac{\Omega_2^2\delta\sqrt{\delta^2-\Omega_2^2}}{\Phi(\delta)}\int_0^{2\pi}\exp\left[-i\delta r\sin(\theta+\phi)\right]d\phi d\delta$$

$$-\frac{F_x}{4\pi^2 G}\int_0^\infty \frac{-\delta^3\left(4\delta^2-3\Omega_2^2-4\sqrt{\left(\delta^2-\Omega_2^2\right)\left(\delta^2-\Omega_1^2\right)}\right)}{\sqrt{\delta^2-\Omega_2^2}\Phi(\delta)}$$

$$\int_0^{2\pi}\cos^2\phi\exp\left[-i\delta r\sin(\theta+\phi)\right]d\phi d\delta \tag{4.10b}$$

$$W(r,\theta) = \frac{-F_x}{4\pi^2 G}\int_0^\infty \frac{-\delta^3\left(4\delta^2-3\Omega_2^2-4\sqrt{\left(\delta^2\Omega_2^2\right)\left(\delta^2-\Omega_2^2\right)}\right)}{\sqrt{\delta^2-\Omega_2^2}\Phi(\delta)}$$

$$\int_0^{2\pi}\sin\phi\cos\phi\exp\left[-i\delta r\sin(\theta+\phi)\right]d\phi d\delta \tag{4.10c}$$

Integrals with respect to ϕ which appear at the end of these equations are given by the following relations in terms of Bessel functions:

$$S_1 = \int_0^{2\pi}\sin\phi\exp\left[-i\delta r\sin(\phi+\theta)\right]d\phi = 2\pi i\cos\theta\ J_1(\delta r) \tag{4.11a}$$

$$S_2 = \int_0^{2\pi}\exp\left[-i\delta r\sin(\phi+\theta)\right]d\phi = 2\pi J_0(\delta r) \tag{4.11b}$$

$$S_3 = \int_0^{2\pi}\sin 2\phi\exp\left[-i\delta r\sin(\phi+\theta)\right]d\phi = 2\pi J_2(\delta r)\sin 2\theta \tag{4.11c}$$

$$S_4 = \int_0^{2\pi}\cos^2\phi\exp\left[-i\delta r\sin(\phi+\theta)\right]d\phi = \pi J_0(\delta r)\cos 2\theta \tag{4.11d}$$

By employing the variables introduced by Eq. 3.14 and substituting for the above integrals, the displacements will take the single integral form:

$$V = -\frac{F_x\cos 2\theta}{4\pi G r}\int_0^\infty \frac{\varepsilon^2\left(2\varepsilon^2-k^2-2\sqrt{\left(\varepsilon^2-k^2\right)\left(\varepsilon^2-h^2\right)}\right)}{F(\varepsilon)}J_1(\varepsilon)\,d\varepsilon \tag{4.12a}$$

$$U = -\frac{-F_x}{4\pi Gr}\left[\int_0^\infty \varepsilon \frac{k^2\sqrt{\varepsilon^2 - k^2}}{F(\varepsilon)} J_O(\varepsilon)\, d\varepsilon - \int_0^\infty \frac{\varepsilon}{\sqrt{\varepsilon^2 - k^2}} J_O(\varepsilon)\, d\varepsilon\right]$$

$$+ \frac{F_x \cos 2\theta}{4\pi Gr}\left[\int_0^\infty \varepsilon \frac{k^2\sqrt{\varepsilon^2 - k^2}}{F(\varepsilon)} J_2(\varepsilon)\, d\varepsilon + \int_0^\infty \frac{\varepsilon}{\sqrt{\varepsilon^2 - k^2}} J_2(\varepsilon)\, d\varepsilon\right] \quad (4.12b)$$

$$W = -\frac{F_x \sin 2\theta}{4\pi Gr}\left[\int_0^\infty \varepsilon \frac{k^2\sqrt{\varepsilon^2 - k^2}}{F(\varepsilon)} J_2(\varepsilon)\, d\varepsilon + \int_0^\infty \frac{\varepsilon}{\sqrt{\varepsilon^2 - k^2}} J_2(\varepsilon)\, d\varepsilon\right] \quad (4.12c)$$

where $F(\varepsilon)$ is the Rayleigh function (see Eq. 3.16).

4.3 Rayleigh Wave Displacements

The free motion of the surface of a half-space in this kind of boundary problem has the same characteristics which were discussed in Sect. 3.4. If the ratio of the transformed shear stress in the x direction and the Rayleigh function is A, then this value can be arbitrary if the surface becomes stress-free at the zero value of the Rayleigh function (given in Sect. 4.3). Thus,

$$A = \frac{\overline{\tau_{yx}}}{F(\kappa)} \quad (4.13)$$

The corresponding displacement in this case can be obtained by substituting the above ratio into Eqs. 4.12a and 4.12b. Completing these substitutions, the Rayleigh wave displacements become

$$V_R = -\frac{A \cos\theta}{Gr}\kappa^2 \left(2\kappa^2 - k^2 - 2\sqrt{(\kappa^2 - k^2)(\kappa^2 - h^2)}\right) J_1(\kappa) \quad (4.14a)$$

$$U_R = -\frac{Ak^2}{2Gr}\left(\kappa\sqrt{\kappa^2 - k^2}\,(J_O(\kappa) - \cos 2\theta\ J_2(\kappa)\right) \quad (4.14b)$$

$$W_R = -\frac{Ak^2}{2Gr}\kappa\sqrt{\kappa^2 - k^2} J_2(\kappa) \sin 2\theta \quad (4.14c)$$

4.4 Evaluation of Displacements

In order to manipulate the contour integration, it will be much easier to replace the Bessel functions in Eqs. 4.12a and 4.12b by their integral representation. These integrals for J_0 and J_1 have been shown in Eq. 3.25 and are

$$J_O(\varepsilon) = -\frac{1}{\pi}\int_0^\infty [\exp(i\varepsilon\cosh u) - \exp(-i\varepsilon\cosh u)]\,du \tag{4.15a}$$

$$J_1(\varepsilon) = -\frac{1}{\pi}\int_0^\infty [\exp(i\varepsilon\cosh u) + \exp(-i\varepsilon\cosh u)]\cosh u\,du \tag{4.15b}$$

Using these two equations by means of recurrence formula, $J_2(\varepsilon)$ will be

$$J_2(\varepsilon) = \frac{1}{\pi}\int_0^\infty [\exp(i\varepsilon\cosh u) - \exp(-i\varepsilon\cosh u)]\cosh 2u\,du \tag{4.15c}$$

By replacing the Bessel functions by the right-hand side of the above equations, Eqs. 4.12a and 4.12b become

$$V=\frac{F_x\cos\theta}{2\pi^2 Gr}\int_0^\infty \cosh u\int_{-\infty}^\infty \varepsilon^2\frac{2\varepsilon^2-k^2-2\sqrt{(\varepsilon^2-k^2)(\varepsilon^2-h^2)}}{(2\varepsilon^2-k^2)^2-4\varepsilon^2\sqrt{(\varepsilon^2-k^2)(\varepsilon^2-h^2)}}\exp(i\varepsilon\cosh u)\,d\varepsilon du \tag{4.16a}$$

$$U = \frac{iF_x}{2\pi^2 Gr}\Bigg[\int_0^\infty du\int_0^\infty \varepsilon\frac{k^2\sqrt{\varepsilon^2-k^2}}{F(\varepsilon)}\exp(i\varepsilon\cosh u)\,d\varepsilon$$

$$-\int_0^\infty du\int_{-\infty}^\infty \varepsilon\frac{k^2\sqrt{\varepsilon^2-k^2}}{F(\varepsilon)}\exp(i\varepsilon\cosh u)\,d\varepsilon\Bigg] + \frac{iF_x\cos 2\theta}{4\pi^2 Gr}$$

$$\Bigg[\int_0^\infty \cosh 2u\int_{-\infty}^\infty \varepsilon\frac{k^2\sqrt{\varepsilon^2-k^2}}{F(\varepsilon)}\exp(i\varepsilon\cosh u)\,d\varepsilon\,du + \int_0^\infty \cosh 2u$$

$$\int_{-\infty}^\infty \varepsilon\frac{1}{\sqrt{\varepsilon^2-k^2}}\exp(i\varepsilon\cosh u)\,d\varepsilon\,du\Bigg] \tag{4.16b}$$

$$W = \frac{-iF_x\sin 2\theta}{4\pi^2 Gr}\Bigg[\int_0^\infty \cosh 2u\int_{-\infty}^\infty \varepsilon\frac{k^2\sqrt{\varepsilon^2-k^2}}{F(\varepsilon)}\exp(i\varepsilon\cosh u)\,d\varepsilon\,du$$

$$+\int_0^\infty \cosh 2u\int_{-\infty}^\infty \varepsilon\frac{1}{\sqrt{\varepsilon^2-k^2}}\exp(i\varepsilon\cosh u)\,d\varepsilon\,du\Bigg] \tag{4.16c}$$

Infinite integrals with respect to ε will be given by performing the $(-i\, d/dx)$ operation on integrals I_2, I_3, and I_4 and then replacing x by $cosh(u)$. These three integrals that were evaluated by Lamb (1904) using a contour integration approach are presented in Appendix B. After simplifications, these operations then yield

$$V = \frac{F_x \cos\theta}{2\pi^2 Gr} \int_0^{\infty} \left[2\pi H_1 \kappa \sin(\kappa \cosh u) + 4k^2 \int_h^{\kappa} \varepsilon^2 \frac{(2\varepsilon^2 - k^2)\sqrt{\varepsilon^2 - k^2}\sqrt{\varepsilon^2 - h^2}}{F(\varepsilon) f(\varepsilon)} \right.$$

$$\left. \exp(-i\varepsilon \cosh u)\, d\varepsilon \right] * \cosh u du \tag{4.17a}$$

$$U = \frac{iF_x}{4\pi^2 Gr} \left\{ \int_0^{\infty} \left[2\pi \kappa H_3 \sin(\kappa \cosh u) - 2k^2 \int_h^{k} \varepsilon \frac{(2\varepsilon^2 - k^2)^2 \beta}{F(\varepsilon) f(\varepsilon)} \right. \right.$$

$$\left. \exp(-i\varepsilon \cosh u)\, d\varepsilon - 2k^2\, \mathbb{C}P \int_k^{\infty} \varepsilon \frac{\sqrt{\varepsilon^2 - k^2}}{F(\varepsilon)} \exp(-i\varepsilon \cosh u)\, d\varepsilon \right] du$$

$$\left. - \int_0^{\infty} \pi k H_1^{(2)}(k \cosh u)\, du \right\} + \frac{iF_x \cos 2\theta}{4\pi^2 Gr} \left\{ \int_0^{\infty} \left[2\pi \kappa H_3 \sin(\kappa \cosh u) - 2k^2 \right. \right.$$

$$\left. \int_h^{k} \varepsilon \frac{(2\varepsilon^2 - k^2)^2 \beta}{F(\varepsilon) f(\varepsilon)} \exp(-i\varepsilon \cosh)\, d\varepsilon - 2k^2\, \mathbb{C}P \int_k^{\infty} \varepsilon \frac{\sqrt{\varepsilon^2 - k^2}}{F(\varepsilon)} \exp(-i\varepsilon \cosh u)\, d\varepsilon \right]$$

$$\left. \cosh 2u du + \int_0^{\infty} \pi k H_1^{(2)}(k \cosh u) \cosh 2u du \right\} \tag{4.17b}$$

$$W = \frac{iF_x \sin\theta}{4\pi^2 Gr} \left\{ \int_0^{\infty} \left[2\pi H_3 \kappa \sin(k \cosh u) - 2k^2 \int_h^{k} \varepsilon \frac{(2\varepsilon^2 - k^2)^2}{F(\varepsilon) f(\varepsilon)} \sqrt{\varepsilon^2 - k^2} \right. \right.$$

$$\left. \exp(-i\varepsilon \cosh u)\, d\varepsilon - 2k^2\, \mathbb{C}P \int_0^{\infty} \varepsilon \frac{\sqrt{\varepsilon^2 - k^2}}{F(\varepsilon)} \exp(-i\varepsilon \cosh u)\, d\varepsilon \right] \cosh 2u du$$

$$\left. + \int_0^{\infty} \pi k H_1^{(2)}(k \cosh u) \cosh u 2u\, du \right\} \tag{4.17c}$$

To simplify these displacements, let us determine the following eight integrals Q_1 to Q_8, which appear in the displacement equations:

$$Q_1 = \int_0^{\infty} 2\pi \kappa H_1 \sin(\kappa \cosh u) \cosh u\, du \tag{4.18a}$$

This integral has been evaluated before in terms of the Bessel function of the second kind, first order in Eq. 3.28 as

$$Q_1 = -\pi^2 \kappa H_1 y_1 (\kappa) \tag{4.18b}$$

$$Q_2 = \int_0^\infty 2\pi\kappa H_3 \sin (\kappa \cosh u)\, du \tag{4.19a}$$

The above integral is similar to integral (3.29); thus

$$Q_2 = \pi^2 \kappa H_3 J_0 (\kappa) \tag{4.19b}$$

$$Q_3 = \int_0^\infty \exp (-i\varepsilon \cosh u)\, du = -i\frac{\pi}{2} H_0^{(2)} (\varepsilon) \tag{4.20}$$

(See Eq. 3.30).

$$Q_4 = \int_0^\infty \exp (-i\varepsilon \cosh u) \cosh u\, du = -\frac{\pi}{2} H_1^{(2)} (\varepsilon) \tag{4.21}$$

(See Eq. 3.31).

$$Q_5 = \int_0^\infty 2\pi\kappa H_3 \sin (\kappa \cosh u) \cosh 2u du = -\int_0^\infty 2\pi\kappa H_3 \sin (\kappa \cosh u) \left(1 - 2\cos^2 hu\right) du$$

$$= -\int_0^\infty 2\pi\kappa H_3 \sin (\kappa \cosh u) + 2\frac{\partial^2}{\partial\kappa^2} \int_0^\infty 2\pi\kappa H_3 \sin (\kappa \cosh u)\, du$$

$$= -\left(1 - 2\frac{\partial^2}{\partial\kappa^2}\right) \int_0^\infty 2\pi\kappa H_3 \sin (\kappa \cosh u)\, du = -\left(1 - \frac{2\partial^2}{\partial\kappa^2}\right)$$

$$Q_2 = \left(1 - \frac{2\partial^2}{\partial\kappa^2}\right) \pi^2 \kappa H_3 J_O (\kappa)$$

$$= -\pi^2 \kappa H_3 \left(J_O (\kappa) - 2\frac{\partial}{\partial\kappa} J_1 (\kappa)\right) \tag{4.22a}$$

From recurrence formulae for Bessel functions, $J_2(\kappa)$ can be presented as $J_2 (\kappa) = J_0 (\kappa) - 2\frac{\partial}{\partial\kappa} - J_1 (\kappa)$. Considering the last result, Q_5 can then be written as

$$Q_5 = -\pi^2\kappa \; H_3 J_2 (\kappa) \tag{4.22b}$$

$$Q_6 = \int_0^\infty \exp(-i\varepsilon \cosh u) \cosh 2u \, du = \left(1 - \frac{\partial^2}{\partial\varepsilon^2}\right) \int_0^\infty \exp(-i\varepsilon \cosh u) \, du \tag{4.23a}$$

The last integral was given by Q_3; thus

$$Q_6 = \left(1 - \frac{\partial^2}{\partial\varepsilon^2}\right) Q3 = \left(1 - \frac{\partial^2}{\partial\varepsilon^2}\right) i\frac{\pi}{2} H_0^{(2)}(\varepsilon) = i\frac{\pi}{2}\left(H_0^{(2)}(\varepsilon) + 2\frac{\partial}{\partial\varepsilon} H_1^{(2)}(\varepsilon)\right) \tag{4.23a}$$

From a recurrence formula, we have

$$2\frac{\partial}{\partial\varepsilon} H_1^{(2)}(\varepsilon) = H_0^{(2)}(\varepsilon) - H_2^{(2)}(\varepsilon) = 2H_0^{(2)}(\varepsilon) - \frac{2}{\varepsilon} H_1^{(2)}(\varepsilon)$$

By substituting the derivative of the Hankel function of the second kind from the above relation into the previous one, J6 becomes

$$Q_6 = i\frac{\pi}{2}\left(3H_0^{(2)}(\varepsilon) - \frac{2}{\varepsilon} H_1^{(2)}(\varepsilon)\right) \tag{4.23b}$$

$$Q_7 = \int_0^\infty \pi k \; H_1^{(2)}(k \cosh u) \, du \tag{4.24a}$$

Introducing $z = cosh(u)$, then

$$Q_7 = \pi \; k \int_1^\infty H_1^{(2)}(kz) \frac{dz}{\sqrt{z^2 - 1}} \tag{4.24b}$$

Separating $H_1^{(2)}$ into its Bessel function components of the first and the second kind results in

$$Q_7 = \pi k \int_1^\infty J_1(kz) \frac{dz}{\sqrt{z^2 - 1}} - i\pi k \int_1^\infty y_1(kz) \frac{1}{\sqrt{z^2 - 1}} dz \tag{4.24c}$$

These two integrals are similar to the Hankel transform of the following functions, which are given in the Tables of Integral Transforms by Bateman and Erdelyi (1954) (pages 24 and 102):

$$f(x) = \begin{cases} x^{-1/2}(x^2-1)^{-1/2}, & 1 < x < \infty \\ 0 & 0 < x < 1 \end{cases} \tag{4.25}$$

The Hankel transforms of this function with respect to J_1 and Y_1 are

$$\int_0^\infty f(x) J_1(kx)(xk)^{1/2} dx = \int_1^\infty \frac{k^{1/2} J_1(kz)}{(z^2-1)^{1/2}} dz = \frac{1}{2}\pi\, k^{1/2} J_{1/2}\left(\frac{1}{2}k\right) y_{1/2}\left(\frac{1}{2}k\right) \tag{4.26}$$

$$\int_0^\infty f(x) Y_1(kx)(xk)^{1/2} dx = \int_1^\infty \frac{k^{1/2} Y_1(kz)}{(z^2-1)^{1/2}} dz$$

$$= \frac{1}{4}\pi k^{1/2} \left\{ \left[J_{1/2}\left(\frac{1}{2}k\right) \right]^2 - \left[Y_{1/2}\left(\frac{1}{2}k\right) \right]^2 \right\} \tag{4.27}$$

By substituting for the components of Q_7 in (4.24c) from the above equations, Q_7 becomes

$$Q_7 = -1/2\pi^2 k J_{1/2}(k/2)\, Y_{1/2}(k/2) - i\frac{\pi^2 k}{4} \left\{ \left[J_{1/2}(k/2) \right]^2 - \left[Y_{1/2}(k/2) \right]^2 \right\} \tag{4.28}$$

$$Q_8 = \int_0^\infty \pi k H_1^{(2)}(k \cosh u) \cosh 2u du$$

After separating $cosh(2u)$,

$$\int_0^\infty \pi k H_1^{(2)}(k \cosh u)\left(2\cosh^2 u - 1\right) du = 2\pi k \int_0^\infty H_1^{(2)}(k \cosh u) \cosh^2 u du -$$
$$\int_0^\infty \pi H_1^{(2)}(k \cosh u)\, du = -2\pi k \frac{\partial}{\partial k} \int_0^\infty H_0^{(2)}(k \cosh u) \cosh u\, du$$
$$- \int_0^\infty \pi k H_1^{(2)}(k \cosh u)\, du \tag{4.29}$$

The second integral is identical to Q_7, except for the negative sign, and the first integral can be written by substituting $z = cosh(u)$

$$-2\pi k \frac{\partial}{\partial k} \left[\int_0^\infty J_0(kz) \frac{z}{\sqrt{z^2-1}} dz - i \int_0^\infty y_0(kz) \frac{z}{\sqrt{z2-1}} dz \right] \tag{4.30}$$

These integrals are also given by integrals (28) and (33) on pages 25 and 103 in the Tables of Integral Transforms by Bateman and Erdelyi (1954). After substituting for all these integrals and simplifying, Q_8 becomes

$$Q_8 = -2\Gamma(1/2)\,k\pi^{1/2}\left[-\left(\frac{\sin k}{k}+\frac{\cos k}{k^2}\right)+i\left(\frac{\sin k}{k^2}-\frac{\cos k}{k}\right)\right]+ \\ 1/2\pi^2 k\left\{J_{1/2}(k/2)\,y_{1/2}\Big(k/2+i/2\left[J_{1/2}^2(k/2)-y_{1/2}^2(k/2)\right]\right\} \quad (4.31)$$

Thus, the displacement functions can be simplified by substituting all eight integrals into Eqs. 4.17a–4.17c, which yields

$$V = \frac{F_x\cos\theta}{2\pi^2 Gr}\left[-\pi^2\kappa H_1 Y_1(\kappa) - 4k^2\int_h^k \varepsilon^2\frac{(2\varepsilon^2-k^2)}{F(\varepsilon)\,f(\varepsilon)}\frac{\pi}{2}H_1^{(2)}(\varepsilon)d\varepsilon\right] \quad (4.32a)$$

$$\mathrm{U} = \tfrac{\mathrm{iF}\cos 2\theta}{4\pi^2\mathrm{Gr}}\Big\{\pi^2\kappa\mathrm{H_3J_0}(\kappa) + 2\mathrm{k}^2\int_h^k \varepsilon\frac{(2\varepsilon^2-\mathrm{k}^2)}{\mathrm{F}(\varepsilon)\,\mathrm{f}(\varepsilon)}\frac{\pi}{2}\mathrm{iH}_0^{(2)}(\varepsilon)\,\mathrm{d}\varepsilon \\ + 2\mathrm{k}^2\,\mathbb{C}\mathrm{P}\int_k^\infty \varepsilon\frac{\sqrt{\varepsilon^2-\mathrm{k}^2}}{\mathrm{F}(\varepsilon)}\frac{\pi}{2}\mathrm{iH}_0^{(2)}(\varepsilon)\,\mathrm{d}(\varepsilon) + 1/2\pi^2\mathrm{k}\Big[\mathrm{J}_{1/2}(\mathrm{k}/2)\,\mathrm{Y}_{1/2}(\mathrm{k}/2) + \mathrm{i}/2 \\ \Big(\mathrm{J}_{1/2}^2(\mathrm{k}/2) - \mathrm{Y}_{1/2}^2\Big(\mathrm{k}/2)\Big]\Big\} + \tfrac{\mathrm{iF_x}\cos 2\theta}{4\pi^2\mathrm{Gr}}\Big\{-\pi^2\kappa\mathrm{H_3J_2}(\kappa) - 2\mathrm{k}^2\int_h^k \varepsilon \\ \tfrac{(2\varepsilon^2-\mathrm{k}^2)^2\sqrt{\varepsilon^2-\mathrm{k}^2}}{\mathrm{F}(\varepsilon)\mathrm{f}(\varepsilon)}\tfrac{\mathrm{i}\pi}{2}\Big(3\mathrm{H}_0^{(2)}(\varepsilon) - \tfrac{2}{\varepsilon}\mathrm{H}_1^{(2)}\Big(\varepsilon)\Big)\,\mathrm{d}\varepsilon - 2\mathrm{k}^2\mathbb{C}\mathrm{P}\int_k^\infty \varepsilon\frac{\sqrt{\varepsilon^2-\mathrm{k}^2}}{\mathrm{F}(\varepsilon)}\mathrm{i}\frac{\pi}{2} \\ \Big(3\mathrm{H}_0^{(2)} - \tfrac{2}{\varepsilon}\mathrm{H}_1^{(2)}(\varepsilon)\Big)\,\mathrm{d}\varepsilon - 2\Gamma\Big(1/2\Big)\mathrm{k}\pi^{1/2}\left[-\left(\tfrac{\sin\mathrm{k}}{\mathrm{k}}+\tfrac{\cos\mathrm{k}}{\mathrm{k}^2}\right)+i\left(\tfrac{\sin\mathrm{k}}{\mathrm{k}^2}-\tfrac{\cos\mathrm{k}}{\mathrm{k}}\right)\right] \\ + 1/2\pi^2\mathrm{k}\Big[\mathrm{J}_{1/2}(\mathrm{k}/2) + \mathrm{i}/2\Big(\mathrm{J}_{1/2}^2\Big(\mathrm{k}/2\Big) - \mathrm{Y}_{1/2}^2\Big(\mathrm{k}/2\Big)\Big)\Big]\Big\} \quad (4.32b)$$

$$\mathrm{W} = -\tfrac{\mathrm{iF_x}\sin 2\theta}{4\pi^2\mathrm{Gr}}\Big\{-\pi^2\kappa\mathrm{H_3J_2}(\kappa) - 2\mathrm{k}^2\int_h^k \varepsilon^2\frac{(2\varepsilon^2-\mathrm{k}^2)^2\sqrt{\varepsilon^2-\mathrm{k}^2}}{\mathrm{F}(\varepsilon)\,\mathrm{f}(\varepsilon)}\frac{\mathrm{i}\pi}{2}\Big(3\mathrm{H}_0^{(2)}(\varepsilon) - \\ - \tfrac{2}{\varepsilon}\mathrm{H}_1^{(2)}(\varepsilon)\Big)\mathrm{d}\varepsilon - 2\mathrm{k}^2\,\mathbb{C}\mathrm{P}\int_k^\infty \varepsilon\frac{\sqrt{\varepsilon^2-\mathrm{k}^2}}{\mathrm{F}(\varepsilon)}\tfrac{\mathrm{i}\pi}{2}\Big(3\mathrm{H}_0^{(2)}(\varepsilon) - \tfrac{2}{\varepsilon}\mathrm{H}_1^{(2)}(\varepsilon)\Big)\mathrm{d}\varepsilon \\ - 2\Gamma(1/2)\,\mathrm{k}\pi^{1/2}\left[-\tfrac{\sin\mathrm{k}}{\mathrm{k}} + \tfrac{\cos\mathrm{k}}{\mathrm{k}^2} + \mathrm{i}\left(\tfrac{\sin\mathrm{k}}{\mathrm{k}^2} - \tfrac{\cos\mathrm{k}}{\mathrm{k}}\right)\right] \\ + 1/2\pi^2\mathrm{k}\Big[\mathrm{J}_{1/2}(\mathrm{k}/2)\;\mathrm{Y}_{1/2}\Big(\mathrm{k}/2\Big) + \mathrm{i}/2\Big(\mathrm{J}_{1/2}^2\Big(\mathrm{k}/2\Big) - \mathrm{Y}_{1-2}^2\Big(\mathrm{k}/2\Big)\Big)\Big]\Big\} \quad (4.32c)$$

The last boundary conditions of the problem are that there should not be any converging or standing waves and that vibrations must be transmitted through the medium by a system of waves diverging from the concentrated point force. Since all of the integrals in the above equations consist of $H_n^{(2)}(\varepsilon)$ terms and these functions can be expressed in the exponential form exp(−iε), they represent diverging waves. The only non-diverging terms in the above equations are residues of integrals, which are in terms of the Bessel functions of Y1, J0, and J2. In order to eliminate these waves, a portion of free waves can be added to the above displacements, without any effect to the boundary conditions. The general solution for free waves is given in Eqs. 4.14a–4.14c. After evaluating the arbitrary constant for the last condition by equating Eqs. 4.14a–4.14c to the standing waves in (4.32a–4.32c), the Rayleigh wave displacements then become

$$V_R = \frac{F_x \cos\theta}{2Gr}\left[i\kappa H_1 J_1(\kappa)\right] \tag{4.33a}$$

$$U_R = \frac{-F_x i\kappa H_3}{4Gr}\left[J_0(\kappa) - J_2(\kappa)\cos 2\theta\right] \tag{4.33b}$$

$$W_R = \frac{-F_x i\kappa H_3}{4Gr} J_2(\kappa)\sin 2\theta \tag{4.33c}$$

Parameters H_1 and H_2 are presented in Appendix B, and the necessary value of A is

$$A = -\frac{iF_x}{2F^1(\kappa)} \tag{4.34}$$

By superimposing these free waves on the displacement equations and introducing

$$\eta = \varepsilon / k \tag{4.35}$$

the displacement equations finally become

$$V = \frac{F_x \cos\theta}{2Gr\pi}\Big[\kappa H_1 \pi \; i H_1^{(2)}(\kappa) + 2k\int_y^1 \eta^2 \frac{\left(2\eta^2 - 1\right)\sqrt{\left(\eta^2 - \gamma^2\right)\left(1 - \eta^2\right)}}{F(\eta)\, f(\eta)} H_1^{(2)} \times (k\eta)\, d\eta \tag{4.36a}$$

$$U = -\frac{F_x}{4Gr\pi}\left[k\int_{\gamma}^{l}\eta\frac{(2\eta^2-1)^2\sqrt{1-\eta^2}}{F(\eta)f(\eta)}iH_0^{(2)}(k\eta) + k\,\mathbb{C}P\int_{l}^{\infty}\eta\frac{\sqrt{\eta^2-1}}{F(\eta)}\right.$$
$$H_0^{(2)}(k\eta)\,d\eta - 1/2\pi kiJ_{1/2}\left(k/2\right)Y_{1/2}\left(k/2\right) + 1/4\pi k\left(J_{1/2}^2\left(k/2\right) - Y_{1/2}^2\right.$$
$$\left.(k/2)\right] + \frac{F_x\cos 2\theta}{4Gr\pi}\left[3k\int_{\gamma}^{l}\eta\frac{(2\eta^2-1)^2\sqrt{1-\eta^2}}{F(\eta)f(\eta)}iH_0^{(2)}(k\eta)\,d\eta\right.$$
$$-2\int_{\gamma}^{1}\frac{(2\eta^2-1)^2\sqrt{1-\eta^2}}{F(\eta)f(\eta)}iH_1^{(2)}(k\eta)\,d\eta + 3k\,\mathbb{C}P\int_{l}^{\infty}\eta\frac{\sqrt{\eta^2-1}}{F(\eta)}H_0^{(2)}(k\eta)\,d\eta$$
$$-2\,\mathbb{C}P\int_{1}^{\infty}\frac{\sqrt{\eta^2-1}}{F(\eta)}H_1^{(2)}(k\eta)\,d\eta + 2ki\left(\frac{\sin k}{k} + \frac{\cos k}{k^2}\right) + 2k\left(\frac{\sin k}{k^2} - \frac{\cos k}{k}\right)$$
$$\left. + 1/2\pi kiJ_{1/2}(k/2)\,Y_{1/2}\left(k/2\right) - 1/4\pi k\left(J_{1/2}^2\left(k/2\right) - Y_{1/2}^2\left(k/2\right)\right)\right] \tag{4.36b}$$

$$W = -\frac{F_x\sin 2\theta}{4Gr\pi}\left[3k\int_{\gamma}^{l}\eta\frac{(2\eta^2-1)^2\sqrt{1-\eta^2}}{F(\eta)f(\eta)}iH_0^{(2)}(k\eta)\,d\eta\right.$$
$$-2\int_{\gamma}^{l}\frac{(2\eta^2-1)^2\sqrt{1-\eta^2}}{F(\eta)f(\eta)}iH_1^{(2)}(k\eta)\,d\eta + 3k\,\mathbb{C}P\int_{1}^{\infty}\eta\frac{\sqrt{\eta^2-1}}{F(\eta)}H_0^{(2)}(k\eta)\,d\eta$$
$$-2\mathbb{C}P\int_{1}^{\infty}\frac{\sqrt{\eta^2-1}}{F(\eta)}H_1^{(2)}(k\eta)\,d\eta + 2ki\left(\frac{\sin k}{k} + \frac{\cos k}{k^2}\right) + 2k\left(\frac{\sin k}{k^2} - \frac{\cos k}{k}\right)$$
$$\left.+1/2\pi kiJ_{1/2}(k/2)\,Y_{1/2}\left(k/2\right) - 1/4\pi k\left(J_{1/2}^2\left(k/2\right) - Y_{1/2}^2\left(k/2\right)\right)\right] \tag{4.36c}$$

As a consequence of the above relations, the amplitude of a displacement on the surface of a half-space can be written in the form

$$V = \frac{F}{Gr}\cos\theta\,(v_3 + iv_4) \tag{4.37a}$$

$$U = \frac{F}{Gr}(u_{31} + iu_{41}) + \frac{F}{Gr}\cos 2\theta\,(u_{32} + iu_{42}) \tag{4.37b}$$

and finally

$$W = \frac{F}{Gr}\sin 2\theta\,((w_3 + iw_4)) \tag{4.37c}$$

where v_3, v_4, u_{31}, u_{41}, u_{32}, u_{42}, w_3, and w_4 are real values that can be evaluated by applying appropriate methods of integration, which were discussed in Sect. 4.6. On the other hand, a displacement in the vertical direction, in this case, is similar to a displacement in the x direction for a vertical point force which was calculated before. Also, from Eqs. 4.36b and 4.36c, it is obvious that

$$w_3 = -u_{32} \tag{4.38}$$

$$w_4 = -u_{42} \tag{4.39}$$

Results that are obtained here will be satisfied by all the conditions discussed by Chi-Chang Chao (1960).

4.5 Numerical Integration of Displacements

To obtain numerical results for a displacement of a particular value of the frequency factor (k), all integrals in displacements should be evaluated by means of the appropriate numerical technique. In the following section, these methods will be introduced for each integral, which are listed below:

$$\mathrm{B}_1 = \int_{\gamma}^{l} \eta^2 \frac{\left(2\eta^2-1\right)^2\left(\eta^2-\gamma^2\right)^{1/2}\left(\eta^2-1\right)^{1/2}}{\left(2\eta^2-1\right)^4 - 16\eta^4\left(\eta^2-\gamma^2\right)\left(\eta^2 1\right)} \mathrm{H}_1^{(2)}(\mathrm{k}\eta)\,d\eta \tag{4.40a}$$

$$\mathrm{B}_2 = \int_{\gamma}^{l} \eta \frac{\left(2\eta^2-1\right)^2\sqrt{1-\eta^2}}{\mathrm{F}(\eta)\,\mathrm{f}(\eta)} \mathrm{H}_0^{(2)}(\mathrm{k}\eta)\,d\eta \tag{4.40b}$$

$$\mathrm{B}_3 = \int_{\gamma}^{l} \frac{\left(2\eta^2-1\right)^2\sqrt{1-\eta^2}}{\mathrm{F}(\eta)\,\mathrm{f}(\eta)} \mathrm{H}_1^{(2)}(\mathrm{k}\eta)\,d\eta \tag{4.40c}$$

$$\mathrm{B}_4 = \mathbb{C}\mathrm{P}\int_{l}^{\infty} \eta \frac{\left(\eta^2-1\right)^{1/2}}{\mathrm{F}(\eta)} \mathrm{H}_0^{(2)}(\mathrm{k}\eta)\,d\eta \tag{4.40d}$$

$$\mathrm{B}_5 = \mathbb{C}\mathrm{P}\int_{1}^{\infty} \frac{\left(\eta^2-1\right)^{1/2}}{\mathrm{F}(\eta)} \mathrm{H}_1^{(2)}(\mathrm{k}\eta)\,d\eta \tag{4.40e}$$

*Evaluation of B*1

This integral is exactly the same as integral $A3$, which was discussed in Sect. 4.6, so please see the evaluation for A_3.

*Evaluation of B*2 *and B*3

These integrals are evaluated by the Gaussian quadrature method, and it was found that 20 abscissa points were sufficient to enable integration in the frequency factor interval.

Evaluation of B4 and B5

The evaluation procedure for these two Cauchy's principal valuea was exactly the same as the procedure for integral A_1 in Sect. 3.6.

4.6 Results and Discussion

The eight displacement functions of v_3, v_4, u_{31}, u_{41}, u_{32}, u_{42}, w_3, and w_4 are plotted against the variations of the frequency factor for the different Poisson ratios; 0.0, 0.25, 0.31, and 0.5 in Figures 4.2, 4.3, 4.4, 4.5, 4.6, 4.7, 4.8, 4.9, 4.10, 4.11, 4.12, and 4.13. This data is also listed in Tables 4.1, 4.2, 4.3, and 4.4. These curves do not start from very low values of the frequency factor because the values of u_{32}, u_{42} are very high in the low range. Consequently, these tables must be used whenever these functions are needed for the low values of the frequency factor.

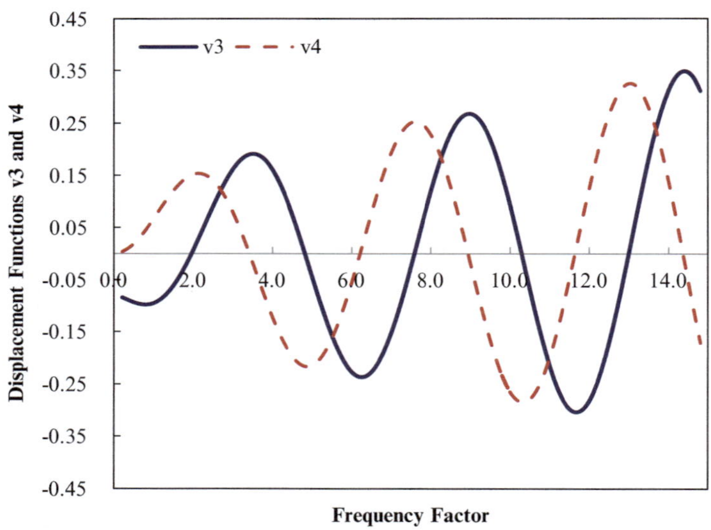

Fig. 4.2 Values of displacement functions v_3 and v_4 versus frequency factor for horizontal harmonic point force on the surface of an elastic half-space with Poisson ratio of 0.0

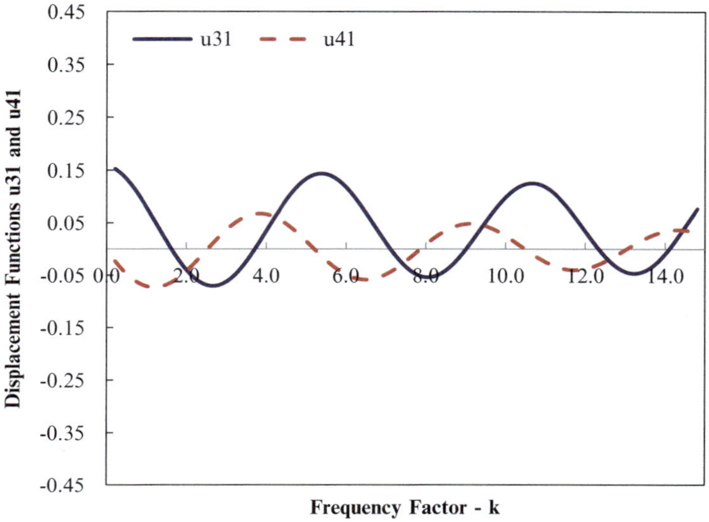

Fig. 4.3 Values of displacement functions u_{31} and u_{41} versus frequency factor for horizontal harmonic point force on the surface of an elastic half-space with Poisson ratio of 0.0

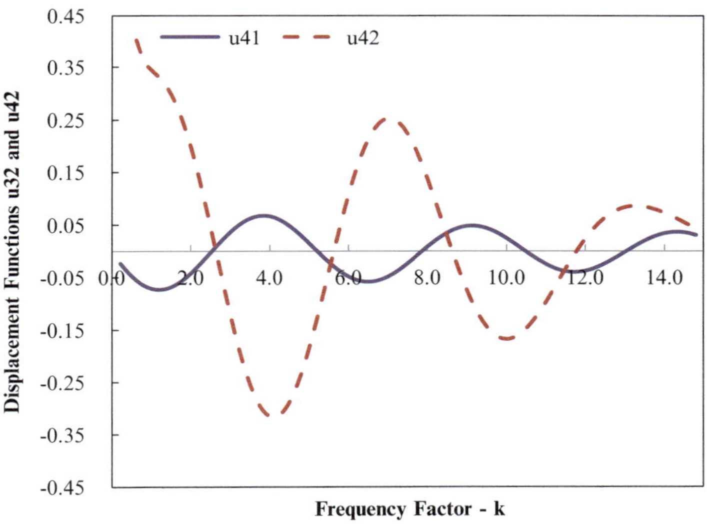

Fig. 4.4 Values of displacement functions u_{32} and u_{42} versus frequency factor for horizontal harmonic point force on the surface of an elastic half-space with Poisson ratio of 0.0

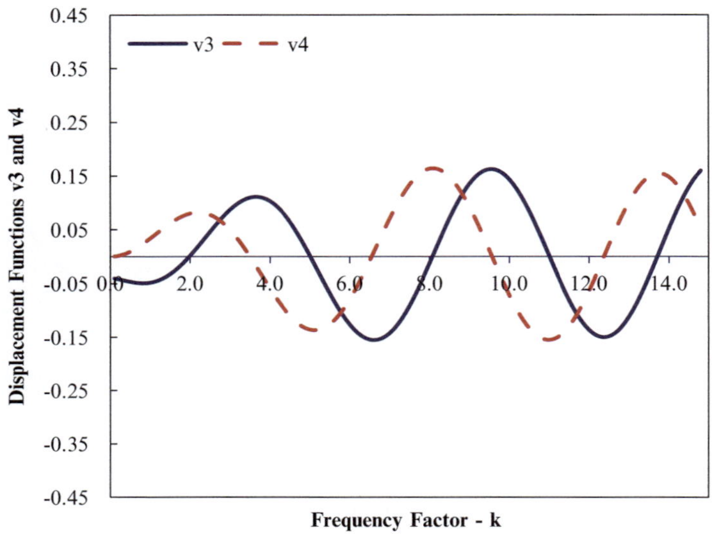

Fig. 4.5 Values of displacement functions v_3 and v_4 versus frequency factor for horizontal harmonic point force on the surface of an elastic half-space with Poisson ratio of 0.25

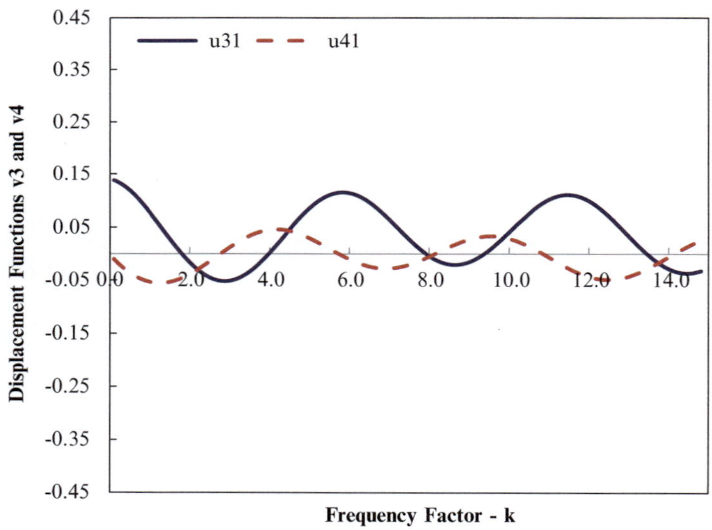

Fig. 4.6 Values of displacement functions u_{31} and u_{41} versus frequency factor for horizontal harmonic point force on the surface of an elastic half-space with Poisson ratio of 0.25

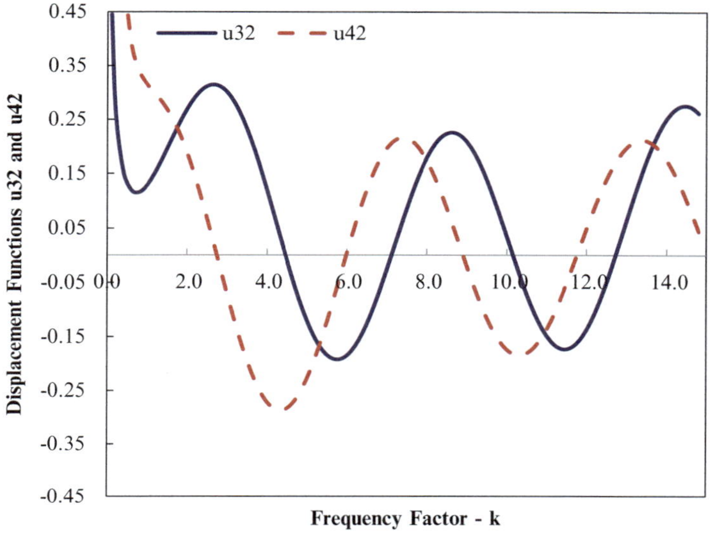

Fig. 4.7 Values of displacement functions u_{32} and u_{42} versus frequency factor for horizontal harmonic point force on the surface of an elastic half-space with Poisson ratio of 0.25

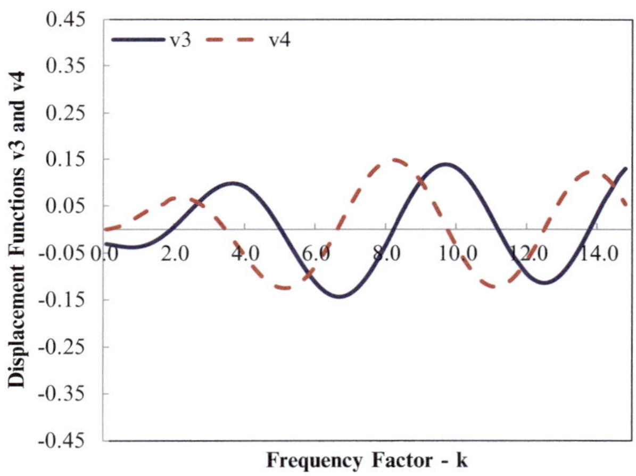

Fig. 4.8 Values of displacement functions v_3 and v_4 versus frequency factor for horizontal harmonic point force on the surface of an elastic half-space with Poisson ratio of 0.31

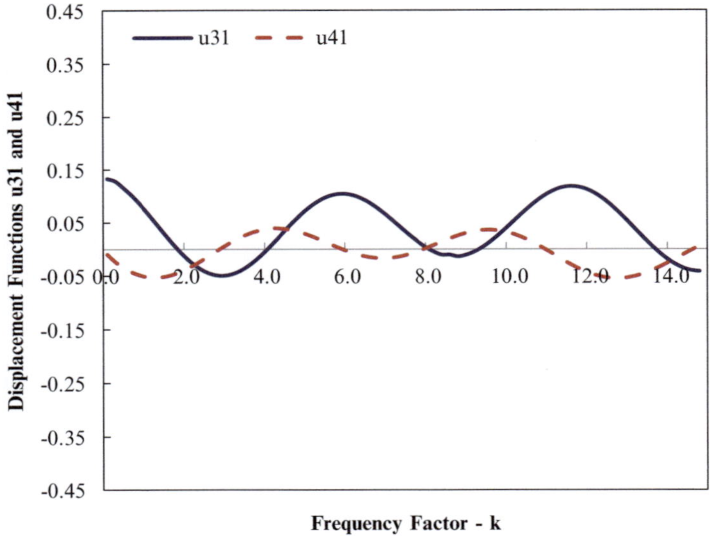

Fig. 4.9 Values of displacement functions u_{31} and u_{41} versus frequency factor for horizontal harmonic point force on the surface of an elastic half-space with Poisson ratio of 0.31

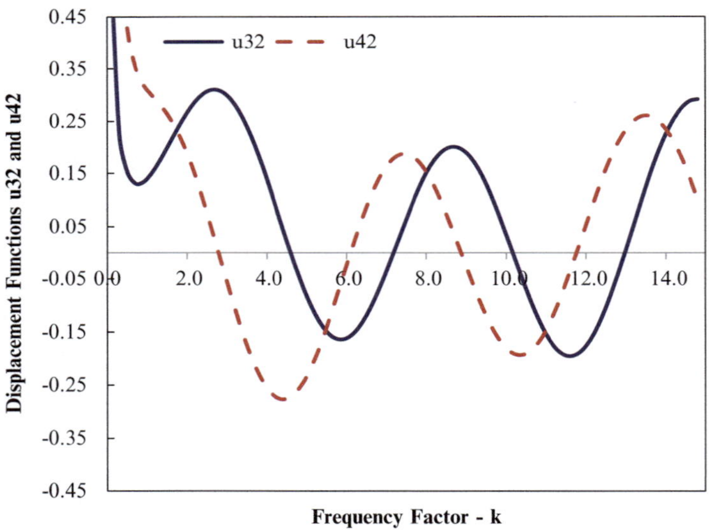

Fig. 4.10 Values of displacement functions u_{32} and u_{42} versus frequency factor for horizontal harmonic point force on the surface of an elastic half-space with Poisson ratio of 0.31

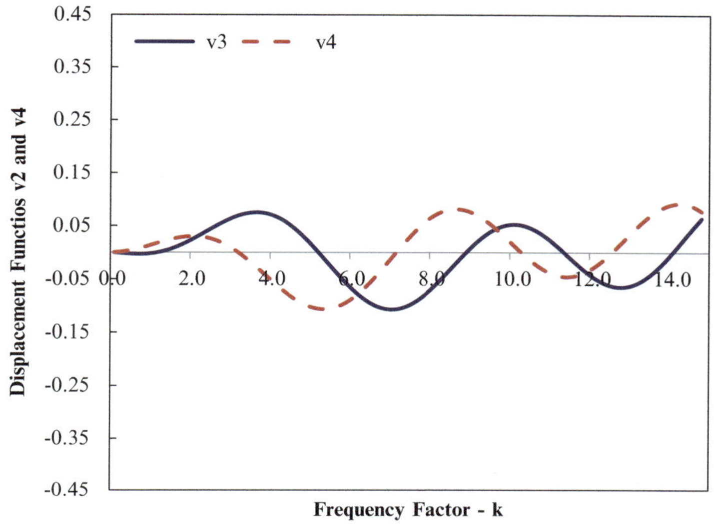

Fig. 4.11 Values of displacement functions v_3 and v_4 versus frequency factor for horizontal harmonic point force on the surface of an elastic half-space with Poisson ratio of 0.5

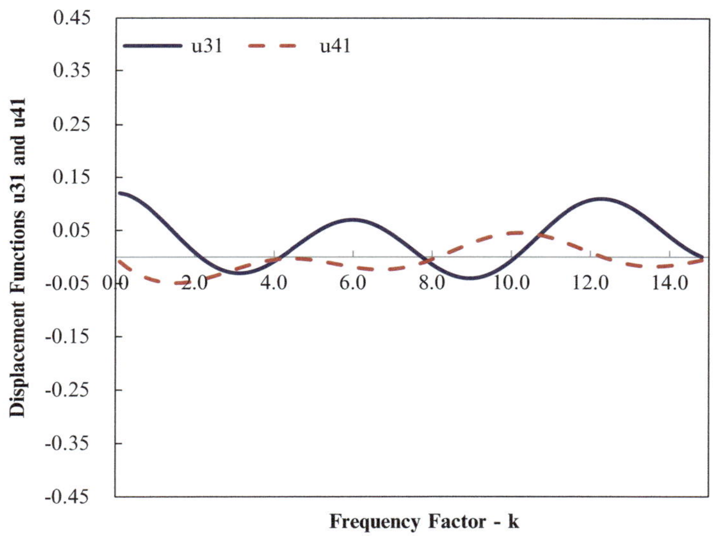

Fig. 4.12 Values of displacement functions u_{31} and u_{41} versus frequency factor for horizontal harmonic point force on the surface of an elastic half-space with Poisson ratio of 0.5

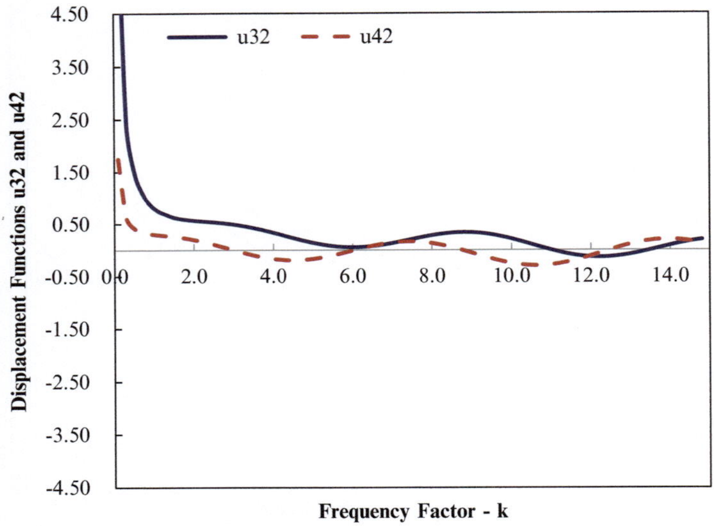

Fig. 4.13 Values of displacement functions u_{32} and u_{42} versus frequency factor for horizontal harmonic point force on the surface of an elastic half-space with Poisson ratio of 0.5

Table 4.1 Values of displacement functions for different frequency factors for a horizontal harmonic force

	Poisson ratio $\nu = 0$					
k	v_3	v_4	u_{31}	u_{41}	u_{32}	u_{42}
0.21	−0.084051	0.003421	0.152661	−0.023449	0.200329	0.931883
0.41	−0.090727	0.012784	0.140515	−0.041006	0.110732	0.517178
0.61	−0.096092	0.027381	0.123463	−0.055058	0.084548	0.402338
0.81	−0.097985	0.046084	0.102408	−0.065309	0.089170	0.361555
1.01	−0.094927	0.067429	0.078427	−0.071445	0.112934	0.343434
1.21	−0.086009	0.089711	0.052741	−0.073298	0.148850	0.329044
1.41	−0.070877	0.111089	0.026660	−0.070896	0.191464	0.309828
1.61	−0.049710	0.129715	0.001515	−0.064480	0.236023	0.281801
1.81	−0.023200	0.143852	−0.021413	−0.054496	0.278264	0.243513
2.01	0.007504	0.151994	−0.040954	−0.041573	0.314416	0.195171
2.21	0.040867	0.152970	−0.056118	−0.026489	0.341296	0.138144
2.41	0.075064	0.146033	−0.066142	−0.010128	0.356406	0.074642
2.61	0.108096	0.130919	−0.070534	0.006574	0.378036	0.007440
2.81	0.137915	0.107880	−0.069098	0.022677	0.345324	−0.060353
3.01	0.162555	0.077688	−0.061945	0.037289	0.318293	−0.125526
3.21	0.180260	0.041601	−0.049486	0.049620	0.277830	−0.184963
3.41	0.189604	0.001305	−0.032413	0.059020	0.225635	−0.235833
3.61	0.189586	−0.041177	−0.011656	0.065012	0.164115	−0.275750
3.81	0.179714	−0.083594	0.011659	0.067322	0.096253	−0.302894
4.01	0.160047	−0.123592	0.036279	0.065886	0.025435	−0.316106
4.21	0.131216	−0.158859	0.060886	0.060845	−0.044742	−0.314939
4.41	0.094401	−0.187254	0.084168	0.052549	−0.110664	−0.299666
4.61	0.051288	−0.206945	0.104892	0.041515	−0.168910	−0.271252
4.81	0.003982	−0.216523	0.211969	0.028410	−0.216439	−0.231291
5.00	−0.042645	−0.215435	0.134004	0.014738	−0.249393	−0.184563
5.20	−0.091009	−0.203260	0.141645	−0.000149	−0.269500	−0.128570
5.40	−0.136039	−0.180061	0.143771	−0.014720	−0.273637	−0.068348
5.60	−0.175192	−0.146744	0.140313	−0.028175	−0.261614	−0.006816
5.80	−0.206175	−0.104806	0.131505	−0.039788	−0.234104	0.053127
6.00	−0.227079	−0.056264	0.117872	−0.048955	−0.192613	0.108752
6.20	−0.236496	−0.003556	0.100195	−0.055219	−0.139394	0.157631
6.40	−0.233612	0.050581	0.079475	−0.058298	−0.077331	0.197743
6.60	−0.218268	0.103262	0.056868	−0.058091	−0.009778	0.227566
6.80	−0.190985	0.151604	0.033630	−0.054686	0.059622	0.246128
7.00	−0.152951	0.192893	0.011040	−0.048349	0.127135	0.253036
7.20	−0.105973	0.224736	−0.009665	−0.039503	0.189145	0.248472
7.40	−0.052387	0.245197	−0.027362	−0.028703	0.242346	0.233154
7.60	0.005052	0.252921	−0.041100	−0.016602	0.283927	0.208281
7.80	0.063321	0.247218	−0.050155	−0.003910	0.311726	0.175448
8.00	0.119283	0.228115	−0.054070	0.008646	0.324346	0.136547
8.20	0.169860	0.196372	−0.052674	0.020362	0.321230	0.093660
8.40	0.212204	0.153452	−0.046096	0.030601	0.302695	0.048943
8.60	0.243857	0.101461	−0.034756	0.038822	0.269907	0.004516

(continued)

Table 4.1 (continued)

	Poisson ratio ν = 0					
k	v_3	v_4	u_{31}	u_{41}	u_{32}	u_{42}
8.80	0.262892	0.043039	−0.019337	0.044612	0.224817	−0.037637
9.00	0.268033	−0.018770	−0.000748	0.047705	0.170051	−0.075762
9.20	0.258735	−0.080676	0.019925	0.047996	0.108763	−0.108407
9.40	0.235228	−0.139323	0.041484	0.045541	0.044454	−0.134472
9.60	0.198520	−0.191465	0.062689	0.040549	−0.019218	−0.153244
9.80	0.150355	−0.234150	0.082321	0.033370	−0.078651	−0.164401
10.00	0.093134	−0.264883	0.099258	0.024472	−0.130495	−0.168004
10.20	0.029795	−0.281774	0.112534	0.014405	−0.171846	−0.164463
10.40	−0.036338	−0.283649	0.121397	0.003779	−0.200404	−0.154489
10.60	−0.101727	−0.270126	0.125349	−0.006784	−0.214607	−0.139036
10.80	−0.162811	−0.241656	0.124178	−0.016675	−0.213718	−0.119227
11.00	−0.216197	−0.199504	0.117965	−0.025340	−0.197867	−0.096290
11.20	−0.258848	−0.145703	0.107086	−0.032309	−0.168041	−0.071479
11.40	−0.288256	−0.082954	0.092186	−0.037216	−0.126034	−0.046010
11.60	−0.302581	−0.014487	0.074145	−0.039825	−0.074342	−0.021006
11.80	−0.300771	0.056097	0.054028	−0.040034	−0.016022	0.002560
12.00	−0.282624	0.125021	0.033024	−0.037882	0.045481	0.023892
12.20	−0.248820	0.188530	0.012378	−0.033543	0.106531	0.042393
12.40	−0.200893	0.243095	−0.006681	0.027312	0.163509	0.057677
12.60	−0.141169	0.285609	−0.023007	−0.019587	0.213021	0.069557
12.80	−0.072648	0.313560	−0.035611	−0.010843	0.252096	0.078031
13.00	0.001144	0.325175	−0.043711	−0.001604	0.278355	0.083254
13.20	0.076335	0.319531	−0.046786	0.007586	0.290160	0.085505
13.40	0.148914	0.296616	−0.044602	0.016198	0.286702	0.085147
13.60	0.214940	0.257345	−0.037231	0.023746	0.266065	0.082591
13.80	0.270757	0.203527	−0.025050	0.029812	0.235220	0.078259
14.00	0.313194	0.137779	−0.008722	0.034070	0.109981	0.072551
14.20	0.339745	0.063402	0.010840	0.036303	0.134906	0.065819
14.40	0.348710	−0.015786	0.032524	0.036413	0.073157	0.058353
14.60	0.339301	−0.095650	0.055077	0.034423	0.008325	0.050366
14.80	0.311698	−0.171948	0.077179	0.030475	−0.055774	0.041998

Table 4.2 Values of displacement functions for different frequency factors for a horizontal harmonic force

	Poisson ratio ν = 0.25					
k	v_3	v_4	u_{31}	u_{41}	u_{32}	u_{42}
0.10	−0.040490	0.000390	0.137505	−0.010322	0.444946	1.830622
0.20	−0.041894	0.001553	0.134282	−0.018280	0.264026	0.941330
0.40	−0.045315	0.006103	0.124657	−0.031345	0.152499	0.509052
0.60	−0.048218	0.013327	0.111521	−0.041505	0.119219	0.386395
0.80	−0.049500	0.022706	0.095556	−0.048964	0.115740	0.338285
1.00	−0.048394	0.033560	0.077475	−0.053721	0.128751	0.314570
1.20	−0.044399	0.045081	0.058053	−0.055772	0.151794	0.297488
1.40	−0.037269	0.056389	0.038109	−0.055171	0.180567	0.279126
1.60	−0.027013	0.066576	0.018477	−0.052054	0.211675	0.255653
1.80	−0.013879	0.074768	−0.000023	−0.046649	0.242217	0.225653
2.00	0.001658	0.080174	−0.016625	−0.039267	0.269679	0.187859
2.20	0.018926	0.082138	−0.030648	−0.030296	0.291907	0.143647
2.40	0.037093	0.080182	−0.041523	−0.020182	0.307144	0.093890
2.60	0.055215	0.074041	−0.048822	−0.009413	0.314066	0.040225
2.80	0.072286	0.063681	−0.052273	0.001505	0.311814	−0.015383
3.00	0.087300	0.049315	−0.051776	0.012069	0.300019	−0.070778
3.20	0.099304	0.031396	−0.047404	0.021806	0.278810	−0.123747
3.40	0.107458	0.010603	−0.039398	0.030289	0.248793	−0.172133
3.60	0.111086	−0.012192	−0.028460	0.037161	0.211024	−0.213954
3.80	0.109717	−0.035961	−0.014231	0.042148	0.166950	−0.247496
4.00	0.103124	−0.059575	0.001735	0.045072	0.118340	−0.271396
4.20	0.091339	−0.081854	0.018997	0.045857	0.067202	−0.284710
4.40	0.074664	−0.101636	0.036767	0.044533	0.015681	−0.286951
4.60	0.053665	−0.117833	0.054239	0.041230	−0.034039	−0.278112
4.80	0.029150	−0.129494	0.070635	0.036171	−0.079841	−0.258667
5.00	0.002138	−0.135857	0.085230	0.029664	−0.119776	−0.229543
5.20	−0.026191	−0.136395	0.097395	0.022077	−0.152155	−0.192075
5.40	−0.054547	−0.130850	0.106617	0.013827	−0.175629	−0.147948
5.60	−0.081595	−0.119255	0.112527	0.005353	−0.189255	−0.099115
5.80	−0.106021	−0.101941	0.114916	−0.002903	−0.192540	−0.047709
6.00	−0.126595	0.079526	0.113738	−0.010519	−0.185464	0.004054
6.20	−0.142240	−0.052896	0.109116	−0.017114	−0.168484	0.053974
6.40	−0.152086	−0.023164	0.101335	−0.022367	−0.142507	0.099963
6.60	−0.145552	0.008376	0.090823	−0.026031	−0.108855	0.140138
6.80	−0.152219	0.040311	0.078134	−0.027947	−0.069194	0.172897
7.00	−0.142179	0.071177	0.063918	−0.028047	−0.025465	0.196992
7.20	−0.125718	0.099526	0.048891	−0.026361	0.020218	0.211574
7.40	−0.103467	0.124001	0.033799	−0.023010	0.065657	0.246226
7.60	−0.076334	0.143403	0.019386	−0.018202	0.108684	0.210973
7.80	−0.045521	0.156752	0.006354	−0.012219	0.147254	0.196277
8.00	−0.012364	0.163339	−0.004667	−0.005405	0.179548	0.173008
8.20	0.021624	0.162767	−0.013151	0.001857	0.204053	0.142394
8.40	0.054881	0.154974	−0.018698	0.009162	0.219637	0.105971

(continued)

Table 4.2 (continued)

	Poisson ratio ν = 0.25					
k	v_3	v_4	u_{31}	u_{41}	u_{32}	u_{42}
8.60	0.085853	0.140243	−0.021053	0.016103	0.225597	0.065498
8.80	0.133076	0.119191	−0.020118	0.022291	0.221696	0.022886
9.00	0.135245	0.092748	−0.015955	0.027378	0.208168	−0.019897
9.20	0.151285	0.062111	−0.008785	0.031072	0.185710	−0.060907
9.40	0.160399	0.028694	0.001030	0.033151	0.155442	−0.098315
9.60	0.162116	−0.005941	0.012998	0.033475	0.118858	−0.130479
9.80	0.156318	0.040155	0.026526	0.031991	0.077753	−0.156017
10.00	0.143249	−0.072315	0.040945	0.028739	0.034134	−0.173863
10.20	0.123506	−0.100873	0.055545	0.023844	−0.009873	−0.183303
10.40	0.098017	−0.124440	0.068611	0.017516	−0.052425	−0.184004
10.60	0.067999	−0.141859	0.082453	0.010035	−0.090565	−0.176019
10.80	0.034899	−0.152262	0.093439	0.001737	−0.123315	−0.159779
11.00	0.000329	−0.155119	0.102029	−0.006999	−0.148770	−0.136064
11.20	−0.034015	−0.150266	0.107797	−0.015774	−0.165669	−0.105966
11.40	−0.066431	−0.137915	0.110452	−0.024184	−0.173155	−0.070835
11.60	−0.095299	−0.118653	0.109850	−0.031844	−0.170814	−0.032214
11.80	−0.119160	−0.093414	0.106002	−0.038402	−0.158692	0.008226
12.00	−0.136788	−0.063435	0.099070	−0.043559	−0.137293	0.048765
12.20	−0.147259	−0.030203	0.089361	−0.047080	−0.107552	0.087704
12.40	−0.149992	0.004619	0.077312	−0.048808	−0.070789	0.123438
12.60	−0.144790	0.039271	0.063465	−0.048667	−0.028646	0.154517
12.80	−0.131849	0.071981	0.048445	−0.046667	0.016993	0.179710
13.00	−0.111756	0.101057	0.032927	−0.042901	0.064090	0.198044
13.20	−0.085461	0.124966	0.017605	−0.037541	0.110551	0.208840
13.40	−0.054241	0.142415	0.003158	−0.030829	0.154323	0.211738
13.60	−0.019633	0.152419	−0.009778	−0.023063	0.193492	0.206695
13.80	0.016630	0.154354	−0.020647	−0.014581	0.226366	0.193986
14.00	0.052712	0.147991	−0.028990	−0.005756	0.251560	0.174181
14.20	0.086754	0.133512	−0.034470	0.003048	0.268050	0.148113
14.40	0.166972	0.111511	−0.036887	0.011465	0.275221	0.116840
14.60	0.141746	0.082967	−0.036184	0.019159	0.272893	0.081593
14.80	0.159697	0.049203	−0.032448	0.025836	0.261318	0.043724

Table 4.3 Values of displacement functions for different frequency factors for a horizontal harmonic force

	Poisson ratio ν = 0.31					
k	v_3	v_4	u_{31}	u_{41}	u_{32}	u_{42}
0.10	−0.030812	0.000320	0.133133	−0.009960	0.545292	1.812119
0.30	−0.033332	0.002843	0.128231	−0.024062	0.224901	0.642562
0.50	−0.036027	0.007711	0.115541	−0.034912	0.152572	0.248860
0.70	−0.037787	0.014576	0.101849	−0.043181	0.131352	0.353971
0.90	−0.037856	0.022944	0.085811	−0.048987	0.133336	0.320820
1.10	−0.035712	0.032198	0.068118	−0.052354	0.148586	0.301377
1.30	−0.031047	0.041639	0.049497	−0.053322	0.171717	0.284317
1.50	−0.023758	0.050524	0.030703	−0.051983	0.198980	0.264295
1.70	−0.013847	0.053110	0.012488	−0.048498	0.227381	0.238734
1.90	−0.001908	0.063699	−0.004424	−0.043098	0.254385	0.206617
2.10	0.011886	0.066683	−0.019367	−0.036078	0.277824	0.167944
2.30	0.026816	0.066577	−0.031762	−0.027791	0.295886	0.123431
2.50	0.042143	0.063057	−0.041141	−0.018630	0.307130	0.074322
2.70	0.057055	0.055981	−0.047162	−0.009018	0.310522	0.022239
2.90	0.070706	0.045405	−0.049631	0.000620	0.305448	−0.030944
3.10	0.082269	0.031586	−0.048501	0.009864	0.291733	−0.083224
3.30	0.090977	0.014977	−0.043881	0.018327	0.269634	−0.132580
3.50	0.096174	−0.003791	−0.036026	0.025664	0.239823	−0.177076
3.70	0.097349	−0.023937	−0.025327	0.031595	0.203351	−0.214960
3.90	0.094176	−0.044564	−0.012293	0.035907	0.161593	0.244747
4.10	0.086531	−0.064705	0.002474	0.039472	0.116189	−0.265287
4.30	0.074513	−0.083369	0.018302	0.039247	0.068959	−0.275824
4.50	0.058443	−0.099959	0.034484	0.038272	0.021822	−0.276020
4.70	0.038855	−0.112483	0.050305	0.035672	−0.023292	−0.265979
4.90	0.016481	−0.121282	0.065079	0.031648	−0.064533	−0.246238
5.00	0.004524	−0.123953	0.071872	0.029182	−0.083167	−0.233009
5.20	−0.020310	−0.125558	0.083915	0.023532	−0.115517	−0.233601
5.40	−0.045472	−0.121968	0.093499	0.017207	−0.140266	−0.161515
5.60	−0.069841	−0.113139	0.100245	0.010567	−0.156459	−0.117491
5.80	−0.092297	−0.099272	0.103911	0.003979	−0.163528	−0.070459
6.00	−0.111777	−0.080810	0.104399	−0.002199	−0.161312	−0.022459
6.20	−0.127318	−0.058426	0.101758	−0.007639	−0.150069	0.024454
6.40	−0.138124	−0.032995	0.096182	−0.012055	−0.130463	0.068295
6.60	−0.143590	−0.005555	0.087997	−0.015221	−0.103535	0.107239
6.80	−0.143347	0.022738	0.077650	−0.016976	−0.070652	0.139695
7.00	−0.137282	0.050661	0.065683	−0.017238	−0.033449	0.164377
7.20	−0.125545	0.076983	0.052709	−0.016003	0.006247	0.180353
7.40	−0.108553	0.100523	0.039384	−0.013347	0.046503	0.187085
7.60	−0.086972	0.120207	0.026375	−0.009420	0.085368	0.184448
7.80	−0.061690	0.135119	0.014331	−0.004440	0.120961	0.172728
8.00	−0.033778	0.144552	0.003849	0.001319	0.151558	0.152610
8.20	−0.004445	0.148040	−0.004550	0.007542	0.175672	0.125141
8.40	0.025023	0.145384	−0.010443	0.013885	0.192120	0.091381
8.60	0.053314	0.136670	−0.010353	0.019996	0.200080	0.053841

(continued)

Table 4.3 (continued)

	Poisson ratio ν = 0.31					
k	v_3	v_4	u_{31}	u_{41}	u_{32}	u_{42}
8.80	0.079161	0.122263	−0.013649	0.025530	0.199125	0.013409
9.00	0.101397	0.102797	−0.010768	0.030167	0.189246	−0.027726
9.20	0.119014	0.079147	−0.005005	0.033626	0.170850	−0.067660
9.40	0.131213	0.052389	0.003387	0.035684	0.144744	−0.104557
9.60	0.137441	0.023755	0.014032	0.036171	0.112096	−0.136724
9.80	0.137424	−0.005431	0.026447	0.035008	0.074384	−0.162687
10.00	0.131184	−0.033808	0.040064	0.032181	0.033325	−0.181247
10.20	0.119035	−0.060047	0.054258	0.027755	−0.009199	−0.191534
10.40	0.101576	−0.082915	0.068374	0.021875	−0.051227	−0.193038
10.60	0.079662	−0.101333	0.081758	0.014749	−0.090813	−0.185627
10.80	0.054369	−0.114432	0.093785	0.006647	−0.126108	−0.169549
11.00	0.026938	−0.121594	0.103892	−0.002116	−0.155448	−0.145420
11.20	−0.001275	−0.122487	0.111598	−0.011192	−0.177425	−0.114191
11.40	−0.028873	−0.117082	0.116529	−0.020218	−0.190955	−0.077107
11.60	−0.054478	−0.105658	0.118432	−0.028833	−0.195321	−0.035648
11.80	−0.074802	−0.088793	0.117190	−0.036690	−0.190212	0.008527
12.00	−0.094709	−0.067335	0.112823	−0.043477	−0.175726	0.053660
12.20	−0.107271	−0.042367	0.105487	−0.048929	−0.152374	0.067962
12.40	−0.113821	0.015155	0.095469	−0.052835	−0.121051	0.139686
12.60	−0.113980	0.012911	0.083170	−0.055055	−0.082994	0.177203
12.80	−0.107685	0.040384	0.069089	−0.055519	−0.039730	0.209059
13.00	−0.095196	0.065830	0.053799	−0.054233	0.006998	0.234037
13.20	−0.077083	0.087998	0.037921	−0.051269	0.055306	0.251199
13.40	−0.054204	0.105389	0.022095	−0.046777	0.103257	0.259923
13.60	−0.027670	0.117315	0.006954	−0.040962	0.148942	0.259915
13.80	0.001211	0.122948	−0.006907	−0.034030	0.190567	0.251221
14.00	0.030992	0.121865	−0.018954	−0.026423	0.226531	0.234218
14.20	0.060156	0.113969	−0.028740	−0.018309	0.255492	0.209589
14.40	0.084190	0.099497	−0.035920	−0.010062	0.276426	0.178293
14.60	0.110658	0.079020	−0.040269	−0.002000	0.288665	0.141520
14.80	0.129272	0.053414	−0.041685	0.005580	0.291924	0.100639

Table 4.4 Values of displacement functions for different frequency factors for a horizontal harmonic force

	Poisson ratio ν = 0.50					
k	v_3	v_4	u_{31}	u_{41}	u_{32}	u_{42}
0.10	−0.000267	0.000156	0.120195	−0.008670	7.269025	1.753200
0.30	−0.001403	0.001387	0.116736	−0.020686	2.462999	0.624366
0.50	−0.002541	0.003755	0.109549	−0.029955	1.496344	0.418666
0.70	−0.003114	0.007083	0.099601	−0.037233	1.090098	0.345353
0.90	−0.002750	0.011113	0.087518	−0.042748	0.875317	0.312239
1.10	−0.001189	0.015525	0.073877	−0.046601	0.749618	0.293124
1.30	0.001725	0.019953	0.059253	−0.048872	0.682472	0.277558
1.50	0.006040	0.024008	0.044226	−0.049650	0.624074	0.260629
1.70	0.011703	0.027297	0.029369	−0.049050	0.593217	0.239922
1.90	0.018557	0.029446	0.015232	−0.047217	0.572841	0.214389
2.10	0.026355	0.030125	0.002327	−0.044325	0.558146	0.183839
2.30	0.034765	0.029068	−0.008896	−0.040573	0.545706	0.148680
2.50	0.043393	0.026073	−0.018056	−0.036180	0.533032	0.109755
2.70	0.051800	0.021058	−0.024865	−0.031377	0.548336	0.068223
2.90	0.059520	0.014024	−0.029130	−0.026400	0.500396	0.025459
3.10	0.066090	0.005086	−0.030766	−0.021478	0.478477	−0.017031
3.30	0.071072	−0.005537	−0.029764	−0.016828	0.452278	−0.057702
3.50	0.074075	−0.017527	−0.026341	−0.012643	0.421876	−0.095042
3.70	0.074777	−0.030478	−0.020638	−0.009088	0.387687	−0.127649
3.90	0.072948	−0.043915	−0.013002	−0.006291	0.350416	−0.154296
4.10	0.068459	−0.057308	−0.003828	−0.004338	0.311002	−0.173990
4.30	0.061297	−0.070105	0.006429	−0.003272	0.270562	−0.186015
4.50	0.051569	−0.081746	0.017274	−0.003092	0.230331	−0.189969
4.70	0.039499	−0.091701	0.028192	−0.003749	0.191598	−0.185781
4.90	0.025424	−0.099486	0.038672	−0.005155	0.155643	−0.173717
5.00	0.017769	−0.102432	0.043594	−0.006101	0.139088	−0.164906
5.20	0.001537	−0.106222	0.052516	−0.008384	0.109522	−0.142240
5.40	−0.015451	−0.107000	0.059869	−0.011044	0.085491	−0.113747
5.60	−0.032589	−0.104624	0.065312	−0.013890	0.067783	−0.080713
5.80	−0.049242	−0.099079	0.068589	−0.016723	0.056943	−0.044617
6.00	−0.064782	−0.090484	0.069539	−0.019339	0.053245	−0.007073
6.20	−0.078608	−0.079085	0.068106	−0.021544	0.056676	0.030240
6.40	−0.090178	−0.065253	0.064340	−0.023160	0.066931	0.065647
6.60	−0.099034	−0.049464	0.058393	−0.024037	0.083422	0.097544
6.80	−0.104820	−0.032286	0.050516	−0.024053	0.105303	0.124470
7.00	−0.107306	−0.014351	0.041045	−0.023127	0.131496	0.145162
7.20	−0.106392	0.003669	0.030387	−0.021220	0.160739	0.158613
7.40	−0.102123	0.021094	0.019007	−0.018337	0.191634	0.164113
7.60	−0.094680	0.037260	0.007402	−0.014523	0.222703	0.161276
7.80	−0.084381	0.051557	−0.003914	−0.009884	0.252453	0.150064
8.00	−0.071666	0.063446	−0.014436	−0.004537	0.279433	0.130783
8.20	−0.057077	0.072494	−0.023688	0.001347	0.302297	0.104076
8.40	−0.041237	0.078387	−0.031241	0.007575	0.319860	0.070896
8.60	−0.024821	0.080946	−0.036738	0.013938	0.331145	0.032469

(continued)

Table 4.4 (continued)

	Poisson ratio ν = 0.50					
k	v_3	v_4	u_{31}	u_{41}	u_{32}	u_{42}
8.80	−0.008532	0.080139	−0.039897	0.020213	0.335429	−0.009758
9.00	0.006940	0.076080	−0.040536	0.026179	0.332273	−0.054170
9.20	0.020939	0.069027	−0.038574	0.031622	0.321537	−0.099054
9.40	0.032879	0.059370	−0.034034	0.036347	0.303394	−0.142664
9.60	0.042267	0.047614	−0.027045	0.040184	0.278319	−0.183294
9.80	0.048731	0.034360	−0.017837	0.042996	0.247073	−0.219347
10.00	0.052031	0.020271	−0.006729	0.044685	0.210671	−0.249396
10.20	0.052074	0.006051	0.005883	0.045196	0.170346	−0.272243
10.40	0.048920	−0.007594	0.019539	0.044518	0.127492	−0.286970
10.60	0.042780	−0.019983	0.033740	0.042687	0.083613	−0.292968
10.80	0.034002	−0.030498	0.047963	0.039780	0.040263	−0.289970
11.00	0.023066	−0.038607	0.061684	0.035916	−0.001024	−0.278054
11.20	0.010550	−0.043890	0.074400	0.031247	−0.038788	−0.257642
11.40	−0.002883	−0.046064	0.085651	0.025956	−0.071708	−0.229482
11.60	−0.016527	−0.044991	0.095035	0.020243	−0.098651	−0.194620
11.80	−0.029656	−0.040691	0.102226	0.014323	−0.118722	−0.154353
12.00	−0.041563	−0.033339	0.106987	0.008413	−0.131295	−0.110179
12.20	−0.051594	−0.023260	0.109178	0.002721	−0.136040	−0.063738
12.40	−0.059173	−0.010916	0.108762	−0.002556	−0.132936	−0.016743
12.60	−0.063837	0.033117	0.105809	−0.007244	−0.122266	0.029083
12.80	−0.065253	0.018171	0.100483	−0.011202	−0.104610	0.072078
13.00	−0.063241	0.033515	0.093047	−0.014320	−0.080815	0.110708
13.20	−0.057777	0.018395	0.083840	−0.016530	−0.051961	0.143623
13.40	−0.049005	0.062061	0.073270	−0.017803	−0.019314	0.169710
13.60	−0.037224	0.073805	0.061792	−0.018152	0.015725	0.188136
13.80	−0.022884	0.082994	0.049894	−0.017631	0.051681	0.198377
14.00	−0.006561	0.089103	0.089103	−0.016328	0.087065	0.200237
14.20	0.011061	0.091735	0.026804	−0.014366	0.120432	0.193851
14.40	0.029227	0.090647	0.016548	−0.011891	0.150447	0.179678
14.60	0.047137	0.085759	0.007706	−0.009069	0.175932	0.158477
14.80	0.063989	0.077164	0.000614	−0.006076	0.195918	0.131276

Chapter 5
Dynamics of a Rigid Foundation on the Surface of an Elastic Half-Space

Abstract This chapter is about the response of a rigid rectangular foundation block resting on an elastic half-space. It is determined by considering first the displacement functions for any position on the surface of an unloaded half-space due to a harmonic point force. The influence of the foundation has been taken into account by assuming a relaxed condition at the interface, i.e., a uniform displacement under the foundation and that the sum of the point forces must equal the total applied force. The vertical, horizontal, and rocking motions have been considered, and numerical values for the in-phase and the quadrature components of the displacement functions are presented for a Poissons ratio of 0.25.

In addition, dynamic response of coupled horizontal and rocking motion of rectangular foundations resting on the surface of an elastic half-space medium when subjected to harmonic horizontal force and moment is considered, and numerical results for a number of cases are provided. Then, a mathematical procedure for analysis of the interaction between two foundations resting on the surface of an elastic half-space medium is presented. The analysis was extended to study the interactions between active and passive as well as two active foundations.

Keywords Dynamic rigid foundation • Elastic half-space response • Massless rigid base • Horizontal and rocking vibration • Two massive base

The response of a rigid rectangular foundation block resting on an elastic half-space has been determined by considering first the displacement functions for any position on the surface of an unloaded half-space due to a harmonic point force. The influence of the foundation has been taken into account by assuming a relaxed condition at the interface, i.e., a uniform displacement under the foundation and that the sum of the point forces must equal the total applied force. The three motions of vertical, horizontal, and rocking have been considered, and numerical values for the in-phase and the quadrature components of the displacement functions are presented for a Poisson ratio of 0.25. The effect of the mass and inertia of the foundation can be allowed for by an impedance matching technique. Response

H.R. Hamidzadeh et al., *Wave Propagation in Solid and Porous Half-Space Media*,
DOI 10.1007/978-1-4614-9269-6_5, © Springer Science+Business Media New York 2014

curves and nondimensional resonant frequency curves are given for a square and a rectangular foundation for different mass and inertia ratios and for several values of Poisson ratio. These curves are for design purposes and are an addition to similar published curves for circular and infinitely long rectangular foundations. Some of the calculated results have been verified by a laboratory experiment.

In addition, dynamic response of coupled horizontal and rocking motion of rectangular foundations resting on the surface of an elastic half-space medium when subjected to harmonic horizontal force and moment is considered, and numerical results for a number of cases are provided. Moreover, a mathematical procedure for analysis of the interaction between two foundations resting on the surface of an elastic half-space medium is presented. The analysis was extended to study the interactions between active and passive as well as two active foundations.

5.1 Introduction

The vibration of a rigid body on an elastic half-space in the vertical and horizontal directions and in rocking is in essence a three-dimensional wave propagation problem involving a set of dual integral equations (see Awojobi and Grootenhuis 1965). The problem reduces to a one-dimensional case only for torsional motion of a circular body about an axis normal to the surface of the half-space, and a solution in close form is then possible.

For all motions other than the torsional case, a difficulty can arise in determining the stress distribution immediately under the rigid base and in obtaining a pattern of displacement compatible with the prescribed motion and with a stress-free surface condition outside the base. The first attempt at a mathematical formation for the vertical oscillation of a rigid circular body was by Reissner (1936). Since then, attention has been confined in many subsequent publications to obtain numerical solutions for the dynamics of rigid bodies either with a circular base or in the shape of an infinitely long strip. Among the premier investigators that have made significant contribution, one should refer to Quinlan (1953), Sung (1953), Arnold et al. (1955), Luco and Westmann (1971), Veletsos and Wei (1971), Grootenhuis (1970), Clemmet (1974), Bycroft (1977), Luco and Westmann (1972), Hamidzadeh (1978), Hamidzadeh and Grootenhuis (1981), Rucker (1982), and Wolf (1985). For more extensive surveys of the literature, see the publications by Hamidzadeh (1978), Triantafyllidis and Prange (1988, 1989), Estorff and Kausel (1989), Hamidzadeh (2010), Verruijt (2010), Lou et al. (2011), Shahi and Noorzad (2011), and Pitilakis et al. (2013).

Over the years, only a few attempts have been made to obtain a numerical solution valid over a large frequency range for the dynamics of a rigid body with a rectangular base of finite dimensions. There are, of course, the other approaches for the prediction of the resonant frequency of a foundation based on the so-called subgrade reaction for the soil, as summarized in the books by Barkan (1962) and by Richart et al. (1970), but these approaches are really limited to the particular shapes of the foundation and to the soil conditions for which measurements have been taken. The asymmetrical wave propagation problem can be studied by considering first a rigid massless base that is resting over a certain rectangular area on the surface of the half-space with a prescribed motion or an oscillating force. Kobori and co-workers (1966a, b, 1970, 1971) considered the dynamic response of a rectangular base for vertical, horizontal, and rocking motion by assuming a single uniform stress distribution under the base for the first two motions and a linear stress distribution for rocking motion. This tends to give an overestimation of the response of a rigid body subjected to a dynamic force for vertical and horizontal motion. Elorduy et al. (1967) introduced a numerical technique for calculating the vertical and rocking displacements of a massless rectangular base by dividing the area into a number of square subregions, each subjected to a point force at the center. The displacement anywhere on the surface of a half-space due to a single point force was obtained using Lamb's approach (1904), and they used the numerical solutions given by Pekeris (1955a,b). It should be noted that these results do not agree closely at all for low values of the frequency factor with the approximate solutions given by Barkan (1962) for the vertical displacement function due to harmonic point forcing. In Elorduy et al.'s (1967) work, the total displacement of a corner of a subregion due to a point force at the center of every subregion was determined by the superpositioning method. The condition for compatibility of displacements over the entire area gave rise to a number of equations in terms of unknown local point forces but with $n + 1$ unknowns. However, summing of all the point forces and equating to the applied force gave an overall solution. Wong and Luco (1976) have improved on the approach developed by Thomson and Kobori (1963) and Kobori et al. (1966a, b) by applying an unknown and different but still uniform stress distribution on each subregion into which the massless base of arbitrary shape has been divided. This produced discontinuities in the stress at the boundaries of adjacent subregions, but by satisfying the compatibility of the overall mean stress, the displacement functions could be evaluated. The effect on the accuracy of the method by increasing the number of subregions from 1 to 64 was tested for the vertical motion of a square base, and these results are to be compared with computer values using the approach provided in this chapter. Wong and Luco (1976) have presented numerical solutions for vertical, horizontal, and the rocking motion for square and rectangular bases. A somewhat approximate solution for the vertical vibration of a rectangular foundation based on the solution for an equivalent circular body was introduced by Awojobi and Tabiowo (1976). It should be noted that in all of the above approaches, it has either been assumed specifically or otherwise it has been implied that there is a relaxed condition at the interface between the rigid body and the elastic half-space. In an article, Wong and Luco (1978) have given numerical values for the displacement functions for a bonded condition at the interface; the method of solution was similar to that used for the relaxed condition. The difference in the displacement values for the two conditions at the interface is not large and is in fact smaller than the difference in the values predicted by the different authors for the same condition.

Hamidzadeh (1978) and Hamidzadeh and Grootenhuis (1981) developed a numerical technique to solve the dynamic response of a foundation. In their mathematical model, the foundation was subjected to harmonic excitation, and it was rested with relaxed conditions on the surface of a homogeneous elastic half-space. In their work, the three independent vibrational modes of vertical, horizontal, and rocking for the foundation were analyzed, and the dynamic responses of a foundation block were then obtained by using an impedance matching technique. In their analysis, they utilized the method which was developed by Elorduy et al. (1967) for obtaining the response of a massless rectangular rigid base subjected to a harmonic force; however, the method was refined by considering calculation of displacements for all corners of each subregion due to the point forces applied at the center of all the subregions. This provided for compatibility of displacement at the boundaries of adjacent subregions. It should be noted that the presented solution was based on new solutions to the Lamb's problems for determining displacements on the surface of an elastic half-space excited by vertical and horizontal point force on the surface. Their solution to the Lamb's problem was independent of the method given by Pekeris (1955a, b). The details of Hamidzadeh's solution for the Lamb's problem (1978) were presented in Chaps. 2, 3, and 4.

Rucker (1982) presented an approximate solution to determine compliance functions for a rigid massless plate resting on an elastic or viscoelastic semi-infinite medium excited by forces and moments for all degrees of freedom. In his method, Rucker used Green's function approach to evaluate vertical, horizontal, rocking, and torsional compliance functions for rectangular plates with different side ratios. Wolf (1997) developed lumped parameter models for the dynamics of rigid foundations on the homogeneous elastic half-space medium, as well as homogeneous layer fixed at its base. Based on his analysis, he presented the frequency-independent coefficients for the mass-spring-dashpot model for different values of dimension ratios and Poisson ratios for all translational and rotational motions. Hamidzadeh (2010) presented a comprehensive literature review on the dynamic response of rigid foundations subjected to harmonic loads. He presented a methodology that can be utilized with some degree of confidence to estimate dynamic responses for a number of problems in the field of soil-foundation interaction. Pitilakis et al. (2013) provided an approximate linearization method for determining the dynamic impedance coefficients of rigid surface footings by considering the nonlinear behavior of soil. Their method was based on subdivision of the soil under the footing into a number of horizontal layers with different shear modulus and damping ratio.

In this chapter, the method of calculation is to be derived for a base with a square or a rectangular area; nevertheless, this approach can be applied to any arbitrary shape of contact area by suitable subdivision. To complete the analysis, design curves similar to those published by Awojobi and Grootenhuis (1965) for circular and infinitely long rectangular foundations (1965) are provided for vertical, horizontal, and rocking vibrations of square bases resting on an elastic half-space medium. It should be noted that in the presented analysis, the effects of embedment of a foundation into the soil will not be considered nor the effects of energy dissipation into heat. Loss of vibratory energy is predominately by radiation into the infinite half-space. Moreover, in the early part of this chapter, the horizontal and rocking motions of a foundation block are treated as independent of each other. There are practical conditions where these modes can operate almost entirely in an uncoupled manner, such as the horizontal motion of a wide slab foundation or the rocking of a tall narrow block. The presented results are furthermore of value for comparison with other published solutions. An extension of this work to the coupled horizontal and rocking motion of a foundation is to be presented in the latter part of this chapter.

5.2 Method of Analysis for a Massless Base

The area of a massless base of sides $2c$ and $2d$ is divided into a number of subregions of either a square or a rectangular shape, as shown in Fig. 5.1.

A point force which is proportional to the mean stress but of unknown amplitude is applied at the center of the jth subregion, $j = 1$ to n. A vertical force, F_j, will give rise to displacements in the vertical direction and to rocking motions about the orthogonal axes in the x and y directions. Please note that the equations for rocking motion will be given only about one axis. Also, a tangential force, P_j, will produce a horizontal displacement. The point forces at the center of the jth subregion can be related to the amplitude of the total applied forces F and P and the moment M by the following equations:

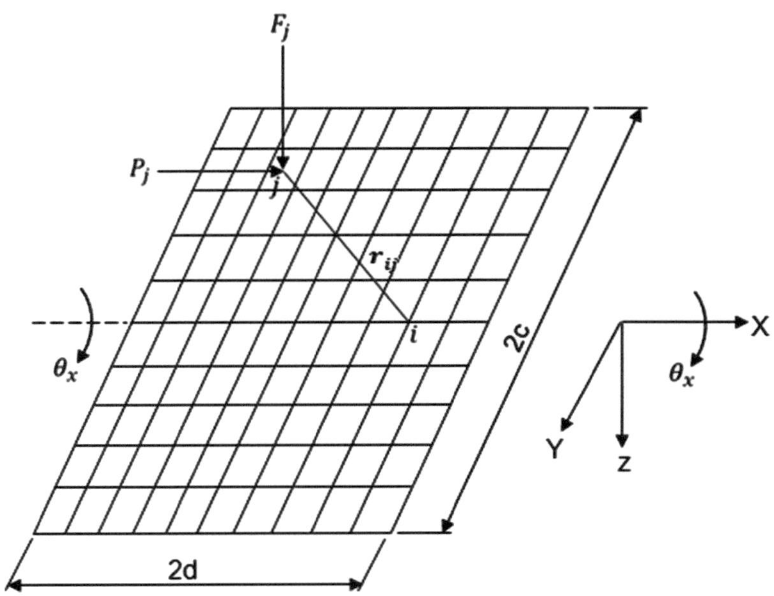

Fig. 5.1 Division of a rectangular base into subregions and definition of the coordinates

Vertical

$$F_j = a_j \frac{\overline{F}}{A} A_j \tag{5.1a}$$

Horizontal

$$P_j = b_j \frac{\overline{P}}{A} A_j \tag{5.1b}$$

Rocking

$$F_j = c_j \frac{\overline{M}}{I_x} y_j A_j \tag{5.1c}$$

where:

A_j—area of the *j*th subregion.
A—total area of the base.
a_j, b_j, and c_j—are unknown coefficients for the three motions.
I_x—is the second moment of area of the base about the *x*-axis.
Y_j—is the coordinate of the *j*th center from the *x*-axis.

Lamb's approach (1904) for determining the displacement on the surface of an infinite half-space due to a vertical point force has been adapted by Hamidzadeh (1978) for the estimation of the response at the *i*th corner, a distance r_{ij} from the point force, applied at the center of the *j*th subregion for the three motions under consideration. The coordinates have been defined in Fig. 5.1. The displacements are, respectively,

Vertical

$$U_{zi} = \frac{\overline{F}}{Gd}\left[\sum_{j=1}^{n} \frac{A_j a_j}{A\beta_{ij}} (V_z)_{ij}\right] \tag{5.2a}$$

Horizontal

$$U_{xi} = \frac{\overline{P}}{Gd}\left[\sum_{j=1}^{n} \frac{A_j b_j}{A\beta_{ij}} (H_x)_{ij}\right] \tag{5.2b}$$

Rocking

$$\theta_{xi} = \frac{\overline{M}}{Gd}\left[\sum_{j=1}^{n} \frac{A_j c_j y_j}{I_x \beta_{ij} y_j} (V_z)_{ij}\right] \tag{5.2c}$$

where, in addition to the symbols defined already,

V_z and H_x are nondimensional complex functions for the response, dependent on the frequency of the applied force and on the shear velocity and the Poisson ratio of the medium. In the above equations, G is the shear modulus of the medium, d is one half of the width of the base, and $\beta_{ij} = r_{ij}/d$.

The integral expressions in terms of the Rayleigh function for the six general response functions were presented in Chaps. 3 and 4. For this analysis, only two complex response functions, V_z and H_x, are used in the above equations, but the remainders are required as well for a general solution of the displacements throughout the half-space. Please note that the represented integrals were simplified by contour integration for the singularity of a Rayleigh pole and numerical values for the complete integrals were computed in the last two chapters.

In order to satisfy the condition of rigidity of the base, the compatibility of the displacements under the base requires that

Vertical

$$U_{zi} = U_z \tag{5.3a}$$

Horizontal

$$U_{xi} = U_x \tag{5.3b}$$

Rocking

$$\theta_{xi} = \theta_x \tag{5.3c}$$

A further condition for equilibrium at the interface under the base is for the sum of the forces at the center of the subregions to be equal to the total applied force, and in view of Eq. 5.1a–5.1c

Vertical

$$A = \sum_{j=1}^{n} a_j A_j \tag{5.4a}$$

Horizontal

$$A = \sum_{j=1}^{n} b_j A_j \tag{5.4b}$$

Rocking

$$I_x = \sum_{j=1}^{n} c_j y_j^2 A_j \tag{5.4c}$$

The above sets of companion equations have been solved numerically for a wide range of frequency of the applied excitation. The displacement for each of the motions considered can be expressed conveniently in terms of two nondimensional displacement functions, one in phase with the motion (F_1) and the other in quadrature (F_2) such that

Vertical

$$U_z = \frac{\overline{F}}{Gc}(F_{V1} + iF_{V2}) \tag{5.5a}$$

Horizontal

$$U_x = \frac{\overline{P}}{Gc}(F_{H1} + iF_{H2}) \tag{5.5b}$$

Rocking

$$\theta_x = \frac{\overline{M}}{Gc^3}(F_{R1} + iF_{R2}) \tag{5.5c}$$

The numerical values of the displacement functions F_1 and F_2, for the different uncoupled motions with subscripts V for vertical, H for horizontal, and R for rocking, have been computed for a square base and for a rectangular shape with a side ratio of ½ ($d/c = 0.5$). Each shape was divided into 64 subregions, which was about the maximum number to be accommodated on the available computer. The subregions for the rectangular shape had the same side ratio as the base. The results are given in Figs. 5.2–5.7 for the square and rectangular bases with side ratio of 0.5.

The abscissa is the frequency factor ($a = \omega d/c_2$), where ω is the excitation frequency in rad/s, d is half the width of the base, and c_2 is the velocity of shear waves in the medium. The displacement functions have been calculated for a frequency factor up to 1.8 where necessary for the subsequent computation of the resonant response of a foundation block. No values have been calculated for the frequency factor of zero, because of a singularity in the computing process. The results presented in Figs. 5.2–5.7 are for a Poisson ratio of 0.25 which is a reasonable value to take for certain types of soil. Similar sets of curves have been computed for values of Poisson ratio of 0, 0.31, and 0.5 (given by Hamidzadeh (1978)).

The in-phase function F_1 gives a measure of the stiffness of the elastic half-space, whereas the quadrature function F_2 is dependent upon the amount of energy radiated into the half-space. The amplitude response at resonance of a massive foundation block is therefore controlled solely by the quadrature function.

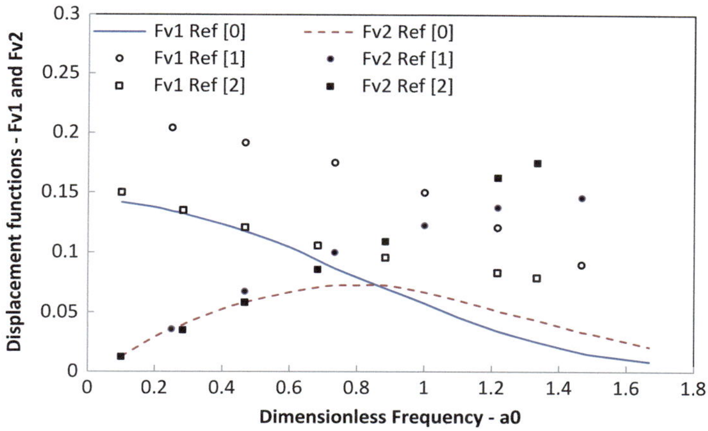

Fig. 5.2 Non dimensional displacement functions F_{V1} and F_{V2} versus frequency factor for vertical motion of a square base resting on an elastic half-space with a Poisson ratio of 0.25

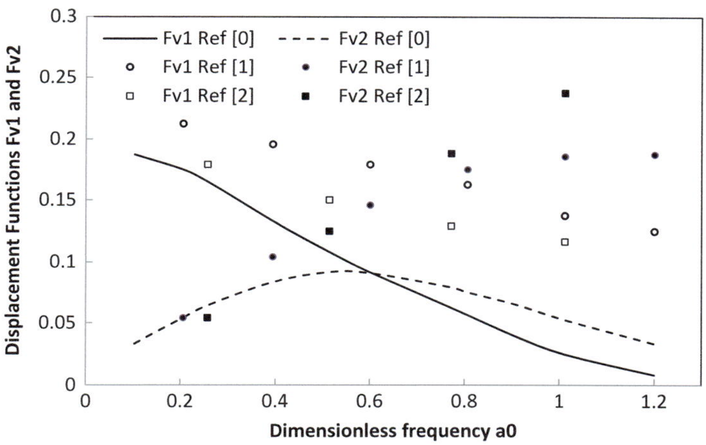

Fig. 5.3 Non dimensional displacement functions F_{V1} and F_{V2} versus frequency factor for vertical motion of a rectangular base with side ratio of 1/2 resting on an elastic half-space with a Poisson ratio of 0.25

5.2.1 Comparison

Kobori et al. (1966a, b) have also calculated the displacement functions for a Poisson ratio of 0.25, and their results have been included in Figs. 5.2–5.4. The values are generally much higher, in particular for F_2, because the assumption of stresses under the entire base gives an overestimation of the amplitude of response for vertical and horizontal motions. Only for the rocking motion, Figs. 5.6 and 5.7, are the values for F_1 at the higher frequency factors somewhat lower than those obtained in the

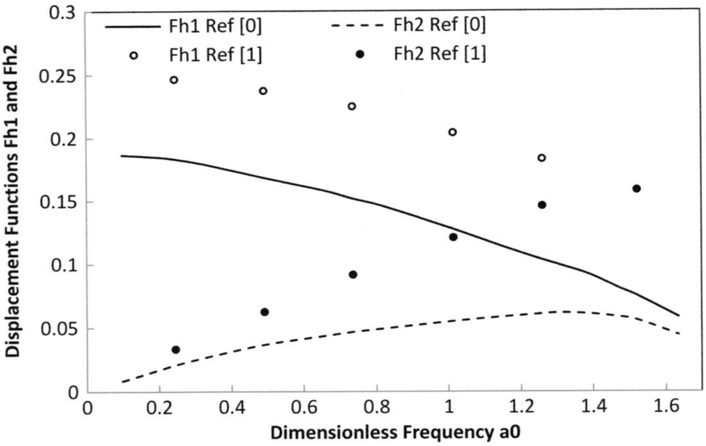

Fig. 5.4 Non dimensional displacement functions F_{H1} and F_{H2} versus frequency factor for horizontal motion of a square base resting on an elastic half-space with a Poisson ratio of 0.25

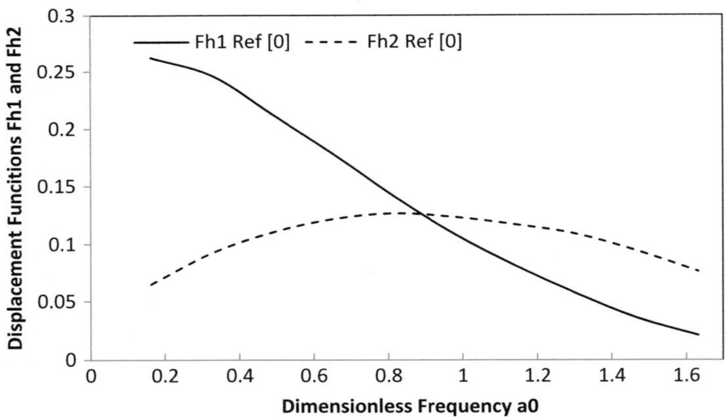

Fig. 5.5 Non dimensional displacement functions F_{H1} and F_{H2} versus frequency factor for horizontal motion of a rectangular base with side ratio of 1/2 resting on an elastic half-space with a Poisson ratio of 0.25

present analysis. The results by Elorduy et al. (1967) are also included and again the values for F_2 are much higher. This was to be expected because in the present analysis the displacement at the four corners of each subregion has been computed, and thereby a more accurate compatibility throughout the interface area has been obtained.

In the analysis by Wong and Luco (1976), a uniform but different stress distribution was imposed on the medium for each subregion which gave rise to a discontinuity at the boundary of adjacent subregions. Hence, it is to be expected

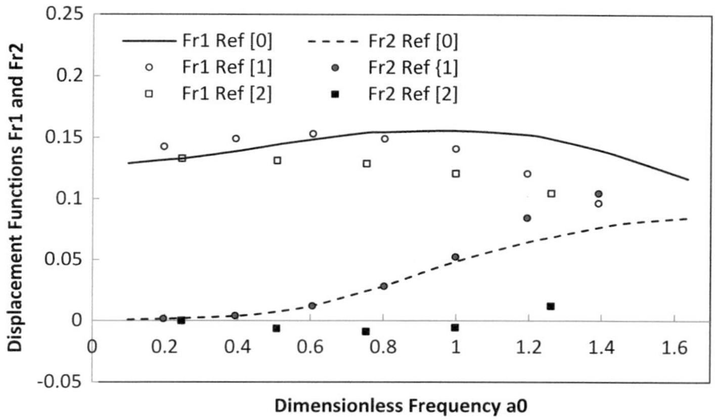

Fig. 5.6 Non dimensional displacement functions F_{R1} and F_{R2} versus frequency factor for rocking motion of a square base resting on an elastic half-space with a Poisson ratio of 0.25

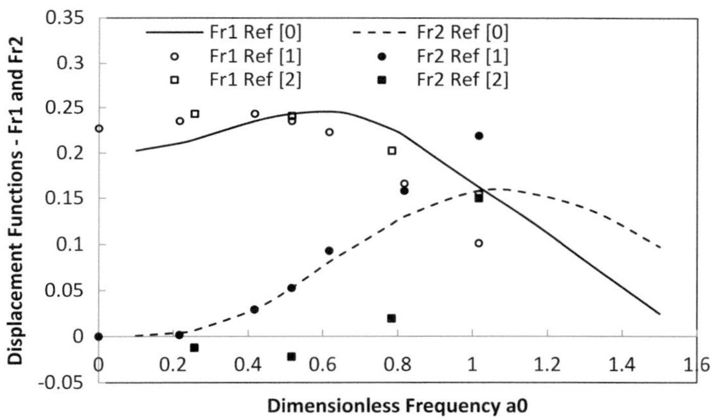

Fig. 5.7 Non dimensional displacement functions F_{R1} and F_{R2} versus frequency factor for rocking motion of a rectangular base with side ratio of 1/2 resting on an elastic half-space with a Poisson ratio of 0.25

that the greater the number of subregions into which the base area has been divided, the closer the values for F_1 and F_2 should approach those predicted by the present solution. They have given curves for F_1 and F_2 versus the frequency factor for the vertical motion of a square base and for a Poisson ratio of 1/3 with subdivision increased from 1 to 64. A comparison is given in Table 5.1 for several values of the frequency factor, a, and it can be seen that their values for $n = 64$ begin to approach those presented here. The values for the present work are also for a Poisson ratio of 1/3 and have been obtained by interpolation of the computed values for Poisson ratios of 0.31 and 0.5. Computation for $n > 64$ needs a very large store, and this

Table 5.1 The values of F_{v1} and F_{v2} displacement functions for vertical vibrations of a square footing for Poisson ratio of 0.25

	a/n	0.25	0.50	0.75	1.00	1.25	1.50
Values of—F_{v1}							
Wong and Luco (1976)	1	0.180	0.173	0.154	0.131	0.105	0.075
	4	0.162	0.150	0.128	0.101	0.075	0.05
	16	0.150	0.135	0.113	0.086	0.064	0.045
	64	0.144	0.128	0.105	0.078	0.058	0.041
Hamidzadeh (1978)	64	0.122	0.104	0.079	0.053	0.031	0.013
Values of F_{v2}							
Wong and Luco (1976)	1	0.030	0.064	0.094	0.113	0.128	0.137
	4	0.030	0.063	0.088	0.101	0.108	0.108
	16	0.030	0.061	0.083	0.094	0.096	0.094
	64	0.030	0.060	0.078	0.089	0.090	0.086
Hamidzadeh (1978)	64	0.032	0.055	0.065	0.062	0.050	0.034

was not done. The values obtained by Wong and Luco would seem to be converging satisfactorily at $n = 64$. However, they retained a square shape for the subregions when dividing a rectangular base area for a side ratio of ½, the number of subregions was only 32. In the present work, the shape of the subregions was the same as the base, and a value of $n = 64$ could be maintained. A further comparison of the other motions for a square and for a rectangular base has shown a similar trend, the exception being at the lower values for a, for a square base with a rocking motion when the values for F_2 were slightly lower than those predicted here. The agreement for all the cases considered in the two approaches is closest for the F_1 function, and this should give similar values for the resonant frequency factor of a massive foundation, whereas there are greater differences for the quadrature function F_2 which can affect the amplitude response at resonance.

5.3 Dynamic Response of a Massive Foundation

To evaluate the dynamic response of a foundation block, the displacement functions and the impedance matching technique will be used. The assembled systems (C) of a mass, m, or of inertia of I and a half-space for the three motions are shown in Fig. 5.8 and can be split into components of mass or inertia (A) and medium (B). The axis of rocking has been taken through the center point O, in the base of the block. The impedance of the rigid body is defined the rigid body is defined as the ratio of the applied force divided by the resultant velocity. Since the applied force and moment are harmonic, the displacements given by Eqs. 5.5a–5.5c can be transformed into velocities. By adding the component impedances, as was shown by Hamidzadeh (1978), the nondimensional amplitude of vibration for each mode can be expressed by the following expressions:

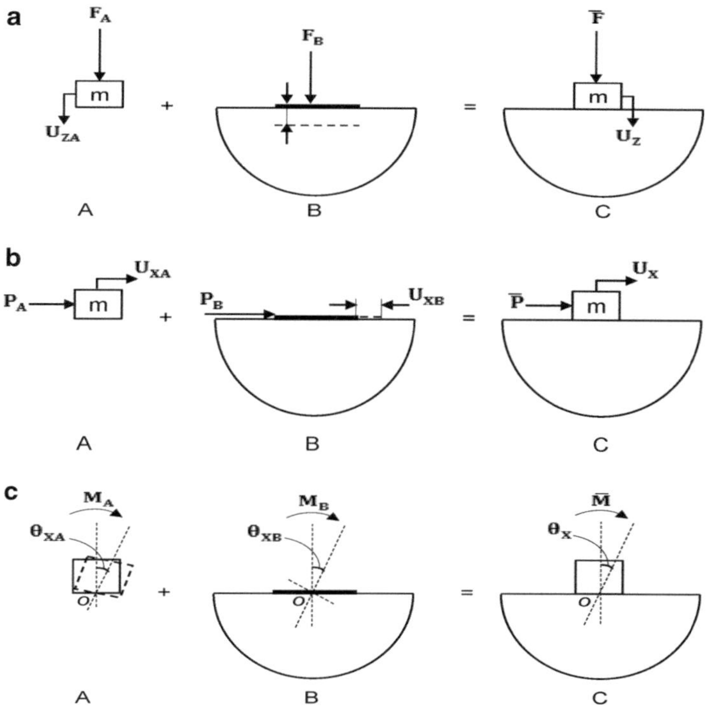

Fig. 5.8 Addition of a foundation block (A) to the elastic half-space (B) to form the complete system (C) for the (**a**) vertical, (**b**) horizontal, and (**c**) rocking motions

Vertical

$$\frac{|U_z|\,Gc}{\overline{F}} = \left[\frac{F_{v1}^2 + F_{v2}^2}{\left(1 + a^2 b F_{v1}\right)^2 + \left(a^2 b F_{v2}\right)^2}\right]^{1/2} \tag{5.6a}$$

Horizontal

$$\frac{|U_x|\,Gc}{\overline{P}} = \left[\frac{F_{H1}^2 + F_{H2}^2}{\left(1 + a^2 b F_{H1}\right)^2 + \left(a^2 b F_{H2}\right)^2}\right]^{1/2} \tag{5.6b}$$

Rocking

$$\frac{|\theta_x|\,Gc^3}{\overline{M}} = \left[\frac{F_{R1}^2 + F_{R2}^2}{\left(1 + a^2 b' F_{R1}\right)^2 + \left(a^2 b' F_{R2}\right)^2}\right]^{1/2} \tag{5.6c}$$

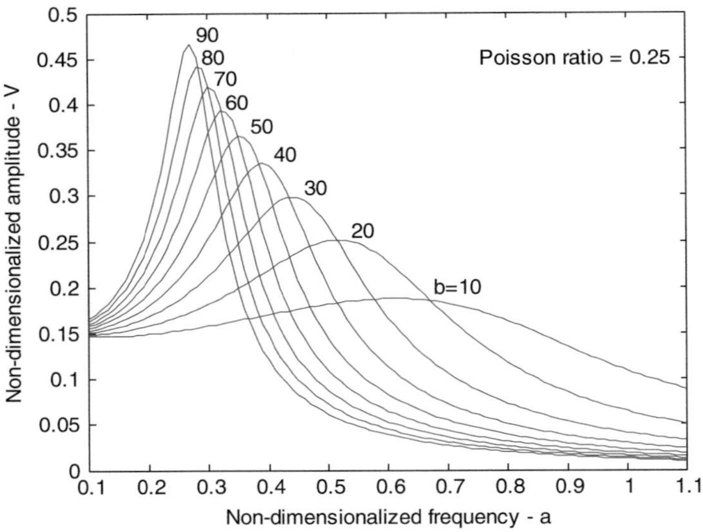

Fig. 5.9 Nondimensional frequency responses for vertical vibration of square bases with different mass ratios

Where the parameter a is the frequency factor, ρ is the density of the medium, and b and b' are the mass and inertia ratios defined as $b = \frac{m}{pcd^2}; \quad b' = \frac{I}{pc^2d^3}$

The amplitude factors have been plotted in Figs. 5.9, 5.10, and 5.11 as a function of the frequency factor "a" for different constant mass or inertia ratios for square bases and a Poisson ratio of 0.25. It can be seen from these frequency response curves that the lower the mass or inertia ratio, the lower the maximum value of the amplitude factor and the higher the value of the frequency factor at which this maximum occurs. There is a limit to this for the vertical and the horizontal modes. The frequency response curve becomes nearly flat for very low values of the mass or inertia ratios. This means that for a very wide range of frequency factors, the amplitude factor is approximately constant.

It may seem rather surprising at first that a foundation block with a high mass or inertia ratio will respond more vigorously at resonance than another block with lower values for the same ratio, when each is exposed to the same disturbing forces or moments. The energy radiated into the infinite half-space will be less in proportion to the kinetic energy of the vibrating system for the higher values of the mass or inertia ratio than for the lower values, and hence there will be less effective damping at resonance. For very low values of the mass ratio for vertical and for horizontal vibration, all the energy is radiated into the half-space, and there is then no resonant response.

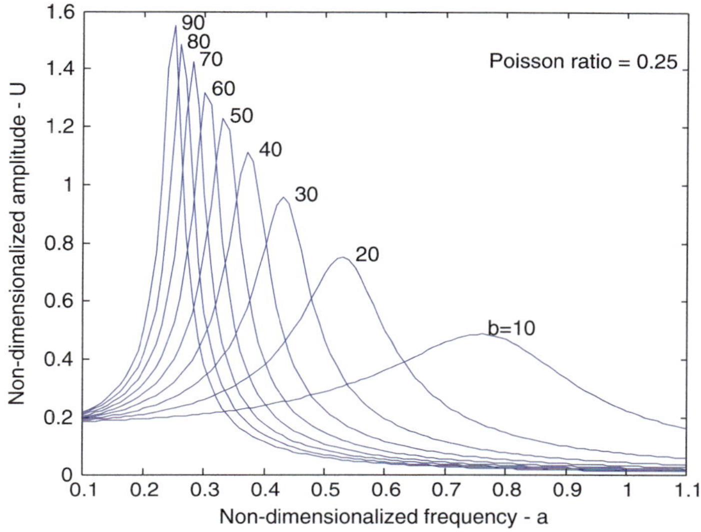

Fig. 5.10 Nondimensional frequency responses for horizontal vibration of square bases with different mass ratios

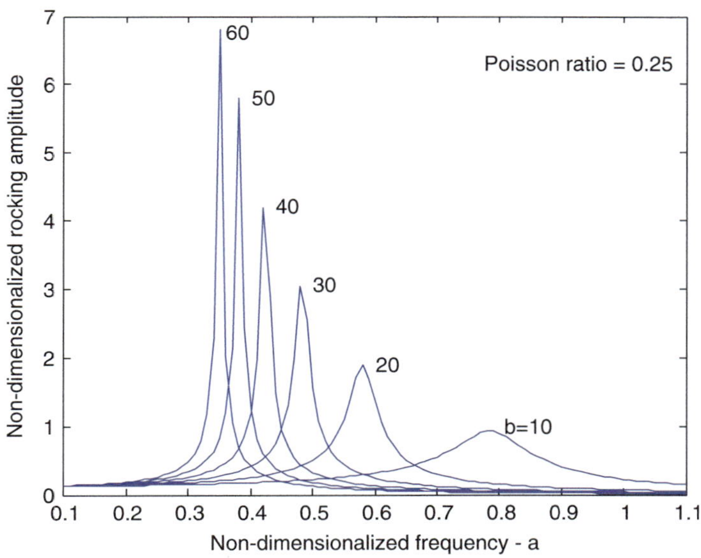

Fig. 5.11 Nondimensional frequency responses for rocking vibration of square bases with different mass ratios

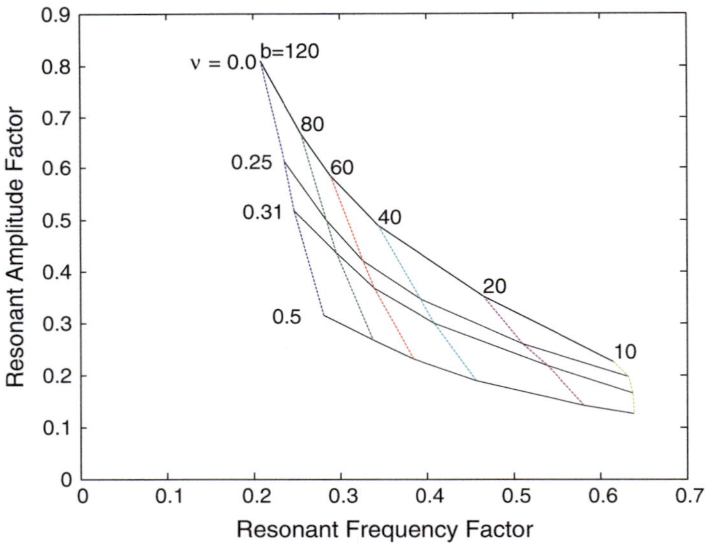

Fig. 5.12 Design curves for vertical oscillation of square bases with different mass and Poisson ratios

Another important parameter which has a significant effect on the dynamic response is the Poisson ratio. The previously established numerical results have shown that the resonant amplitude factor will decrease when the Poisson ratio of the medium is increased. The effects of mass ratio and of Poisson ratio on the frequency factor at resonance and the amplitude factor for the three modes are shown in Figs. 5.12, 5.13, and 5.14. These results play an important part in the design of a foundation block. The variation of the resonant frequency factor and the amplitude factor can be explained physically by the change in stiffness which accompanies a variation in Poisson ratio.

The static displacement factors, i.e., for a zero value of the frequency factor, have been estimated by extrapolation of the curves or F_1 presented in Figs. 5.2–5.7 and for other values of the Poisson ratio from Hamidzadeh (1978). These static values have been included on the design curves, Figs. 5.12–5.14, and it can be seen that the curves for each Poisson ratio tend toward these static values at the higher resonant frequency factors for the lower values of the mass or inertia ratios. There is therefore no dynamic response in any of the three motions considered for a foundation with a low mass or inertia ratio.

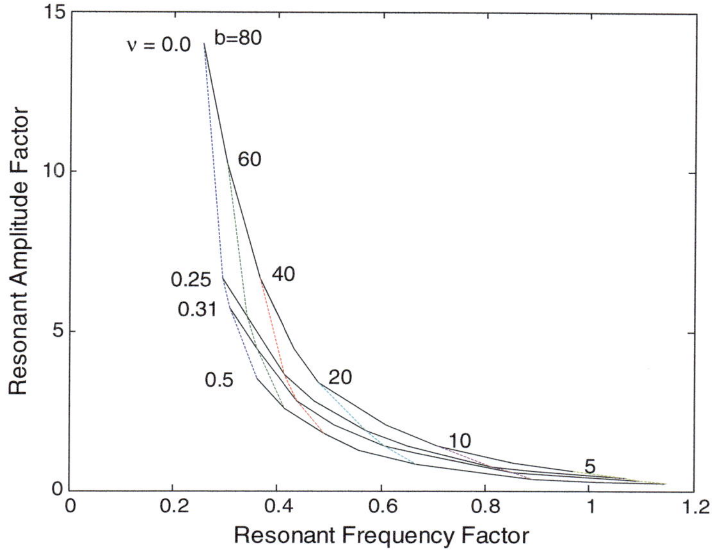

Fig. 5.13 Design curves for horizontal oscillation of square bases with different mass and Poisson ratios

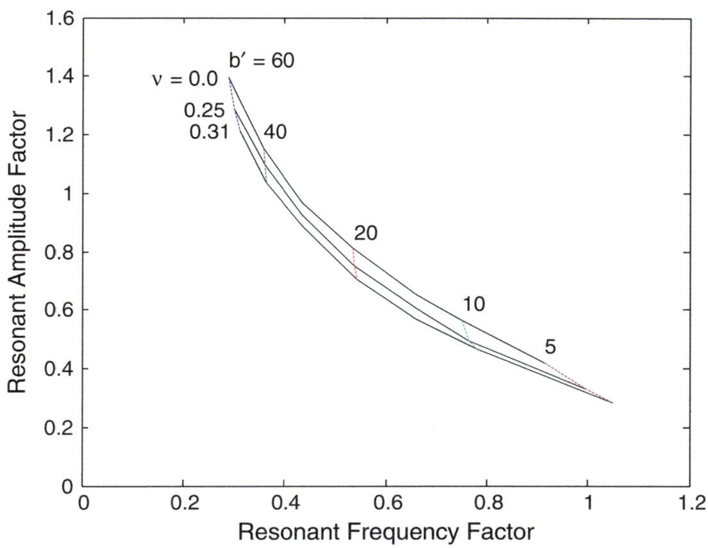

Fig. 5.14 Design curves for rocking oscillation of square bases with different Inertia and Poisson ratios

5.4 Experimental Verification

To verify some of the computed results, experimental work has been carried out. A laboratory model of an elastic half-space was built with sheets of sponge rubber 1.2 m by 0.9 m and 0.18 m total thickness laid on a bed of Plasticine to act as an energy sink with more Plasticine around the sides. The shear modulus and Poisson ratio under the dynamic loading were determined experimentally with a circular block for which the dynamic response can be predicted with confidence (see Awojobi and Grootenhuis (1965)). The dynamic responses for vertical vibration of different rectangular bases (side ratios of 1 and 1/2) were measured. In Fig. 5.15, b, the variation of the mass ratio with respect to the resonant frequency factor for a square and a rectangular base has been compared with the present theory, the theoretical results of Elorduy et al. (1967) and Wong and Luco (1976), and the experimental results of Awojobi and Tabiowo (1976). The latter measurements were taken by Tabiowo (1973) on the same half-space model as used in the present work. Some further experimental results are shown on Fig. 5.15 obtained by Girard (1968) for concrete slabs cast directly on the soil, giving very low mass ratios and apparently rather high values for the resonant frequency factor. The comparison shows a reasonable agreement between all these results.

5.5 Discussion and Conclusion

The displacement of a foundation block at resonance in any one of the three motions considered depends largely upon the impedance of the system, which in turn is dependent solely upon the quadrature component (F_2) of the displacement function. The close agreement between the predicted and the measured displacements at resonance and of the resonant frequencies which depend upon the in-phase component (F_1) provides a good basis for confidence in the computed values for the nondimensional displacement functions. The values of F_1 and F_2 presented are for a square base and for a rectangular base with a side ratio of ½ and for a Poisson ratio of 0.25. The range of the frequency factor covered is more than sufficient for the calculation of the frequency response curves for practical values of the mass and inertia ratios. Different shapes and proportions of the base can be accommodated by appropriate subdivision and by interpolation.

The frequency response curves show the usual characteristics of a larger displacement per unit value of the exciting force or moment correspond to larger mass and inertia ratios. This is because of the rather small amount of energy radiated into the half-space with a large mass and inertia ratio compared with the kinetic energy involved during a cycle of oscillation. On the other hand, a large base area lightly loaded will permit extensive radiation into the half-space, and hardly any dynamic magnification will occur at resonance. The frequency response curves become almost flat for mass or inertia ratios less than a certain value, but this minimum

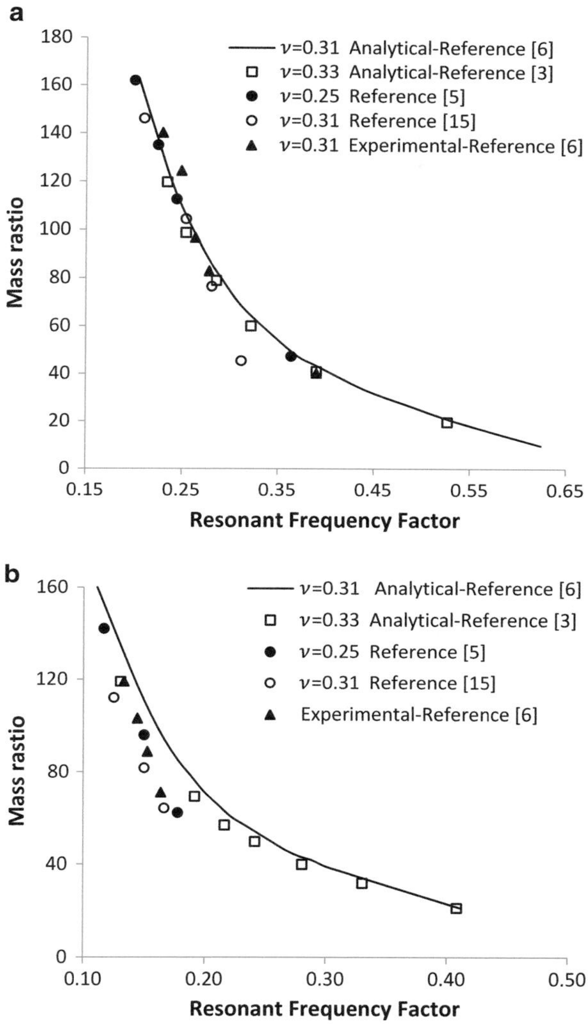

Fig. 5.15 Comparison of experimental results with the theoretical prediction of mass ratios versus the resonant frequency factors for vertical motion of (**a**) square footings on an elastic half-space medium, (**b**) rectangular footings on an elastic half-space medium (side ratio of 0.5)

value varies for the different modes of oscillation. There is no well-defined value of the resonant frequency factor for a low mass ratio for the vertical and horizontal modes, and the design curves become indeterminate for mass ratios less than about 5. The rocking mode is somewhat different, and a resonant frequency factor for an inertia ratio of 2 has been predicted although the magnitude of the displacement factor is very small. It can be seen from the frequency response curves that these are sharper for the rocking mode at low values of the inertia ratio than for the other two modes at low mass ratios. There is less radiation into the half-space during rocking because there is hardly any motion imposed on the medium in the center region of the base along the axis of rocking.

The design curves, Fig. 5.12–5.14, cover a wide range of mass and inertia ratio values and any value of the Poisson ratio between 0 and 0.5 by interpolation. These design curves provide a useful addition to the similar curves for vertical and rocking modes of an infinitely long rectangular foundation and for vertical and torsional modes of a circular foundation, published by Awojobi and Grootenhuis (1965).

The maximum value of the harmonic exciting forces F and P and the moment M is either a constant, or when caused by out of balance of a machine, it has to be calculated for the appropriate value of ω used in the frequency factor.

5.6 Simultaneous Horizontal and Rocking Vibration of Rectangular Footing

As was indicated in the last chapter, neither pure horizontal nor pure rocking motion can exist separately for practical foundations. In a realistic footing vibration problem, it is necessary to consider both motions, as a coupled motion or at least simultaneously. To the author's knowledge, an exact solution to the boundary value problem for the coupled motion is not known. Only an approximate solution is given for the coupled motion of a circular base. Veletsos and Wei (1971) obtained stiffness and damping coefficients for a rigid massless circular base on the surface of an elastic half-space medium. They also compared these with their simultaneous horizontal and rocking vibration results and found a close agreement between these two approaches. Luco and Westmann (1971) presented numerical results for displacement functions for the coupled motion of a rigid massless circular base. Clemmet (1974) improved the horizontal translation mode by allowing the rocking and the horizontal vibration of the circular base by defining new boundary conditions. Krizek et al. (1972) considered this problem for embedded foundations by using the finite lumped parameter model for the half-space. For rectangular footings Karasudhi et al. (1968) and Luco and Westmann (1972) presented an approximate solution for the coupled horizontal and rocking response of an infinitely long rigid foundation resting on the surface of an elastic half-space.

In the field of simultaneous rocking and horizontal motion of a footing, Hsieh (1962) introduced the derivation of the equation of motion. Hall (1967) extended the theory of the pure horizontal and the rocking modes for the simultaneous motion of a circular massive base under a harmonic applied moment. Richart and Whitman (1967) used a similar approach to compare their results with the data which was described for a footing model by Fry (1963), and their prediction was satisfactory. Ratay (1971) derived the solutions for this problem by following the Hsieh (1962) approach for a circular base on the surface of an elastic half-space which was excited by a harmonic horizontal force. Beredugo and Novak (1972) followed Ratay's approach by considering the problem for embedded footings.

The problem of simultaneous rocking and horizontal vibration of a rectangular base has not been considered by any investigator due to lack of a theoretical solution

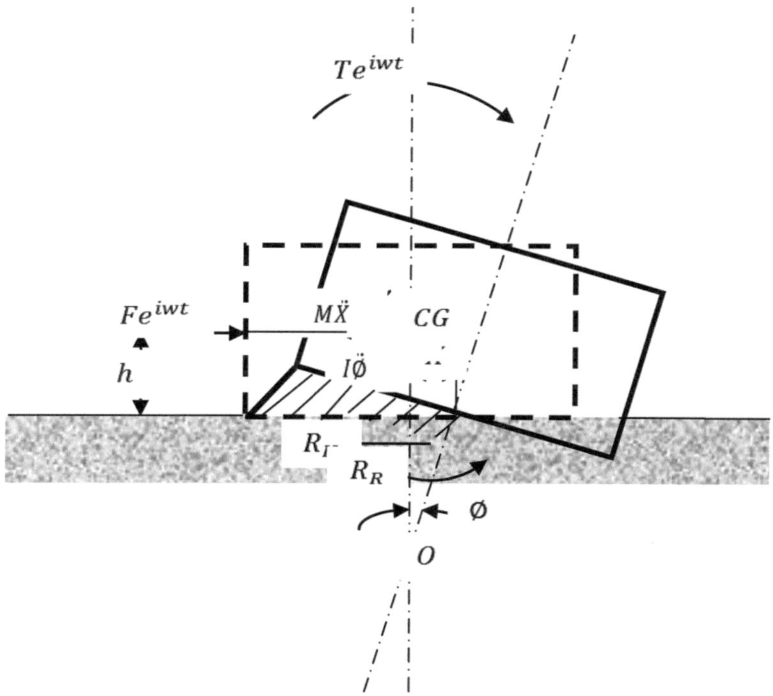

Fig. 5.16 Soil-foundation model for simultaneous rocking and horizontal vibrations

for this type of motion. The purpose of this chapter is to derive Hsieh's equation for the rectangular base. This is done by using the numerical solutions for the horizontal and rocking modes of a rectangular base which were established in Sect. 5.3 of this chapter and by considering the effects of coupling on both pure motions are investigated.

5.6.1 Equation of Motion

The idealized model considered here is a rigid rectangular footing attached to the top of a homogeneous *elastic* half-space medium. This is excited by a harmonic horizontal force (F) through its center of mass and a harmonic torque (T) about its horizontal axis, as shown in Fig. 5.16.

Hall (1967) demonstrated that the horizontal displacement of the contact in terms of the displacement of mass center is

$$x = x_{CG} - h\varnothing \tag{5.7}$$

and the equations of motion are

$$M\ddot{x} + Mh\ddot{\varnothing} + R_H = Fe^{jpt} \tag{5.8a}$$

$$I_{CG}\ddot{\varnothing} + R_R - R_H h = Te^{jpt} \tag{5.8b}$$

where

M is the mass of the footing.
I_{CG} is the mass moment of inertia in rocking about the center of gravity.
h is the height of center of gravity from the surface of the contact.
R_H is the amplitude of the reaction force due to the half-space media acting on the contact.
R_R is the amplitude of reaction torque due to the half-space media.
x is the horizontal displacement of the contact surface.
Ø is the angle of reaction of the base about the center of the mass.

Substituting for the reaction forces in terms of *x and* Ø as they were given in Eqs. 5.5b and 5.5c and also replacing the coefficient by the model notations of the half-space, the differential equations for the harmonic motion become

$$-a^2b\ GCx - a^2b\ G\overline{H}C\overline{\varnothing} + \frac{GCx}{F_{H1} + iF_{H2}} = F \tag{5.9a}$$

$$a^2b'G\ C^2\overline{\varnothing} + \frac{GC^2\overline{\varnothing}}{(F_{R1} + iF_{R2})} - \frac{\overline{H}C\ C_x}{(F_{H1} + iF_{H2})} = T \tag{5.9b}$$

Where

a is the frequency factor $\left(a = \frac{pd}{C_2}\right)$.
b is the mass ratio $\left(\frac{M}{\rho C d^2}\right)$.
b' is the inertia ratio $\left(\frac{I_{CG}}{C^3 d^2}\right)$.
G is the shear modulus.
C is the half length of the rectangular footing.
is the half length of the rectangular footing.
$\overline{H}$ is the ratio of h/C.
$(F_{H1} + iF_{H2})$ is the complex displacement function for the horizontal vibration.
$(F_{R1} + iF_{R2})$ is the complex displacement functions for the rocking vibration.
$\overline{\Phi}$ is equal to $c\ \Phi$.

These equations in matrix form are

$$\begin{bmatrix} -a^2b + 1/\left(F_{H1} + iF_{H2}\right) & -a^2b\overline{H} \\ -\overline{H}/\left(F_{H1} + iF_{H2}\right) & -a^2b' + 1/\left(F_{R1} + iF_{R2}\right) \end{bmatrix} \begin{pmatrix} \frac{GCX}{F} \\ \frac{GC\overline{\varnothing}}{F} \end{pmatrix} = \begin{pmatrix} 1 \\ \mu \end{pmatrix} \tag{5.10}$$

where $\mu = T/(CF)$.

The frequency response of the base can be given by solving the system of the last equation for different values of the frequency factor. In other words, the displacement amplitudes are

$$\left\{ \begin{array}{c} \frac{CGx}{F} \\ \frac{CG\varnothing}{F} \end{array} \right\} = \left[A \right]^{-1} \left\{ \begin{array}{c} 1 \\ \mu \end{array} \right\} \tag{5.11}$$

where $[A]^{-1}$ is the inverse of the coefficient matrix in Eq. 6.4.

5.6.2 Results and Discussions

In order to investigate the important characteristics of simultaneous horizontal and rocking motions of square bases, the frequency response of the base for different sizes and positions of applied force is studied. For a specific base with the following characteristics, $v = 0.31$, $\overline{H} = 1.5$, $\rho_b/p_s = 10.0$, and $\mu = 0$, where ρ_b is the density of the rigid block and ρ_s is the density of the half-space medium. The mass ratio and inertia ratio in terms of the above parameters can be written as

$$b = 8\overline{H}\rho_b/\rho_s = 120 \tag{5.12}$$

$$b' = \frac{1}{3}b\left(1 + \overline{H}^2\right) = 130 \tag{5.13}$$

In Fig. 5.17, nondimensional frequency responses are compared with the resonant conditions for the base when its motion is either pure horizontal translation or pure rocking. The comparing shows a significant coupling effect on both motions. In this figure, curves 1 and 2 represent the horizontal nondimensional amplitude response of the mass center. Also the horizontal resonant amplitude factor considerably increases, while the rocking amplitude factor decreases.

The decreasing of the fundamental frequency factor is due to the flexibility caused by coupling. This means that for each motion the flexibility of the other has decreased from infinity. The increase of resonant amplitude factor for the horizontal motion is also due to the flexibility of the foundation. As it was shown in Fig. 5.16, the axis of rotation is not on the contact surface of the block with the medium. In fact its position varies along the vertical axis through the mass center. Therefore, the amplitude of rocking should be lower than the calculated value for pure rocking. This can be shown to be true by substituting for R_H from Eq. 5.8a, when x = 0, into Eq. 5.8b. The resulting equation of motion for rocking shows that for pure rocking the center of rotation is on the contact surface of the base and the medium. The peak amplitude for the second resonance is much lower than the first one, because of the increase in radiational damping with the frequency factor.

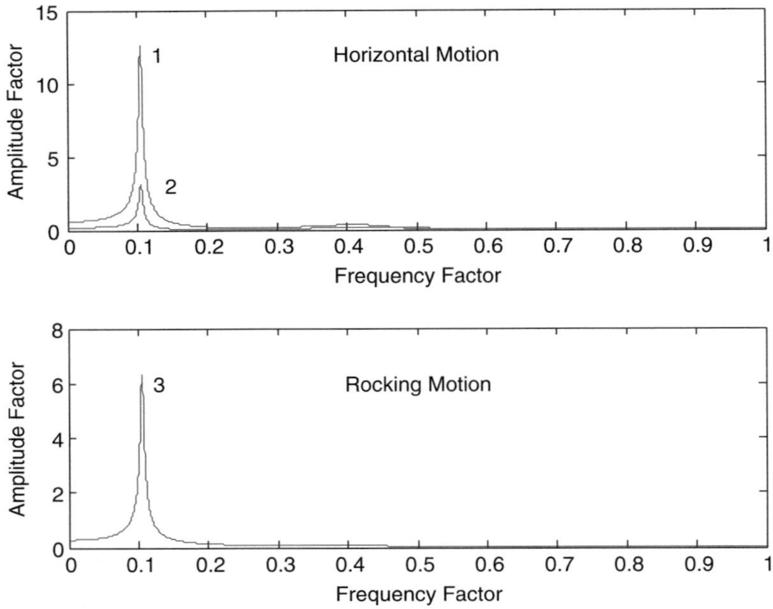

Fig. 5.17 Nondimensional frequency responses for coupled horizontal and rocking motion of rigid square block resting on an elastic half-space medium for $\nu = 0.31, \overline{H} = 1.5, \rho_b/\rho_s = 10, \mu = 1$ (1) GCX_{CG}/F, (2) GCX/F, (3) $GC\ \overline{\Phi}/F$

Another important factor is the location of the applied force. In Fig. 5.18 the effect of this location on the nondimensional frequency response of the rigid square block, with the following specification, is shown: $\upsilon = 0.31, \overline{H} = 1$, and $\rho_b/\rho_s = 10$.

Consequently, the mass and inertia ratios for the square block are 80 and 53.333. In this figure, curves 1, 2, and 3 correspond to values of 1, 0, and −0.5 for μ. These locations are indicated on the block shown in Fig. 5.18. Since the flexibility, mass, and inertia of the block are independent of the location of applied force, the first and second resonant frequency factors do not change. However, the value of applied moment depends on the position of the force, and therefore the nondimensional amplitudes are effectively changed by the variation of the location of the dynamic force.

The two parameters of mass ratio and inertia ratio for the square block are functions of $\overline{H}$ and ρ_b/ρ_s as they are given in Eqs. 5.12 and 5.13. Therefore any variation of $\overline{H}$ should effectively change the frequency responses of simultaneous motion. These equations show that by increasing the $\overline{H}$, values of b and b′ will increase. As a result of these increases, the resonant amplitude factors should also increase, while the first resonant frequency factor decreases. To verify this behavior, for a special case when $\rho_b/\rho_s = 10$ and the exciting horizontal force passes through the mass center of the block, the frequency responses of the motions for three different $\overline{H}$ values of 1.5, 1, and 0.5 were calculated and are given in Fig. 5.19.

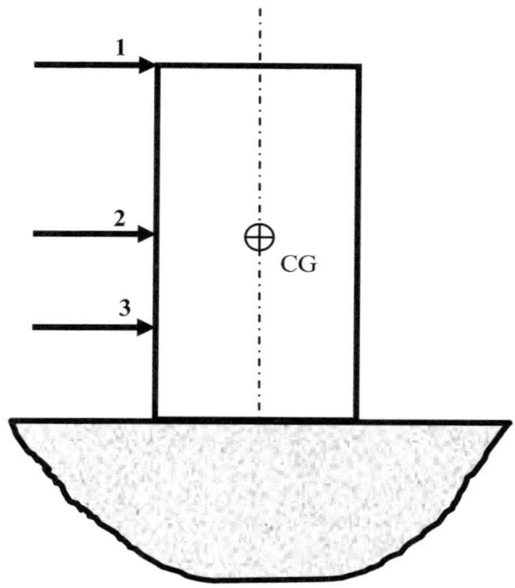

Fig. 5.18 Locations of applied horizontal harmonic forces for simultaneous horizontal and rocking vibration analysis of square block on the half-space medium

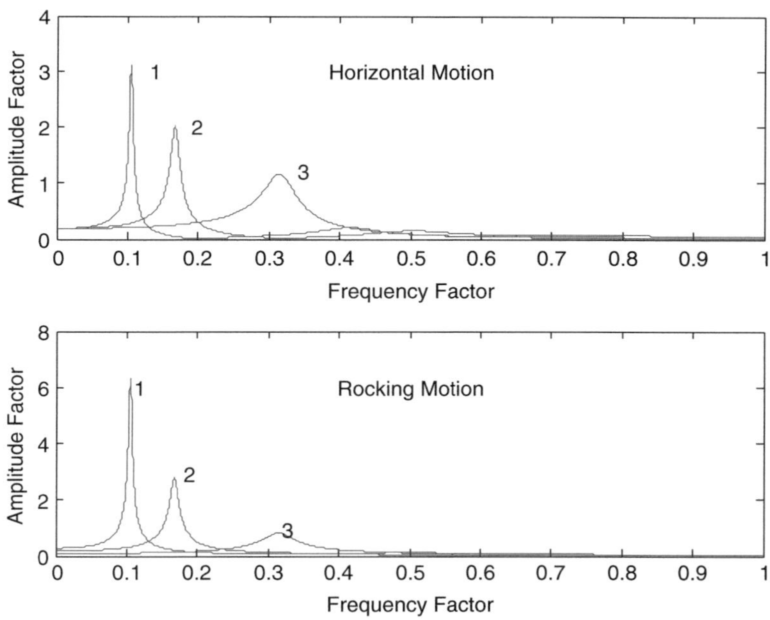

Fig. 5.19 Nondimensional frequency responses for coupled horizontal and rocking motion of rigid square block resting on an elastic half-space medium for $\nu = 0.31, \overline{H} = 1.0, \rho_b/\rho_s = 10$. (1) $\mu = 1$, (2) $\mu = 0.0$, (3) $\mu = -0.5$

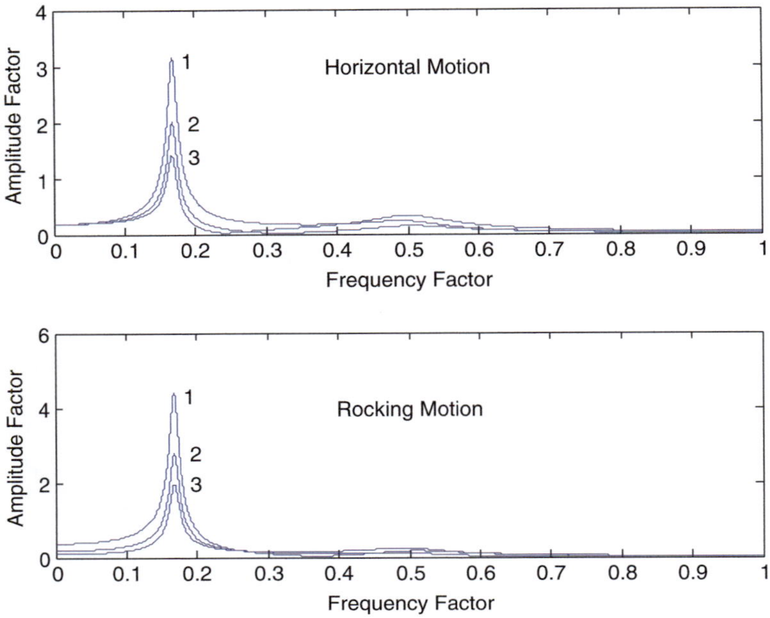

Fig. 5.20 Nondimensional frequency responses for coupled horizontal and rocking motion of rigid square block resting on an elastic half-space medium for $\nu = 0.31$, $\mu = 0$, $\rho_b/\rho_s = 10$. (1) $\overline{H} = 1.5$, (2) $\overline{H} = 1$, (3) $\overline{H} = 0.5$

Finally results have shown that for most cases only the first resonant frequency has any significant effect on the response, even though the motion has two peaks at low frequency factors (Fig. 5.20).

5.7 Response of Two Massive Bases on an Elastic Half-Space Medium

5.7.1 Introduction

The problem of two massive bases on the surface of an elastic half-space is an important problem which has only recently been considered by a few authors. Whitman (1969) was one of the earliest authors. He introduced this problem as a study of the interaction of nearby masses on the ground in his survey of soil dynamics. The first mathematical attempts in this field were made by Richardson (1969) and Warburton et al. (1971, 1972). Their work concerned the vibrations of two circular massive bases which were attached to the surface of an elastic half-space and were excited by harmonic forces.

In general, the displacement of a base is a combination of two components. One component is that of the base itself, and the other component is the displacement of the free surface upon which it rests due to the reaction force on the second base. The results for the second component are tabulated by Richardson (1969) and Richardson et al. (1971). To approximate the displacement of the second massless disc due to the reaction forces in the first base, an averaging procedure was introduced. They assumed that the displacement distribution under the second disc could be approximated by a parabolic function, and consequently they estimated the vibration of the passive disc by evaluating the response of two particular points on the discs.

Another publication was given by Lee and Wesley (1973). In their paper a general method of analysis is discussed where they determine the dynamic response of a group of flexible structure which is bonded in close proximity to an elastic half-space with seismic loading. They have improved on the technique of Richardson et al. (which was reasonable for a distance ratio of more than 10) in that they can find an accurate solution for small separation distances. This improvement was made by representing discrete values of displacements on the free surface of the half-space which were computed by Richardson by a set of continuous function and after integrating them over a circular occupied area of passive bases.

MacCalden and Matthieson gave experimental and theoretical results for the transmission of harmonic vibrations through the half-space media from an active rigid circular foundation to a passive one. They have used the same principle of superposition with the Bycroft solution (1965) for a single disc and the far-field equations for the surface displacements for horizontal, vertical, and rocking modes of vibrations. These can be described by Rayleigh waves.

Clemmet (1974) used the principle of superposition for his single circular mass theory. This is based on Richardsons's theory except that he has included hysteretic damping of the half-space and had also improved the horizontal vibration. He also compared one of his results with MacCalden and Tabiowo (1973). Tabiowo has only given experimental results in this field for vertical vibration of active and passive foundations.

5.7.2 Displacement of a Massless Passive Footing Due to Oscillations of an Active Massless Footing

In this section analyses of the oscillation of a passive foundation due to the oscillation of an active foundation are given. Here it is assumed that the displacement of the passive massless foundation is equal to the average displacement of the area occupied by the passive base, on the surface of an elastic half-space media. The numerical method which is used is the same as for the vibration of a single base, in that the surfaces of the connection of the footings are also divided into a number of square elements. Then, the displacements of the center of each element in the

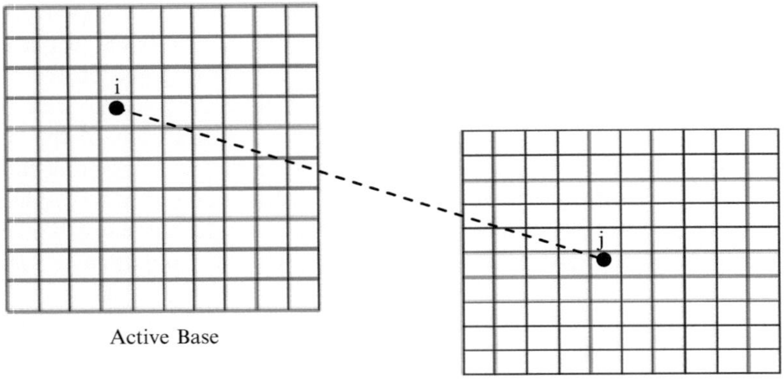

Fig. 5.21 Subregions for active and passive massless footings

passive base are evaluated by superimposing the response of all the forces in the centers of the subregions in the active base. Figure 5.21 shows the necessary form of the elements in the two passive and active bases for both rectangular bases. In the analysis only three modes of vibration, in z and x directions and rotation about y-axis, will be considered. In a previous chapter, values of K_j for different frequency factors have been evaluated for the three mentioned modes of vibration. Using these values, the displacement of the center of each element of the passive base, due to any excitation of the active foundation, can be evaluated by the results developed in Chaps. 3 and 4 for computing displacements of any point due to applied point forces.

The displacement of the ith center of the passive base due to the vertical, horizontal, and rocking modes of the active base can be evaluated by means of a point force, point displacement response theory which was discussed in previous chapters. The amplitude of displacement of the ith center of the passive base in the three directions for different modes is as follows.

For the vertical vibration of the active base,

$$z_{vi} = \sum_{j=1}^{N1} \frac{R_v K_j A_j}{AGr_{ij}} (\overline{v}_3 + i\overline{v}_4)_v \tag{5.14a}$$

$$Y_{vi} = \sum_{j=1}^{N1} \frac{R_v K_j A_j}{AGr_{ij}} (\overline{u}_3 + i\overline{u}_4) \sin\theta_{ij} \tag{5.14b}$$

$$x_{vi} = \sum_{j=1}^{N1} \frac{R_v K_j A_j}{AGr_{ij}} (\overline{u}_3 + i\overline{u}_4) \cos\theta_{ij} \tag{5.14c}$$

where N1 is the number of square subregions in the active base and K_j is the coefficient for the vertical force in the jth center of the active base. And the functions of $\overline{u}_3, \overline{u}_4, \overline{v}_3, and \overline{v}_4$ are given in Chap. 3.

For horizontal excitation in the x direction of the active foundation, displacement of ith center of the passive base

$$z_{Hi} = \sum_{j=1}^{N1} \frac{R_H K_j A_j}{AGr_{ij}} (\overline{v}_3 + i\overline{v}_4)_H \cos\theta_{ij} \tag{5.15a}$$

$$y_{Hi} = \sum_{j=1}^{N1} \frac{R_H K_j A_j}{AGr_{ij}} (\overline{w}_3 + i\overline{w}_4)_H \sin 2\theta_{ij} \tag{5.15b}$$

$$x_{Hi} = \sum_{j=1}^{N1} \frac{R_H K_j A_j}{AGr_{ij}} \left[(\overline{u}_{31} + \overline{u}_{41}) + (\overline{u}_{32} + \overline{u}_{42}) \cos 2\theta_{ij}\right]_H \tag{5.15c}$$

$K_j (j = 1, N1)$ in the above three relations are coefficients for a particular frequency factor for horizontal excitation. Values of $\overline{u}_{31}, \overline{u}_{41}, \overline{u}_{32}$, and $\overline{u}_{42}$ are given in Chap. 4. The three displacement amplitudes due to the rocking vibration of the active base are

$$z_{Ri} = \sum_{j=1}^{N1} \frac{R_R K_j x_i A_j}{I_y G r_{ij}} (\overline{v}_3 + i\overline{v}_4)_v \tag{5.16a}$$

$$y_{Ri} = \sum_{j=1}^{N1} \frac{R_R K_j x_i A_j}{I_y G r_{ij}} (\overline{u}_3 + i\overline{u}_4)_v \sin\theta_{ij} \tag{5.16b}$$

$$x_{Ri} = \sum_{j=1}^{N1} \frac{R_R K_j x_i A_j}{I_y G r_{ij}} (\overline{u}_3 + i\overline{u}_4)_v \cos\theta_{ij} \tag{5.16c}$$

In the last tree relations, $K_j (j = 1, N1)$ are coefficients for rocking vibration at a particular frequency factor. The total displacements of the passive base can be estimated by taking the average of the displacement of the centers of the square elements in the passive base. Therefore, displacements in the three directions of x, y, and z for the passive base are

$$z_p = \frac{1}{N2} \sum_{i=1}^{N2} z_{vi} + z_{Hi} + z_{Ri} \tag{5.17a}$$

$$x_p = \frac{1}{N2} \sum_{i=1}^{N2} x_{vi} + x_{Hi} + x_{Ri} \tag{5.17b}$$

$$y_p = \frac{1}{N2} \sum_{i=1}^{N2} Y_{Vi} + Y_{Hi} + Y_{Ri} \tag{5.17c}$$

where N2 is the number of elements in the passive base.

Rotation of the passive base about the y-axis will be given by subtracting the vertical displacements of the central points from z_p and evaluating the average rotation due to these remaining displacements:

$$\theta_p = \frac{1}{N2}\sum_{i=1}^{N2}\frac{1}{xpp(i) - xop}\left[(z_i + z_{Hi} + z_{Ri}) - z_p\right] \tag{5.17d}$$

Xpp (i) is the "x" coordinate of the ith center of the element in the passive base, and xop is the "x" coordinate of the center of the passive base. Putting the three displacements of z_p, x_p and $D\,\theta_p$ into matrix form,

$$\begin{Bmatrix} z_p \\ x_p \\ D\theta_p \end{Bmatrix} = \begin{bmatrix} (F_1 + iF_2)_{vv} & (F_1 + iF_2)_{Hv} & (F_1 + iF_2)_{Rv} \\ (F_1 + iF_2)_{vH} & (F_1 + iF_2)_{HH} & (F_1 + iF_2)_{RH} \\ (F_1 + iF_2)_{vR} & (F_1 + iF_2)_{HR} & (F_1 + iF_2)_{RR} \end{bmatrix} \begin{Bmatrix} \frac{R_v}{GD_1} \\ \frac{R_H}{GD_1} \\ \frac{R_R}{GD_1^2} \end{Bmatrix} \tag{5.18}$$

where $(F_1 + iF_2)_{\alpha\beta}$ is the complex coefficient for the β-vibration of the passive base due to α-vibration of the active base and D_1 is the dimension of the active base (for circular base $D_1 = R_1$ and for rectangular base $D_1 = C_1$). These complex functions for the vertical vibration of rectangular footing are

$$(F_1 + iF_2)_{vv} = \frac{C_1}{d_1}\frac{1}{N2}\sum_{i=1}^{N2}\sum_{j=1}^{N1}\frac{k_j}{N1\,H\,(I, J)}(\overline{v}_3 + i\overline{v}_4)_v \tag{5.19a}$$

$$(F_1 + iF_2)_{vH} = \frac{C_1}{d_1}\frac{1}{N2}\sum_{i=1}^{N2}\sum_{j=1}^{N1}\frac{k_j}{N1H\,(I, J)}(\overline{u}_3 + i\overline{u}_4)_v \cos\theta_{ij} \tag{5.19b}$$

$$(F_1 + iF_2)_{vR} = \frac{C_1^2}{d_1^2}\frac{1}{N2}\sum_{i=1}^{N2}\frac{1}{(xpp(I) - xop)/d_1} \times \left\{\sum_{j=1}^{N1}\left[\frac{k_j}{N1H\,(I, J)}(\overline{v}_3 + i\overline{v}_4)_v\right] - [(F_1 + iF_2)_{vv}]\right\} \tag{5.19c}$$

In the last three relations, k_j are values which have been evaluated for vertical vibration of a massless base on a half-space medium. Other complex functions due to the horizontal vibrations of rectangular bases are

$$(F_1 + iF_2)_{Hv} = \frac{C_1}{d_1}\frac{1}{N2}\sum_{i=1}^{N2}\sum_{j=1}^{N1}\frac{k_j}{N1H\,(I, J)}(\overline{v}_3 + i\overline{v}_4)_H \cos\theta_{ij} \tag{5.20a}$$

$$(F_1 + iF_2)_{HH} = \frac{C_1}{d_1}\frac{1}{N2}\sum_{i=1}^{N2}\sum_{j=1}^{N1}\frac{k_j}{N1H\,(I, J)} \times \left[(\overline{u}_{31} + i\overline{u}_{41}) + (u_{32} + iu_{42})\cos 2\theta_{ij}\right]_H \tag{5.20b}$$

$$(F_1 + iF_2)_{HR} = \frac{C_1^2}{d_1^2}\frac{1}{N2}\sum_{i=1}^{N2}\frac{1}{(xpp(I) - xop)/d} \times \left\{\sum_{j=1}^{N1}\left[\frac{k_j}{N1H(I,J)}(\overline{v}_3 + i\overline{v}_4)\cos\theta_{ij}\right][(F_1 + iF_2)_{Hv}]\right\} \tag{5.20c}$$

where k_j is coefficient for horizontal vibration of an active base. Finally the complex functions due to a rocking vibration of the active base are

$$(F_1 + iF_2)_{Rv} = \frac{C_1^2}{d_1}\frac{1}{N2}\sum_{i=1}^{N2}\sum_{j=1}^{N1}\frac{k_j x_j}{N1\,(I_Y/A_A)\,H(I,J)}(\overline{v}_3 + i\overline{v}_4)_v \tag{5.21a}$$

$$(F_1 + iF_2)_{RH} = \frac{C_1^2}{d_1}\frac{1}{N2}\sum_{i=1}^{N2}\sum_{j=1}^{N1}\frac{k_j x_j}{N1\,(I_Y/A_A)\,H(I,J)}(\overline{u}_3 + i\overline{u}_4)_v \tag{5.21b}$$

$$(F_1 + iF_2)_{RR} = \frac{C_1^3}{d_1^2}\frac{1}{N2}\sum_{i=1}^{N2}\frac{1}{(xPE(I) - xOP)/d} \times \left\{\sum_{j=1}^{N1}\left[\frac{k_j x_j}{N1\,(I_Y/A_A)\,H(I,J)}(\overline{v}_3 + i\overline{v}_4)_v\right] - [(F_1 + iF_2)_{Rv}]\right\} \tag{5.21c}$$

where A_A is the area of the active base.

Finally the equations of motion for a rigid massless passive base are given in matrix form by

$$[F]_P\{d\}_P = \{R\}_a \tag{5.22}$$

where [F] is the matrix which was introduced in Eq. 5.18, $\{d\}_P$ is the displacement vector for the passive footing, and $\{R\}_a$ is the reaction force vector for the active footing.

5.7.3 Interactions Between Two Massive Bases

In this part the effect of two rigid and massive foundations on the surface of an elastic half-space is considered. For the analysis, only three modes of vibration of masses are studied in two directions, i.e., x and z, and also rocking about the y-axis. Here, it is assumed that the displacements of each of the bases are direct displacements due to their external force plus any displacement which is caused by reaction force at the surface of contact of the other foundation.

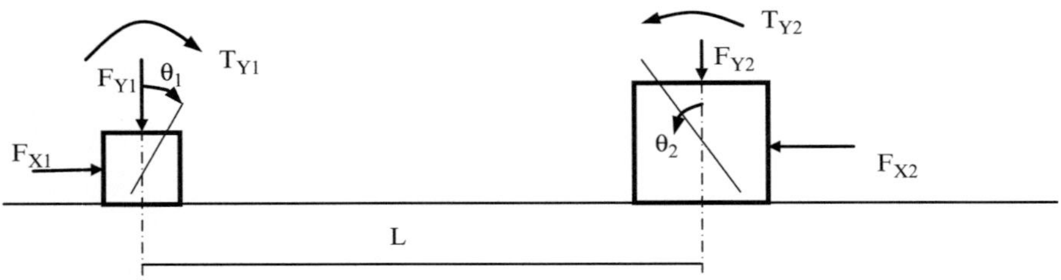

Fig. 5.22 Schematic of rigid bases resting on the surface of an elastic half-space

Figure 5.22 shows rigid bodies resting upon a semi-infinite elastic half-space with finite distance between them. The displacements and the acting forces on the first foundation are defined in terms of a left-handed system of Cartesian coordinate (x_1, y_1 and z_1). The second body is treated in the same way. These coordinate system have their origins situated at the centroid of each of the two bases.

The height of the centroids is given by the products $\overline{H}_1 D_1$ and $\overline{H}_2 D_2$ sphere D is one dimension of the base as it was introduced before. From the geometry of the problem, the relation between the displacement of the centroids and the contact surfaces is

$$U_{gi} = U_i + \overline{H}_i D_i \theta_i \tag{5.23a}$$

$$W_{gi} = w_i \tag{5.23b}$$

where the index g_i indicates the motion of the centroid of the foundation and i indicates the ith base. U_i, w_i, and θ_i are displacements of the rigid interface between ith base and elastic medium.

Equations of motion for the ith block are

$$M_i \frac{d^2}{dt^2} u_{gi} + R_{Hi} = F_{Hi} \tag{5.24a}$$

$$M_i \frac{d^2}{dt^2} w_{gi} + R_{vi} = F_{vi} \tag{5.24b}$$

$$I_i \frac{d^2}{dt^2} \theta_i + R_{Ri} - \overline{H}_i D_i R_{Hi} = T_{Ri} \tag{5.24c}$$

Substituting for u_{gi} and w_{gi} from Eqs. 5.23a and 5.23b into 5.24a and 5.24b gives

$$M_i \ddot{u}_i + M_i \overline{H}_i D_i \ddot{\theta} + R_{Hi} = F_{Hi} \tag{5.25a}$$

$$M_i \ddot{w}_i + R_{vi} = F_{vi} \tag{5.25b}$$

Also substituting for R_{Hi} in Eqs. 5.24c from 5.25a results in the third equation, which is

$$I_i\ddot{\theta}_i + R_{Ri} + \overline{H}_i D_i \left(M_i\ddot{u}_i + \overline{H}_i D_i\theta - F_{Hi}\right) T_{Ri} \tag{5.25c}$$

or

$$\left(I_i + M_i\overline{H}_i^2 D_i^2\right)^{\ddot{\theta}} + \overline{H}_i D_i M_i\ddot{u}_i = T_{Ri} - \overline{H}_i D_i F_{Hi} + R_{Ri} \tag{5.26}$$

where ($i = 1, 2$) for two blocks.

The matrix equation for motion of the two bases can be written as

$$[M]\left\{\ddot{d}\right\} + \{R\} = \{F\} \tag{5.27}$$

where

$$\{d\}^T = \{u_1, w_1, D_1, \theta_1, u_2, w_2, D_2, \theta_2\} \tag{5.28a}$$

$$\{R\}^T = \{R_{H1}, R_{V1}, R_{R1}, R_{H2}, R_{V2}, R_{R2}\} \tag{5.28b}$$

$$F^T = \left\{F_{H1}, F_{V1}, T_{R1} + \overline{H}_1 D_1 F_{H1}, F_{H2}, F_{V2}, T_{R2} + \overline{H}_2 D_2 F_{H2}\right\} \tag{5.28c}$$

$\left\{\ddot{d}\right\}$ is the second derivation of $\{d\}$ with respect to time and $[M]$. is the mass matrix.

$$[M] = \begin{bmatrix} M_1 & 0 & \overline{H}_1 D_1\, M_1 & 0 & 0 & 0 \\ 0 & M_1 & 0 & 0 & 0 & 0 \\ \overline{H}_1 D_1\, M_1 & 0 & I + \overline{H}_1^2 D_1^2 M_1 & 0 & 0 & 0 \\ 0 & 0 & 0 & M_2 & 0 & \overline{H}_2 D_2 M_2 \\ 0 & 0 & 0 & 0 & M_2 & 0 \\ 0 & 0 & 0 & \overline{H}_2 D_2 M_2 & 0 & I + \overline{H}_2^2 D_2^2 M_2 \end{bmatrix} \tag{5.29}$$

As it was indicated the displacements of any base have two main components. The first portion is caused by its own reaction force and the second one is caused by the reaction force of other base. Thus the relation between reaction forces and displacements for rectangular bases are given by

$$\begin{Bmatrix} u_1 \\ w_1 \\ G_1\theta_1 \\ u_2 \\ w_2 \\ C_2\theta_2 \end{Bmatrix} \left[\begin{array}{ccc|c} (F_{H1}+iF_{H2}) & 0 & 0 & \\ 0 & (F_{V1}+iF_{V2}) & 0 & \left[\ F\ \right]_{P_1} \\ 0 & 0 & (F_{R1}+iF_{R2}) & \\ \hline & & & (F_{H1}+iF_{H2}) \quad 0 \quad 0 \\ & \left[\ F\ \right]_{P_2} & & 0 \quad (F_{V1}+iF_{V2}) \quad 0 \\ & & & 0 \quad 0 \quad (F_{R1}+iF_{R2}) \end{array} \right] \begin{Bmatrix} R_{H1} \\ R_{V1} \\ R_{R1} \\ R_{H2} \\ R_{V2} \\ R_{R2} \end{Bmatrix} \tag{5.30}$$

or

$$\{d\} = [A]\{R\} \tag{5.31}$$

where $[A]$ is a complex matrix. Now multiplying Eq. 5.27 by $[A]$ the result is

$$[A][M]\left\{\ddot{d}\right\} + [A]\{R\} = [A]\{F\} \tag{5.32}$$

Since responses is harmonic, thus

$$\left\{\ddot{d}\right\} = -P^2\{d\} \tag{5.33}$$

After simplification Eq. 5.32 becomes

$$-p^2[A][M]\{d\} + \{d\} = [A]\{F\} \tag{5.34}$$

or

$$([I]) - P^2[A][M]\{d\} = [A]\{F\} \tag{5.35}$$

Solving the system of the above equations for $\{d\}$, the displacements of the two bases can be estimated on an elastic half-space medium.

5.7.4 Results and Discussion

In the present chapter, the dynamics of two rectangular bases on the surface of an elastic half-space medium were considered. In the analyses it was understood that there are many important parameters affecting the dynamic responses of the bases such as the shape of each base, their height, mass and inertia ratios, position of applied forces, distance of two bases, Poisson ratio of the medium, and many more. In fact, it is very difficult to consider the effect of all these parameters.

Special attention is paid to the response of two identical square bases. For this purpose, two problems were considered. Firstly, one of the bases was vertically active, while the other one was passive. Secondly, both bases were vertically active with similar harmonic exciting force at the mass center. Analysis of this problem

Table 5.2 Nondimensional displacement amplitudes of the active base subjected to vertical harmonic force for different frequency factors without and with passive base

a	Single active base	Active and passive bases		
	$\frac{\lvert W\rvert Gc}{F_V}$	$\frac{\lvert W_1\rvert Gc}{F_{V1}}$	$\frac{\lvert U_1\rvert Gc}{F_{V1}}$	$\frac{\lvert \theta_1\rvert Gc^2}{F_{V1}}$
0.1	0.17688	0.16635	0.00031	0.00027
0.2	0.32214	0.28663	0.00503	0.00681
0.3	0.32874	0.34512	0.01581	0.03636
0.4	0.10357	0.09996	0.00299	0.01183
0.5	0.05371	0.05322	0.00153	0.01064
0.6	0.03377	0.03394	0.00101	0.01344
0.7	0.02349	0.02368	0.00074	0.02238
0.8	0.01740	0.01746	0.00056	0.03579
0.9	0.01346	0.01344	0.00046	0.01645

Table 5.3 Nondimensional displacement amplitudes of the passive base excited by an active base subjected to vertical harmonic force for different frequency factors

a	Active and passive bases		
	$\frac{\lvert W_2\rvert Gc}{F_{V1}}$	$\frac{\lvert U_2\rvert Gc}{F_{V1}}$	$\frac{\lvert \theta_2\rvert Gc^2}{F_{V1}}$
0.1	0.03285	0.01110	0.00949
0.2	0.10920	0.02400	0.03245
0.3	0.13216	0.03115	0.07161
0.4	0.01266	0.01042	0.04120
0.5	0.00334	0.00647	0.04515
0.6	0.00169	0.00468	0.06206
0.7	0.00088	0.00354	0.10707
0.8	0.00051	0.00273	0.17331
0.9	0.00033	0.00221	0.07940

is based on the response of a single base and also the response of the medium free surface due to the oscillation of the base. As it was mentioned in the present chapter, second displacement components under the bases are averaged displacement of 64 discrete points on the contact surface due to the vibration of the other base.

Computing displacements for these two cases are given for special conditions which are set out below:

$$b_1 = b_2 = 100, b'_1 = b'_2 = 10, \quad \overline{H}_1 = \overline{H}_2 = 1, \quad v = 0.25, \text{and } \overline{d} = \frac{L}{c} = 5$$

where L is the distance between the centers of the bases. For the active and passive bases, the amplitude factors for nine discrete values of the frequency factor are given in Tables 5.2 and 5.3. In the case when both bases are active, the results are given in Table 5.4. The vertical amplitudes of active bases for both cases are compared with the results of an identical single base, when excited with a similar force. These results show that even though the system is excited by a vertical harmonic force, each base has three components of vertical, horizontal, and rocking displacement. The comparison of vertical displacement amplitude with those of single base on the medium shows that vertical displacement is not greatly affected by the presence of a passive base.

Table 5.4 Nondimensional displacement amplitudes of active base subjected to vertical harmonic force for different frequency factors without and with other active bases

a	Single active base	Two active bases		
	$\frac{\lvert W \rvert Gc}{F_V}$	$\frac{\lvert W_1 \rvert Gc}{F_{V1}}$	$\frac{\lvert U_1 \rvert Gc}{F_{V1}}$	$\frac{\lvert \theta_1 \rvert Gc^2}{F_{V1}}$
0.1	0.17688	0.19555	0.01135	0.00971
0.2	0.32214	0.33795	0.02500	0.03381
0.3	0.32874	0.23722	0.02211	0.05084
0.4	0.10357	0.09735	0.01127	0.04459
0.5	0.05371	0.05426	0.00761	0.05310
0.6	0.03377	0.03510	0.00567	0.07516
0.7	0.02349	0.03451	0.00426	0.12872
0.8	0.01740	0.01797	0.00315	0.19982
0.9	0.01346	0.01372	0.00236	0.08501

Also it is shown that the vertical displacement has more contribution in the motion of the active base. The vertical amplitudes of active bases for both cases are compared with the results of an identical single base, when excited with a similar force. These results show that even though the system is excited by a vertical harmonic force, each base has three components of vertical, horizontal, and rocking displacement.

Chapter 6
Experiments on Elastic Half-Space Medium

Abstract The vibration of a rigid foundation resting on the surface of an elastic half-space is in essence a three-dimensional wave propagation problem. Over the years, extensive research has been conducted to obtain analytical and numerical solutions for the dynamics of a rigid foundation with different geometries.

To verify the reported mathematical solutions, it is necessary to conduct experiments on an elastic half-space. This requires either in situ field tests or a simulation of an elastic half-space by the use of a laboratory model. The finite size model needs to be examined for all specified dynamic conditions of the elastic half-space. Also, due to the possibility of the elastic waves being reflected from the boundaries of the laboratory model, it is important to determine the dynamic limitations of the model, such as the size of the footing, the maximum amplitude of the displacement, and the size of the model itself. Prior to obtaining experimental results for the dynamics of a foundation, the knowledge of density, dynamic shear modulus, and the Poisson's ratio of the medium is essential.

Keywords Shear modulus of half-space medium • Dynamic properties of half-space medium • Response of coupled horizontal and rocking vibration

6.1 Introduction

The vibration of a rigid foundation resting on the surface of an elastic half-space is in essence a three-dimensional wave propagation problem. Over the years, extensive research has been conducted to obtain analytical and numerical solutions for the dynamics of a rigid foundation with different geometries. Arnold et al. (1955) considered the four vibrational modes of vertical, horizontal, torsional, and rocking of a circular base on the surface of elastic media. Awojobi and Grootenhuis (1965) used Hankel transform to define the complete dynamic problem by integral equations. They provided analytical solutions for the vertical and torsional oscillations of a circular foundation and the vertical and rocking oscillation of an infinitely long

H.R. Hamidzadeh et al., *Wave Propagation in Solid and Porous Half-Space Media*,
DOI 10.1007/978-1-4614-9269-6_6,

strip. Kobori et al. (1966a, b) studied the dynamic response of a rectangular base for vertical, horizontal, and rocking motions by assuming a single uniform stress distribution under the base for the first two motions and a linear stress distribution for the rocking motion. Elorduy et al. (1967) introduced a numerical technique for calculating the vertical and rocking displacements of a massless rectangular base. Wong and Luco (1967) improved the approach developed by Kobori et al. (1966a, b) by applying a different but still uniform stress distribution to each subregion into which the massless base of an arbitrary shape has been divided. They presented a numerical solution for vertical, horizontal, and rocking motions for square and rectangular bases. Hamidzadeh and Grootenhuis (1981) developed a numerical technique for the vertical, horizontal, and rocking vibrations of an arbitrary-shaped foundation. Karabalis and Beskos (1984) employed the boundary element method and provided a solution to the dynamics of a rectangular foundation in the time domain. Triantafyllidis (1986) formulated the dynamics of a rectangular foundation in terms of integral equations and presented an approximate solution to the influence functions characterizing the dynamic interaction of the foundation and the half-space. Hamidzadeh (1986) provided a solution to the displacements of the free surface of an elastic half-space subjected to a vertical, harmonic force on the surface.

To verify the reported mathematical solutions, it is necessary to conduct experiments on an elastic half-space. This requires either in situ field tests or a simulation of an elastic half-space by the use of a laboratory model. The finite size model needs to be examined for all specified dynamic conditions of the elastic half-space. Also, due to the possibility of the elastic waves being reflected from the boundaries of the laboratory model, it is important to determine the dynamic limitations of the model, such as the size of the footing, the maximum amplitude of the displacement, and the size of the model itself. Prior to obtaining experimental results for the dynamics of a foundation, the knowledge of density, dynamic shear modulus, and the Poisson ratio of the medium is essential.

Finite modeling has been used by a few investigators. Arnold et al. (1955) used 12 sheets of foam rubber as a medium and carried out reliable experiments for the vibrations of circular bases. The test model of Elorduy et al. (1967) consisted of sand in a concrete cylinder with an inside diameter of 2.24 m. A hemisphere of rubber was embedded in the middle of the sand such that its flat surface coincided with the free surface of the sand. Chae's (1967) model was made of sand in a concrete bin. He carried out harmonic tests to determine resonant frequencies of different rigid circular foundations and evaluated the shear modulus based on Sung's theoretical solution (1953). Stokoe and Woods (1972) in their model used sand as an elastic medium and saw dust as a damping medium for the reflected waves. Tabiowo (1973) tested different kinds of rubber to select a suitable material for his medium and found that ten sheets of NAT (natural rubber) was the most suitable material for simulating an elastic half-space medium. Using this model, he carried out vibration tests for both circular and rectangular bases. Luco et al. (2010) investigated the tracking (signal reproduction) capability of a shake table system through a series of broadband and harmonic experiments with different tuning and test amplitudes. Baidya and Mandal (2006) used a Lazan-type mechanical oscillator to collect a number of model block vibration tests carried out on differently prepared layered beds. Choudhury and Subba Rao (2005) determined seismic bearing capacity factors for shallow strip footing using the limit equilibrium approach and pseudo-static method of analysis. It is thus seen that in Choudhury et al. (2005), the pseudo-static approach is being used to determine bearing capacity and settlement and tilt of the foundations subjected to seismic loads in non-liquefying soils. The dynamic nature of the load and other factors which affect the dynamic response are not being accounted for. Also, no guidelines are available for the design of footings in liquefying soil.

6.2 Determination of Shear Modulus for the Medium

During the last decades, considerable attention has been paid to the design of footings subjected to dynamic loads. In all of the theories concerning the dynamic response of foundations, the shear modulus and Poisson ratio play an important role. The effect of other soil dynamic properties such as strain amplitude, effective mean principal stress, void ratio, number of cycles of loading, soil structure, etc., is also important. Numerous field and laboratory testing methods have been developed and are in use for determining the shear modulus of soils.

1. Field methods
   These methods are based on the measurement of the velocities of stress wave propagation through the medium from which the shear modulus can be estimated. They can be classified as:

   (a) Refraction and reflection surveys
   The refraction and reflection survey method is capable of estimating the average dilatation wave velocity C_1. The shear modulus will then be given by

   $$G = C_1^2\,(1 - 2v)\,\rho/\,[2\,(1 - v)] \tag{6.1}$$

   This method cannot be reliable unless the Poisson ratio is known accurately. See Grootenhuis and Awojobi (1965).

   (b) Rayleigh wave length method
   This can be used to measure the velocity of surface waves which are propagating on the surface of media. By neglecting the small difference between the velocity of shear wave and Rayleigh waves, the shear modulus is given by

   $$G = \rho C_1^2 \tag{6.2}$$

   In this type of field test, contributions have been made by many authors such as Jones (1958, 1959) who performed many steady-state vibration surveys.

(c) Direct shear wave measurement
Measurement of direct shear waves between surface and subsurface has been performed in several different types of excitation on the surface, and in each case geophones were placed down a borehole to detect the arrival of the direct shear wave. Erickson, Miller, and Waters (1968) used a horizontal vibrator for excitation. Jolly (1956), Shima et al. (1968), and White, Heaps, and Lawrence (1956) all used a horizontal impulse. Halperin and Frolova (1963) and White and Mannering (1974) used a falling weight. Many other attempts have been made by using explosive sources located at the subsurface. In this method, the shear waves arrive after dilatation waves, and so great care is needed to identify the shear waves. Yutak Ohta et al. (1978) measured the velocity of shear wave with the standard penetration test. In this test an impact at the bottom of a borehole is produced by weight dropping, and the generated wave was recorded on the surface near the borehole.
(d) Cross-hole method
In this method the time for the body wave to travel the horizontal path between two points in the soil is measured. These body waves are generated by a vertical impulse applied at the bottom of one borehole. Then, arrival of compression and shear waves is monitored in a second borehole with the same depth with a vertical velocity transducer. By bearing in mind that the compression wave has a higher velocity than the shear wave and also knowing the distance between the two boreholes, the shear wave velocity can be evaluated.

In practice several different types of sources have been used by many people: Ballard and Leach (1969) used a vertical vibrator, and Dobrin, Lawrence, and Sengbush (1954), McDonal et al. (1958), Swain (1962), and White and Senbush (1953, 1963) all used a falling weight. Riggs (1955) used an explosive source. Stokoe and Woods (1972) have used an impulse rod and hammer which triggers the oscilloscope, and Stokoe and Richard (1974) have developed this method. Beeston and McEvilly (1977) measured the velocity of the shear wave by downhole measurements.
(e) Footing vibration measurements
The calculated response characteristic of the vertical vibration of a rigid circular disc on the surface of an elastic half-space may be used to determine, indirectly, the shear modulus, by measuring the resonant responses of the rigid bodies for known values of mass ratios. This method has been used by a few authors such as Eastwood (1953), Jones (1958), Sung (1953), and Arnold et al. (1955) which have used the foam rubber model for half-space media. Bycroft (1959) and Kanai and Yoshizawa (1961) used the rocking of a rectangular building, where the length is much greater than the width, but this method cannot evaluate both Poisson ratio and the shear modulus independently. Grootenhuis and Awojobi (1965) estimated these two properties by measuring the resonant frequencies of the circular base for each of two close mass ratios. By comparing these two mass ratios with their exact solution, it was possible to estimate both Poisson ratio and the

shear modulus. These methods are very reliable for a pure elastic half-space, but not for a half-space with hysteretic damping. Tabiowo (1973) has carried uut a laboratory version of Grootenhuis and Awojobi's method and a model which was discussed before.

2. Laboratory testing: methods
These tests can be classified in the following manner:

(a) Resonant column test
This technique can be conducted by employing either a solid cylindrical soil specimen or a hollow cylindrical specimen. The specimen is vibrated by either a vertical cyclic load or a cyclic torque. The shear modulus can be determined by measuring the resonant frequency. Then, the elastic properties can be evaluated using the proportionality relations for elastic moduli with density of the medium, measured resonant frequency, and the height of specimen:

$$\mathrm{E\ or\ G} \propto \rho(2f_r h)^2 \tag{6.3}$$

where h is the height of specimen, ρ is the mass density of the soil, and f_r is the lowest resonant frequency. Hardin and Black (1966, 1968), Afifi and Woods (1971), and Cunny and Fry (1973) used solid cylindrical specimen, and Hardin and Drnevich (1972a, b) used a hollow cylindrical one.

(b) Pulse technique
Pulse methods have been used to determine the dynamic properties of soils subjected to small-strain amplitude vibrations. Basically, this technique consists of the measurement of the travel time of a stress pulse between two ends of a soil sample from which the stress wave velocity can be evaluated. From this the velocity, shear modulus, or Young's modulus may be estimated depending on the nature of the stress pulse applied (compression or shear). This technique has been used and further developed by a number of researchers such as Lawrence (1965), Matsukawa and Hunter (1956), and Taylor and Hughes (1965).

(c) Cyclic simple shear test
This method is a modification of the static simple shear test for application to low-frequency cyclic loading. This method provides a reasonable way of determining shear modulus and damping ratio of soil layers by normally using a cylindrical specimen under a cyclic shear test. The following authors have attempted this method: Converse (1962), Theirs and Seed (1968), and Kovacs et al. (1971). This method can also provide liquefaction characteristics. In general it has been shown that the shear strain level in this technique is greater than that in the resonant column tests.

(d) Free vibration method
Shear moduli and damping ratios of soils at very small strains can be found easily using the free vibration of samples excited by a shaking table. The free response of the vibration can give an estimated value of shear modulus using:

$G = 16H^2f_n^2\gamma/g$, where f_n is the natural frequency in cycles per second, H is the thickness of the layer in feet, G is the share modulus lbf/ft^2, g is the acceleration of gravity (32.2 ft/s^2), and γ is the unit weight of soil in lbf/ft^3.

To further study this technique, see De Graft-Johnson (1967) and Kovacs et al. (1971).

6.3 Determination of Dynamic Properties of the Medium

In most of the measuring techniques except for the footing vibration measurement technique, a value for the Poisson ratio cannot be determined accurately. The response of foundations on an elastic medium requires knowledge of the properties of both shear modulus and Poisson ratio of the media. Consequently the method of Grootenhuis and Awojobi (1965) is used. This is based on the accurate solution of a vertical vibration of a circular base on an elastic half-space. They give values for computed nondimensional amplitude and frequency factor in the resonance condition as curves for different mass ratios of the circular base and Poisson ratios of media.

To find the shear modulus, it is sufficient to measure the resonance frequencies of vertical vibration of a circular base with two slightly different mass ratios. In order to find the resonant frequency, the amplitude of vibration was kept constant at various frequencies, and the minimum required force was measured. By assuming that shear modulus is not frequency dependent, then ratios of the frequencies will give the ratio of their corresponding frequency factor. By using this ratio, the mass ratios and the appropriate curves of the proved theory of Grootenhuis and Awojobi, the correct value of the Poisson ratio and frequency factors can be determined, and consequently the value of shear modulus may be estimated from one of the frequency factors by

$$G = \left[\frac{2\pi R f_r}{a_r}\right]^2 \rho \tag{6.4}$$

where R is the radius of circular base, f_r is the resonant frequency, a_r is the resonant frequency factor, and ρ is the density of medium.

The important conditions of this technique are the correct measurement of masses of the two bodies and their resonant frequencies. Accurate measurement is possible provided that appropriate experimental equipment is used. In order to check the linearity of the medium, the same procedure must follow for different mass ratios and different radii of the bases.

If the medium acts as an elastic homogeneous half-space, it is possible to check the shear modulus of the medium by measuring the ratio of the resonant amplitudes of the two different mass ratios. Poisson ratio and the nondimensional amplitude at resonance are evaluated by using the previously mentioned theory of Grootenhuis and Awojobi. Since the determination of the amplitude of vibration is affected a great deal by the nonlinearity of medium and also varies due to hysteretic damping, this method is not recommended.

On the other hand, the shear modulus varies with changes in variation of strain (amplitude of vibration), and so it seems reasonable to use a constant amplitude technique to determine the frequency of resonance. This is due to the fact that any nonlinear characteristics of vibration in a constant force technique are avoided. Using this kind of technique, it is also possible to check if the ratio of minimum force and amplitude of vibration is varying in the resonant condition for a circular base. This was done by changing the amplitude of vibration for two particular bases in resonance and measuring the minimum forces.

6.4 Laboratory Half-Space Medium

The objectives of this section are to outline the procedure for simulating an elastic half-space medium and to investigate the dynamic response of a foundation resting on the surface of the elastic half-space medium when subjected to a harmonic excitation. The vertical and the simultaneous horizontal and rocking vibration modes of a foundation are to be considered experimentally.

The expanded natural rubber sheets selected by Tabiowo are used as the elastic medium, and Plasticine (modeling clay) is utilized as a uniform wave damper around the boundaries of the stuck rubber sheets. To determine the dynamic shear modulus and Poisson ratio of the medium, the experimental technique developed by Grootenhuis and Awojobi (1965) for the determination of soil dynamic properties is employed. Their experimental method is based on the closed form solution for the response of a vertical vibration of a circular base on an elastic half-space medium given by Awojobi and Grootenhuis (1965), where they gave values of nondimensional amplitude and frequency factor at the resonance condition as curves for different mass ratios of circular bases and the Poisson ratio of the media.

A successful laboratory model for a homogeneous elastic half-space has to possess the following important characteristics:

1. The material must, as much as possible, be homogeneous and isotropic with constant elastic moduli for dynamic loading.
2. Poisson ratio must be within the range of stable material.
3. Hysteretic damping of the material should be very low, since it has been neglected in most of the reported theoretical solutions.
4. All generated waves from the source must dissipate before or at least as they reach the boundary of the medium.
5. The surface of the medium should be flat and uniform to assure full contact between the rigid base and the surface of the medium.

Considering the above requirements, ten layers of ($1.194\ \text{m} \times 0.889\ \text{m} \times 0.025\ \text{m}$) expanded natural rubber (NAT), manufactured by Textile Industrial Company

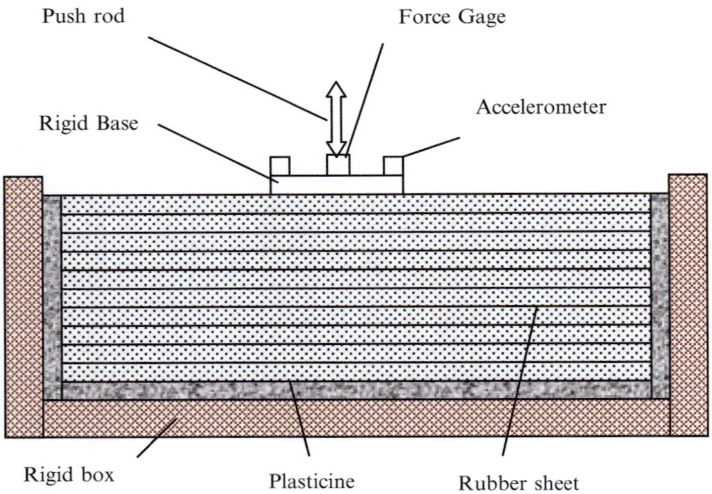

Fig. 6.1 Schematic drawing of laboratory model for an elastic half-space medium

Limited, were stuck and laid on a 19 mm thick base of firmly compacted Plasticine (modeling clay). In addition to the base, Plasticine was used around the stock of rubber sheets to obtain a nearly uniform wave damper. The whole model was enclosed on four sides by a wooden box. The experimental setup is shown in Fig. 6.1.

It should be noted that the natural rubber sheets that were implemented were polymer of isoprene which is obtained from rubber trees. As was discussed by Davey and Payne (1964), this material has good mechanical properties—tensile strength, elasticity, and resilience (very low hysteresis development)—but its resistance to oxidation is not good. The elastic properties of this material in dynamic loading are frequency, strain, and temperature dependent, and they are nearly constant only for a small range of variation of these three parameters.

6.4.1 Apparatus

To determine the dynamic properties of the medium and to perform the resonance tests, different circular, square, and rectangular blocks were used. The harmonic exciting force was produced by transmitting the motion of a shaker to the base by a push rod via a calibrated force gauge. The force gauge was mounted on the center of the base, and the acceleration of the base was measured by two identical accelerometers, which were placed on two symmetrical points on the base. Signals from the accelerometers and the force gauge were amplified by a charge

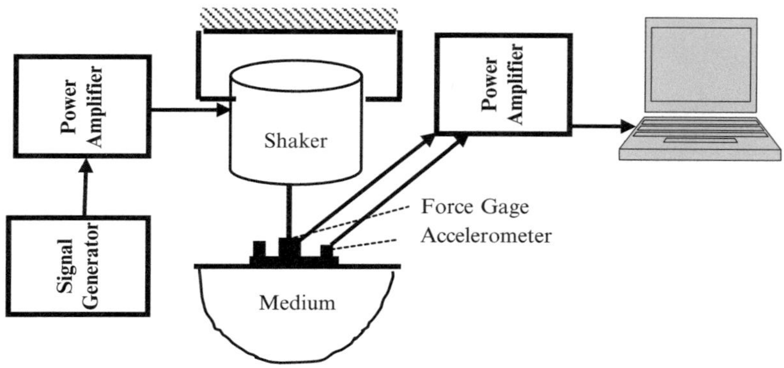

Fig. 6.2 Block diagram of the experimental setup

amplifier and then measured by a digital voltmeter. These signals were also observed on an oscilloscope to ensure that they were harmonic.

The driving signal was generated by a decade oscillator and was amplified by a power amplifier to drive the shaker. The frequency of the amplified signal was checked by a digital frequency meter. A block diagram of the whole measuring system and the experimental setup for the vertical vibration tests is shown in Figs. 6.2 and 6.3.

6.4.2 Static Properties of the Medium

To determine the static elastic properties of shear modulus and the Poisson ratio of the material used for the medium, the following static tests were conducted.

1. Tensile test
   For the tensile test, a specimen with an effective length of 245 mm, width of 43.4 mm, and thickness of 25.3 mm was used. The tensile test was carried out on an Instron machine with tension speed of 0.05 cm/min a the load range of 0–98.10 N. Variations of the axial extension versus applied load and the axial extension versus the lateral contraction were determined to be linear. As a result of this test, values of the elastic modulus and the Poisson ratio for the material were estimated, and consequently shear modulus of the medium was obtained.
2. Double block shear test
   In addition to the above test, the static shear modulus was also estimated by conducting the double block shear test. For this purpose, two equivalent blocks of rubber were stuck to the flat faces of three parallel steel plates using Ciba—Araldite 2000 Epoxy. The two edge face plates were rigidly fixed and a load added to a scale pan which was attached to the bottom of the center face plate.

Fig. 6.3 Experimental setup for measuring vertical vibration response of a square base subjected to a vertical harmonic force while resting on the laboratory model simulating an elastic half-space medium

The displacement of the moving central plate was measured by a dial gauge which was capable of measuring movements as small as 0.01 mm.

The displacement of the center face plate was plotted against the applied force. As it was pointed out by Hull (1937) and Rivlin (1948), it is not possible to have pure shear in the single or double block shear test, because the measured displacement is due to not only shearing but also bending of the rubber. Assuming that shape effects do not arise in shear, Southwell (1944) formula for deflection was used to determine the static value of the shear modulus:

$$\mathrm{GA\,d} = \frac{\mathrm{F}}{2}\mathrm{b}\left(1 + \mathrm{b}^2/36\mathrm{k}^2\right) \tag{6.5}$$

where in this equation *k* is the radius of gyration of the cross-section about the neutral axis of bending, *A* is the cross sectional area of blocks, *b* is the thickness of the rubber block, *F* is the applied force, *d* is the displacement of center face plate, and *G* is the shear modulus. It should be noted that since the cross section of the blocks were rectangular, the gyration radius can be estimated by the following equation:

$$k^2 = a^3c/12ac = a^2/12 \tag{6.6}$$

Thus, the value of shear modulus for the material from this test can be calculated using the following equation:

$$G = \frac{F}{2d}\frac{b\left(1 + b^2/3a^2\right)}{ac} \tag{6.7}$$

The values of *a*, *b*, and *c* were equal to 4.45, 2.26, and 2.43 cm. The obtained experimental results indicated that this static shear deformation was linearly elastic for moderate strains.

Based on these static experiments, Young's modulus, shear modulus, and Poisson ratio of the material were determined to be

$E = 450.9$ kPa (tensile test)
$G = 168.6$ kPa (double block shear test)
$\nu = 0.31$ (tensile test)
$\rho = 336.42$ kg/m^3

It should be noted that the results of these two tests for determination of the shear modulus were in a good agreement with each other.

6.4.3 Dynamic Properties of the Medium

To determine the dynamic response of a foundation on the elastic medium, all the established theoretical solutions require knowledge of both the shear modulus and the Poisson ratio of the medium. Numerous field and laboratory testing methods have been developed and are in use for determining the shear modulus of soils. For extensive surveys of literature, see the publication by Hamidzadeh (1978). In most of the measuring techniques, excluding the footing vibration technique developed by Grootenhuis and Awojobi (1965), a value for the Poisson ratio cannot accurately be determined. According to their experimental method, to estimate the shear modulus and the Poisson ratio of the medium, it is sufficient to measure the resonant frequencies of vertical vibration of two circular bases with the same base diameter but slightly different mass ratios. The mass ratio of the circular base is defined as (b = m/R ^ 3 p), where m is the mass of the base, R is the radius of the base, and p is the density of the medium.

Assuming that the dynamic shear modulus is independent of the exciting frequency, the estimated value of the Poisson ratio can be determined from the paper by Grootenhuis and Awojobi (1965). This can be achieved by finding an appropriate

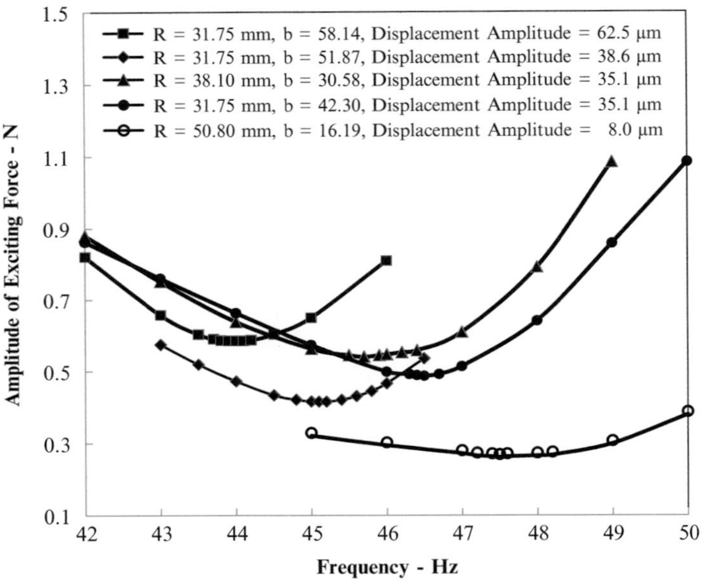

Fig. 6.4 Experimental frequency responses for different circular bases resting on the surface of an elastic half-space with Poisson ratio of $\nu = 0.31$

curve for which the ratio of the resonant frequency factors for the two chosen mass ratios corresponds to the ratio of the measured resonant frequencies. The dynamic shear modulus can then be determined from one of the frequency factor values. The resonant frequency factor for circular base is defined as

$$a_r = \frac{R\omega_r}{\sqrt{(G/\rho)}} \tag{6.8}$$

where ω_r is the resonant frequency and G is the shear modulus of the medium. To determine the resonant frequency of the circular base, the amplitude of vibration was kept constant at various frequencies, and the required forces were measured. Resonant frequency can be defined as the frequency at which the exciting force is at its minimum. This experimental method is more logical than measuring the conventional amplitude frequency response curve because the dynamic shear modulus is sensitive to the amplitude of the vibration. In the conducted experiments, a variety of rigid bases with different radiuses were used, and their frequency responses were measured. These experimental results are shown in Fig. 6.4.

Also experiments were conducted to obtain the frequency response of two bases for three different constant displacement amplitudes. These results are depicted in Fig. 6.5. The presented results were utilized along with technique established by Grootenhuis and Awojobi (1965), to determine Poisson ratio and resonant frequency factor and shear modulus of the medium. The estimated results for both static and dynamic cases are presented in Table 6.1.

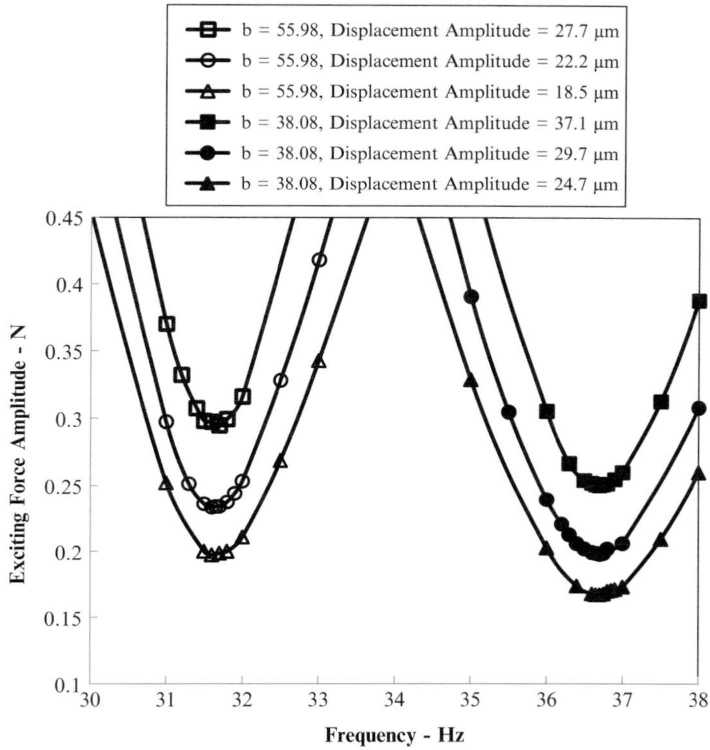

Fig. 6.5 Frequency responses of circular bases resting on an elastic half-space for different constant vibration amplitudes

Table 6.1 Dynamic and static properties of the model for the experimental elastic half-space

Test	Base radius (mm)	Mass ratio	Resonant frequency (Hz)	Resonant frequency factor	Shear modulus (kPa)	Poisson's ratio
Dynamic	31.7	58.15	44.0	0.315	260.97	0.31
	31.7	51.88	45.1	0.325	257.59	0.31
	31.7	42.31	46.5	0.340	251.24	0.31
	38.1	30.75	45.7	0.392	264.48	0.31
	50.8	16.20	47.5	0.392	262.00	0.31
Static					168.65	0.31

It has been shown by other investigators that the shear modulus in the dynamic case is not the same as statically determined. In fact, in the dynamic case, the shear modulus depends on amplitude of vibration, frequency of exciting force, and the depth of measuring point in the medium. Considering these factors, an attempt was made to carry out the tests with very small amplitude of oscillation. Moreover, the variation of resonant frequencies for different bases was within a very small range, thus avoiding any nonlinearity due to the frequency dependency of the shear modulus.

6.5 Experimental Vibration Response of Massive Rectangular and Circular Bases

The experimental frequency response of different rectangular (side ratios of 1.0 and ½) and circular bases is determined for vertical mode of vibration. In these experiments, the amplitude of vibration was kept constant, and the same procedure as described before was followed. By using the previously estimated average dynamic properties of the medium, the nondimensional resonant frequency factors and the mass ratios for the abovementioned cases were evaluated and compared with the previously established results in Figs. 6.6, 6.7, and 6.8. It should be noted that the mass ratio of rectangular and circular bases are defined as

$$b = m/\left(\rho c d^2\right) \qquad \text{for a rectangular footings} \qquad (6.9a)$$

and

$$b = m/\left(\rho R^3\right) \qquad \text{for a circular footings} \qquad (6.9b)$$

where ρ is the mass density of the medium, m is the mass of the base, c and d are the length and width of the rectangular base, and R is the radius of the circular footing. Also, the resonant frequency factor for rectangular and circular bases are defined by

$$a_r = \frac{\omega_r d}{\sqrt{G/\rho}}. \qquad \text{for rectangular footings} \qquad (6.10a)$$

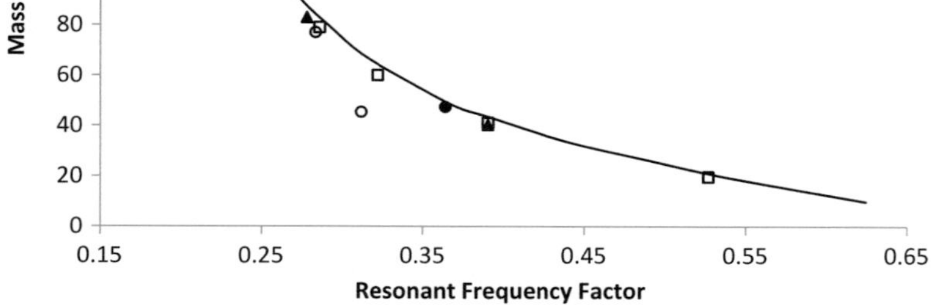

Fig. 6.6 Comparison of experimental results with the theoretical prediction of mass ratio versus the resonant frequency factor for the vertical motion of a square base

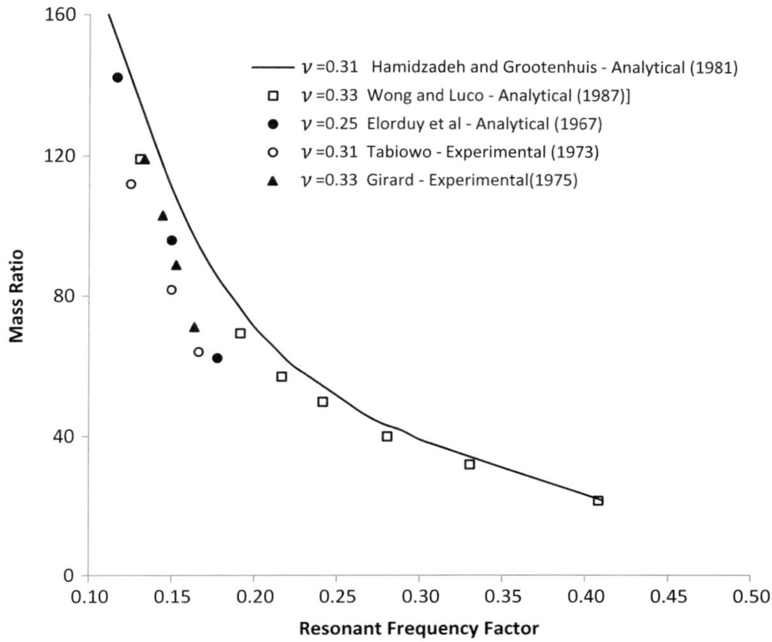

Fig. 6.7 Comparison of experimental results with the theoretical prediction of mass ratio versus the resonant frequency factor for the vertical motion of a rectangular base (side ratio of ½)

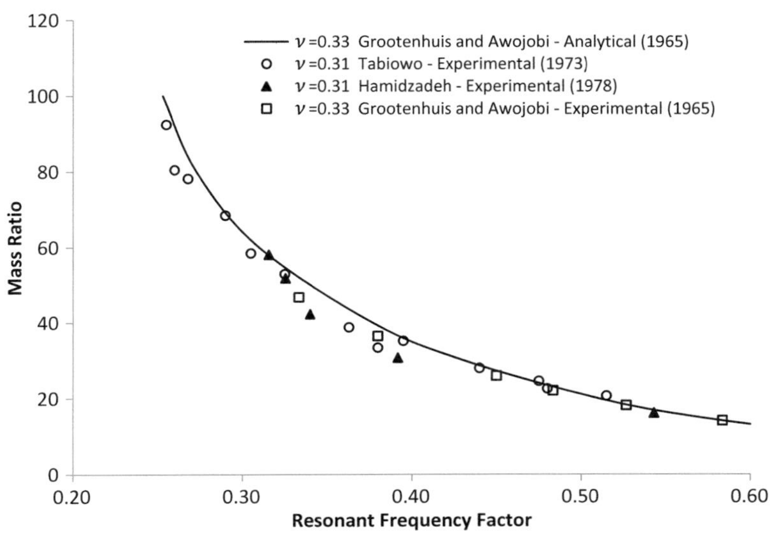

Fig. 6.8 Comparison of experimental results with the theoretical prediction of mass ratio versus the resonant frequency factor for the vertical motion of a circular base

and

$$a_r = \frac{\omega_r R}{\sqrt{G/\rho}}. \qquad \text{for circular footings} \tag{6.10b}$$

The comparison shows that an increase in the mass ratio of the base increases the difference between the experimental and theoretical results. There are two main explanations for these results. First, the stress distribution in practice differs from the stress distribution caused by test rig. As was indicated by Richart and Whitman (1967), it is not possible to have an infinite stresses near the periphery of the footing. They also indicated that as the total load increases, the frequency factor at resonance decreases. Second, the difference for higher mass ratios can be explained by the fact that as the mass increases, the energy through the medium also increases. Thus, a portion of the energy reflects from the boundaries and returns to the base causing an increase in the amplitude and the resonant frequency of the oscillation. In this case, the experimental model cannot act as an elastic half-space medium.

6.6 Experimental Response of Coupled Horizontal and Rocking Vibration

This section presents the experimental study of a coupled horizontal and rocking vibration of a square base. The base setup for this study is depicted in Fig. 6.9.

In this test two accelerometers were used to measure the horizontal displacement of two points on the base. A horizontal harmonic force was applied using a shaker via a force gauge. In this experiment the amplitude of the applied force was kept constant, and the amplitudes of the horizontal displacements were measured. The inertia ratio of a rectangular base is defined as

$$b' = I/\left(\rho c^3 d^2\right) \tag{6.11}$$

where I is the mass moment of inertia about the mass center of the base. The amplitude factor is defined by $X\ cG/F$, where X/F is the receptance of the mass

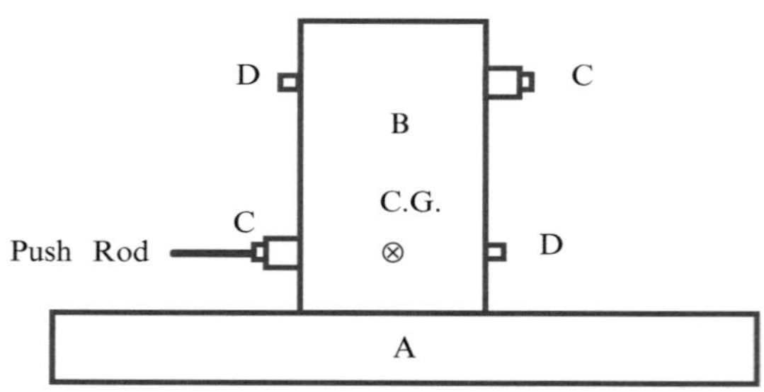

Fig. 6.9 Schematic diagram of the assembly of the base for the measurement of the coupled horizontal and rocking vibration: A—contact base, B—steel block, C—force gauge, D—accelerometer, E—push rod

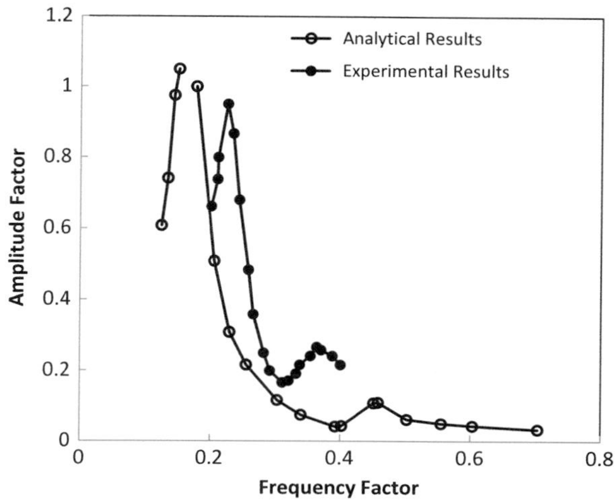

Fig. 6.10 Dimensionless frequency response of coupled horizontal and rocking vibration of a square base

center; X and F are the displacement amplitude of the horizontal vibration and the amplitude of the exciting harmonic force, respectively; and G is the shear modulus of the medium. Figure 6.10 depicts the comparison between the obtained experimental and its respective theoretical results for simultaneous horizontal and rocking motion. The experimental results show that at the first resonant frequency, the contribution of the horizontal vibration is much higher than the rocking.

Furthermore, the nondimensional resonant frequencies for independent horizontal and rocking modes of vibration which were given by Hamidzadeh and Grootenhuis (1981) were compared with these results. The experimental results indicate that the resonant frequencies of the two independent modes are higher than the first resonant frequency obtained in this experiment for simultaneous motion. This can be explained by the fact that in the coupled motion, for each motion, flexibility of the other motion has decreased from infinity to a finite value. Results also reveal that the first experimentally determined resonance has a higher frequency factor and lower amplitude factor than the predicted one.

These differences are due to the imperfection of the finite model and the presence of hysteretic damping in the medium, which was neglected in the presented theory.

Moreover, the experimental results for the second resonance demonstrate that it has a lower frequency and a higher amplitude factors in comparison with the predicted results.

This is due to the fact that the experimental stress distribution for rocking motion is not the same as the theoretically predicted. According to Richart and Whitman (1967), stresses at the periphery of the base are 0; therefore, the reaction moment which is caused by the moment of stress distribution about the rotation axis is less than those expected theoretically. Therefore, the second resonant frequency factor obtained from the experimental results is lower, while its resonant amplitude factor is higher than the theoretical determined values.

6.7 Measurement of Dynamic Properties of Elastic Half-Space Medium Using Square Footings

Reliable dynamic analysis of soil-foundation interaction requires knowledge of dynamic properties of shear modulus, Poisson ratio, and the geometrical damping ratio of the medium. In the case of a machine foundation resting on the surface of ground, it is common to assume a rigid base for the foundation and an elastic half-space model for the ground. To properly evaluate dynamic properties of a medium, a technique based on resonant testing in conjunction with an analytical solution is adopted. In the experimental part of the analysis, vertical frequency responses of different rigid square blocks individually resting on the medium are measured. These results are correlated with frequency responses computed using an established analytical solution. Analyzing this equitability of results can provide estimated values for dynamic properties of the medium. The validity of the proposed method is verified by performing experiments on a laboratory model.

6.7.1 Mathematical Model

This section presents a brief overview of the mathematical procedure for evaluation of dynamic properties of the medium. A schematic model of a rigid base resting on the surface of an elastic half-space is presented in Fig. 6.11. The medium is assumed to have infinite depth and an infinite surface area.

As presented in Chap. 5, the vertical response of a square base on an elastic half-space medium subjected to a vertical harmonic force is given by Eq. 6.5. In this equation f_1 and f_2 are the in-phase and quadrature components of the displacement function. The numerical values of these components, for a wide range of dimensionless frequencies, were presented in the last chapter. These components are functions of the dimensionless frequency "a" and the Poisson ratio of the medium "ν":

$$V = \frac{Gcv}{F} = \left[\frac{{f_1}^2 + {f_2}^2}{\left(1 + b\, a^2\, f_1\right)^2 + \left(b\, a^2\, f_2\right)^2} - \right]^{\frac{1}{2}} \tag{6.12}$$

where "b" is the mass ratio of the square base and was defined by Eq. 6.9a.

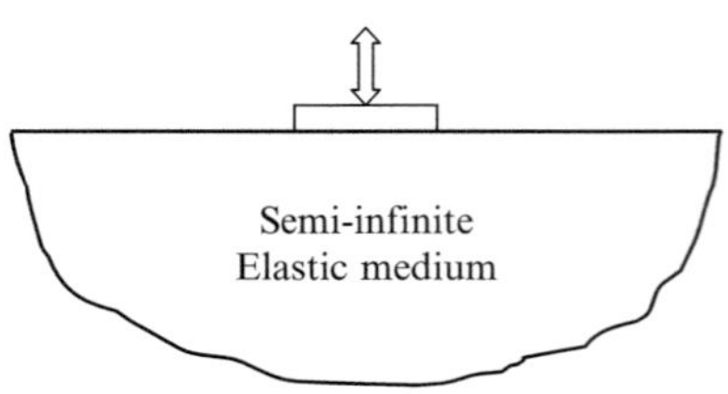

Fig. 6.11 A rigid square base on the surface of an elastic half-space medium

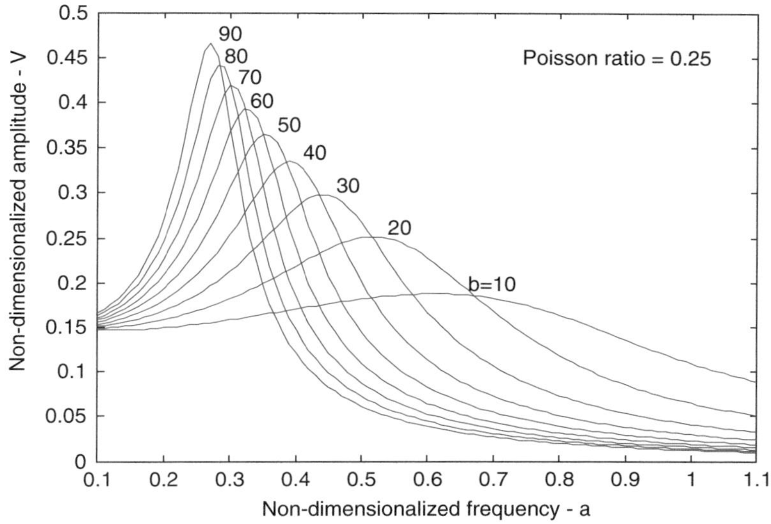

Fig. 6.12 Dimensionless frequency responses for vertical vibration of different square base resting on surface of an elastic half-space

Nondimensional frequency responses of square bases with different mass ratios are computed by sweeping the frequency factor "a" for a wide range using the above equation. These responses for a wide range of mass ratios are presented for a Poisson ratio of 0.25 in Fig. 6.12.

The dimensionless resonant frequencies versus mass ratios and different Poisson ratios are also illustrated in Fig. 6.13. The presented technique for evaluation of dynamic properties requires computation of dimensionless resonant frequencies for mass ratios used in the experiments.

6.7.2 Experimental Results

The experimental frequency responses for different square bases were determined. To determine the resonant frequency of the circular base, the amplitude of vibration is kept constant at various frequencies, and the required forces were measured. Resonance is then defined as the frequency at which the exciting force is minimum. This experimental method is more logical than measuring the conventional amplitude frequency response curve, due to the fact that the dynamic shear modulus is sensitive to strain or the amplitude of the vibration. The finite difference method was applied to the Nyquist representations of the frequency responses to attain a more accurate approximation for the resonant frequencies. The density of the medium is measured to be $\rho = 336.4$ kg/m^3, and the half width of all square bases was $c = 0.0317$ m. The experimental resonant frequencies for different bases are provided in Table 6.2.

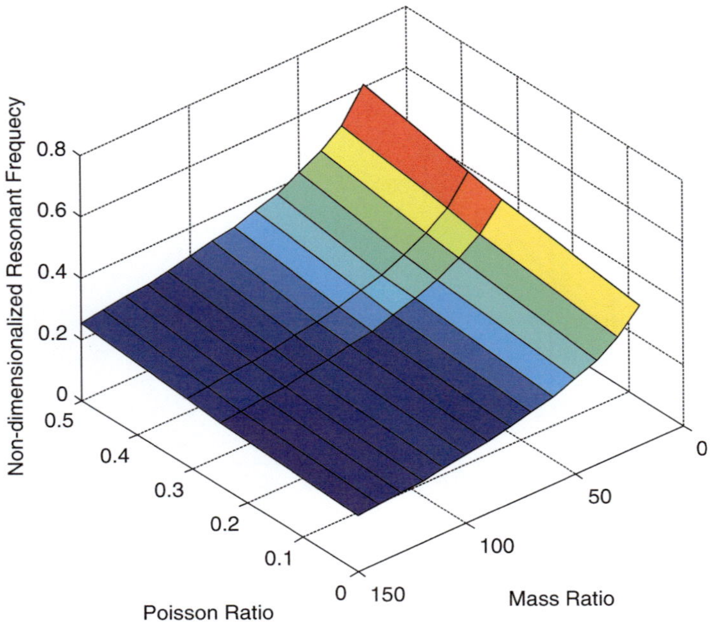

Fig. 6.13 Dimensionless resonant frequencies versus mass and Poisson ratios

Table 6.2 Experimental data for rigid square bases

Test no.	Mass (kg)	Mass ratio	Resonant freq. (Hz)
1	0.226	21	77.14
2	0.506	47	55.43
3	0.732	68	45.14
4	0.824	76.5	43.17
5	0.947	88	41.11
6	1.103	102.5	38.02
7	1.216	113	36.31
8	1.324	123	35.04
9	1.421	132	33.64
10	1.539	143	32.68

Considering that the dynamic shear modulus is independent of the exciting frequency, then the value of Poisson ratio and shear modulus of the medium can be determined using the following procedures:

1. Develop a table of computed dimensionless resonant frequencies for all the experimental mass ratios and different Poisson ratios. Then by using interpolation, estimate values of dimensionless resonant frequencies for each mass ratio for 26 equally spaced Poisson ratios within the elastic range (0.0–0.5).
2. Determine ratios of interpolated dimensionless resonant frequencies for each pair of mass ratios for all 26 Poisson ratios. Then find an appropriate Poisson ratio for

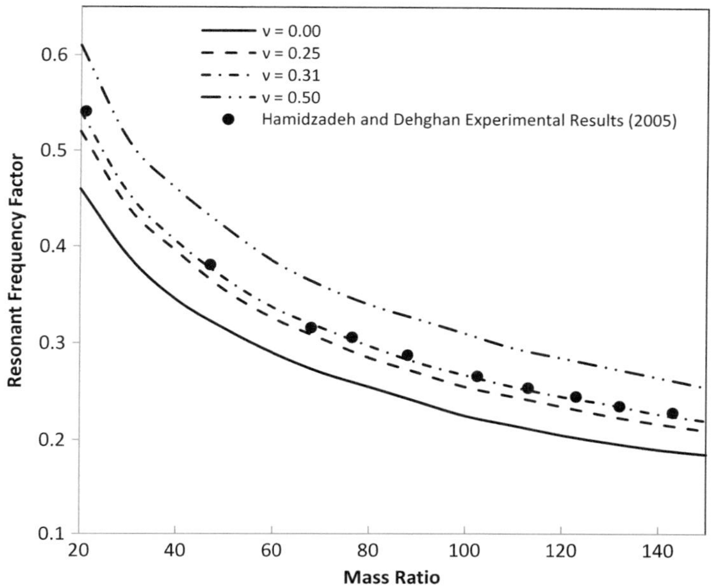

Fig. 6.14 Dimensionless frequency responses for vertical vibration of different square bases on surface of an elastic half-space

each pair by matching the ratio of their resonant frequencies with the ratio of their dimensionless resonant frequencies. The estimated Poisson ratio of the medium can be determined by averaging the Poisson ratios for all pairs of mass ratios.

3. The estimated shear modulus for each experimental mass ratio can be obtained using the corresponding dimensionless resonant frequencies for the evaluated Poisson ratio in the following relation:

$$G = \frac{\omega_r{}^2 c^2 \rho}{a_r{}^2} \tag{6.13}$$

Figure 6.14 illustrates the variation of dimensionless resonant frequencies versus experimental mass ratios for four different Poisson ratios. In addition, dimensionless resonant frequencies for the experimental mass ratios are also indicated in the same figure. It can be concluded that the Poisson ratio and shear modulus of the medium are approximated to be 0.31 and 273,032 Pa, respectively.

Chapter 7
Dynamic Response of a Rigid Foundation Subjected to a Distance Blast

Abstract Blasts and explosions occur in many activities that are either man-made or nature induced. The effect of the blasts could have a residual or devastating effect on the buildings at some distance within the vicinity of the explosion. In this investigation, an analytical solution for the time response of a rigid foundation subjected to a distant blast is considered. The medium is considered to be an elastic half-space. A formal solution to the wave propagations on the medium is obtained by the integral transform method. To achieve numerical results for this case, an effective numerical technique has been developed for calculation of the integrals represented in the inversion of the transformed relations. Time functions for the vertical and radial displacements of the surface of the elastic half-space due to a distant blast load are determined. Mathematical procedures for determination of the dynamic response of the surface of an elastic half-space subjected to the blast along with numerical results for displacements of a rigid foundation are provided.

Keywords Dissipative blast • Surface response to concentrated force

Nomenclature

F_0	Amplitude of the applied blast force
ω	Angular frequency of the applied force
G	Shear modulus of the medium
r	Radial distance from the blast
$u_1 + iu_2$	Complex nondimensional radial displacement function
$v_1 + iv_2$	Complex nondimensional vertical displacement function
u	Radial displacement at any point on the surface
v	Vertical displacement at any point on the surface
∇^2	Laplace operator for Cartesian coordinates
$u(t)$	Displacement of the massless foundation along x-axis
$v(t)$	Displacement of the massless foundation along y-axis

H.R. Hamidzadeh et al., *Wave Propagation in Solid and Porous Half-Space Media*,
DOI 10.1007/978-1-4614-9269-6_7,

$w(t)$	Displacement of the massless foundation along z-axis
t	Time
θ	Angle of radial blast w.r.t. x-axis
E	Young's modulus
υ	Poisson ratio
ε	Linear strain direction
φ_x	Elastic rotation of the massless foundation about x-axis
φ_y	Elastic rotation of the massless foundation about y-axis
φ_z	Elastic rotation of the massless foundation about z-axis

7.1 Introduction

The possible occurrence of extreme dynamic excitation, either natural or man-made, has a major influence on the design of buildings. A primary concern in designing foundations is having the knowledge of how they are expected to respond when subjected to blast loadings in the vicinity of the building. The validity of the mathematical analysis depends entirely on how well the mathematical model simulates the behavior of the real foundation. Over the past decades, the ability to analyze mathematical models for dynamics of foundations has been improved by the use of different analytical and numerical techniques. Nearly all established methods assume that the footing is rigid and the medium is a homogeneous elastic half-space. Extensive efforts have been confined to the development of procedures and computer simulations to tackle some practical problems that arise in this field, while other important problems have been neglected. It should be noted that interactions between foundations for noncircular footings were not treated in a satisfactory manner and significant deficiencies remain in most of the previous analyses.

This paper will discuss some of the issues of dynamics of a rigid foundation subjected to a distant blast load from a practical point of view. Since this topic is quite broad, a brief description of the methodology will be outlined, while details will be given for a few procedures that have proven to be effective and accurate. Special attention is directed to the dynamic response of the surface of the medium due to concentrated dynamic loads and its application in developing a solution to the response of a building founded on a rigid foundation resting on the medium.

7.2 Surface Response Due to Concentrated Forces

In the field of propagation of disturbances on the surface of an elastic half-space, the first mathematical attempt was made by Lamb (1904). He gave integral representations for the vertical and radial displacements of the surface of an elastic half-space due to a concentrated vertical harmonic force. Evaluation of these integrals involves considerable mathematical difficulties, due to the evaluation of a Cauchy principal integral and certain infinite integrals with oscillatory integrands. Nakano (1930) considered the same problem for a normal and tangential force distribution on the surface. Barkan (1962) presented a series solution for the evaluation of integrals for the vertical displacement caused by a vertical force on the surface, which was given by Shekhter (1948). Pekeris (1955a, b) gave a greatly improved solution to this problem when the surface motion is produced by a vertical point load varying with time, like the Heaviside function. Elorduy et al. (1967) developed a solution by applying Duhamel's integral to obtain the harmonic response of the surface of an elastic half-space due to a vertical harmonic point force. Heller and Weiss (1967) studied the far-field ground motion due to an energy source on the surface of the ground.

Among the investigators who considered the three-dimensional problem for a tangential point force, Chao (1960) presented an integral solution to this problem for an applied force varying with time like the Heaviside unit function. Papadopulus (1963) and Aggarwal and Ablow (1967) have presented solutions, in integral expressions, to a class of three-dimensional pulse propagations in an elastic half-space. Johnson (1974) used Green's functions for solving Lamb's problem, and Apsel (1979) employed Green's functions to formulate the procedure for layered media. Kausel (1981) reported an explicit solution for dynamic response of layered media. Davies and Banerjee (1983) used Green's functions to determine responses of the medium due to forces which were harmonic in time with a constant amplitude. The solution was derived from the general analysis for impulsive sources. Kobayashi and Nishimura (1980) utilized the Fourier transform to develop a solution for this problem and expressed the results in terms of the full-space Green's functions, which include infinite integrals of exponential and Bessel's function products. Banerjee and Mamoon (1990) provided a solution for a periodic point force in the interior of a three-dimensional, isotropic elastic half-space by employing the methods of synthesis and superposition. The solution was obtained in the Laplace transform as well as the frequency domain.

Hamidzadeh (1978) presented mathematical procedures for determination of the dynamic response of the surface of an elastic half-space subjected to harmonic loadings and provided numerical results for displacement of any point on the surface in terms of properties of the medium and of the exciting force. The solution was analytically formulated by employing double Fourier transforms and was presented by integral expressions. Hamidzadeh and Grootenhuis (1981) implemented his developed numerical results to address the dynamic response of a rigid foundation that is subjected to harmonic loadings. Holzlohner (1980) considered the same problem and presented numerical results. Hamidzadeh and Chandler (1991) and Hamidzadeh and Chandler (1986) provided dimensionless response for an elastic half-space and compared their results with other available approximate results. Furthermore, based on wave propagation on the surface of ground, Hamidzadeh and Luo (2006) presented an analytical solution to determine the location of an unknown vertical exciting force for surveillance applications.

As depicted in Fig. 7.1, consider a rectangular rigid foundation resting on the surface of an elastic half-space which is excited by a vertical harmonic

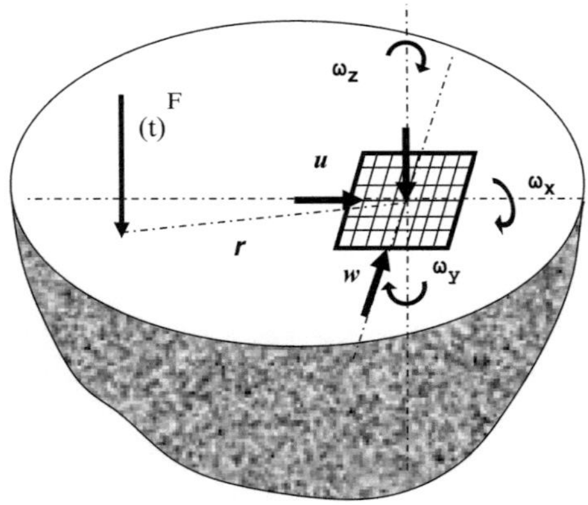

Fig. 7.1 A rectangular rigid foundation on the surface of an elastic half-space subjected to a blast force

force away from the foundation. Under this condition, the rigid foundation will undertake motions with six degrees of freedom. These motions can be determined by considering the radial and vertical displacement time functions for several points under the rigid foundation. As reported by Hamidzadeh (1991, 1986), for a point at a distance of r from the vertical force, its radial and vertical displacement time functions can be expressed by the following equations:

$$u(r) = \frac{F_0}{Gr}(u_1 + iu_2)\,e^{i\omega t} \tag{7.1a}$$

$$v(r) = \frac{F_0}{Gr}(v_1 + iv_2)\,e^{i\omega t} \tag{7.1b}$$

Figure 7.2 presents numerical results computed for the nondimensional displacement functions. The displacements of a point on the surface of the medium depend largely upon in-phase and quadrature components of these nondimensional complex displacement functions. These components are functions of the frequency factor and the Poisson ratio of the medium. The range of frequency factor presented here is more than sufficient for practical purposes of considering near-field displacements. Far-field displacements can be determined using Lamb's equations. In Fig. 7.2, the values of u_1, u_2, v_1, and v_2 are for Poisson ratio of 0.25.

7.3 Governing Equation of Motion

As reported by Ojetola and Hamidzadeh (2012), the model for a vertical blast force can be presented by a complex Fourier series by assuming a large time period such that within each period the response of the rigid foundation will be dissipated due to geometrical damping of the half-space medium.

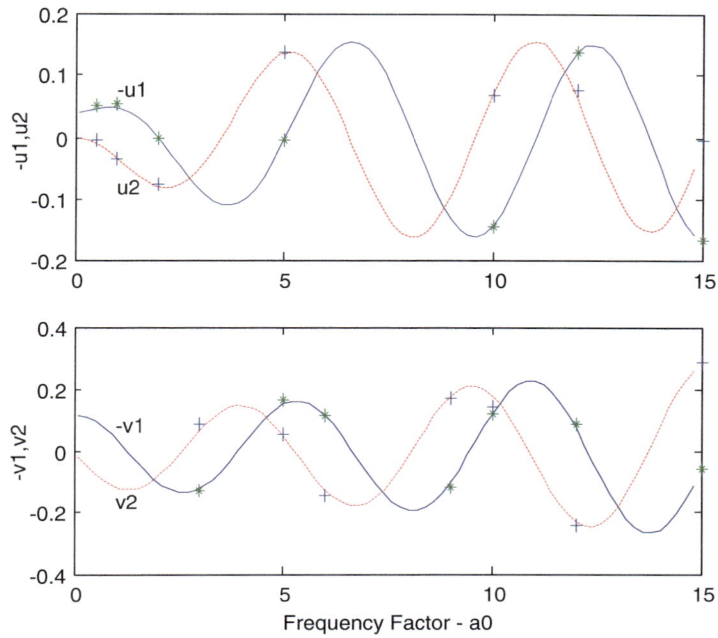

Fig. 7.2 Complex non-dimensional radial and vertical functions versus frequency factor a_0 (Hamidzadeh's—*solid line*; and Holzlohner's—*symbols* Choudhury et al. (2005)

$$f(t) = \sum_{-\infty}^{\infty} \gamma_n e^{\frac{2in\pi t}{T_0}} \tag{7.2}$$

where

$$\gamma_n = \frac{1}{T_0} \int_{-\frac{T_0}{2}}^{\frac{T_0}{2}} f(t) e^{\frac{-2in\pi t}{T_0}} dt$$

$$\gamma_n = \frac{1}{T_0} \left(\frac{-T_0}{2in\pi} \right) e^{\frac{-2in\pi t}{T_0}} \Big]_0^{\Delta T}$$

$$\gamma_n = \frac{1}{2in\pi} \left(1 - e^{\frac{-2in\pi \Delta T}{T_0}} \right)$$

The lowest frequency is given by

$$\omega_0 = \frac{2\pi}{T_0} \tag{7.3}$$

Thus, the blast force can be expressed in terms of the lowest frequency:

$$f(t) = \sum_{-\infty}^{\infty} \gamma_n e^{in\omega_0 t} \tag{7.4}$$

Then, displacements for the *i*th point under the foundation are given by the expression

$$u_{i_n} = \frac{\gamma_n}{Gr_i}(u_{i_1} + u_{i_2})\cos\theta \tag{7.5a}$$

$$v_{i_n} = \frac{\gamma_n}{Gr_i}(v_{i_1} + v_{i_2}) \tag{7.5b}$$

$$w_{i_n} = \frac{\gamma_n}{Gr_i}(u_{i_3} + u_{i_4})\sin\theta \tag{7.5c}$$

Rigid body displacement of the massless foundation can be extracted from displacements of each of the assigned points using the following equations:

$$u_{i_n} = u_{i_0} + y\varphi_{y_n} \tag{7.6a}$$

$$v_{i_n} = v_{i_0} + x\varphi_{z_n} - z\varphi_{x_n} \tag{7.6b}$$

$$w_{i_n} = w_{i_0} - x\varphi_{y_n} \tag{7.6c}$$

Rigid body motion of each point of the massless foundation away from the blast can be written as follows:

$$\begin{Bmatrix} u_1 \\ v_1 \\ w_1 \\ u_2 \\ v_2 \\ w_2 \\ \vdots \\ u_n \\ v_n \\ w_n \end{Bmatrix} = \begin{bmatrix} 1 & 0 & 0 & 0 & y_1 & 0 \\ 0 & 1 & 0 & -z_1 & 0 & x_1 \\ 0 & 0 & 1 & 0 & -x_1 & 0 \\ 1 & 0 & 0 & 0 & y_2 & 0 \\ 0 & 1 & 0 & -z_2 & 0 & x_2 \\ 0 & 0 & 1 & 0 & -x_2 & 0 \\ \vdots & \vdots & \vdots & \vdots & \vdots & \vdots \\ 1 & 0 & 0 & 0 & y_n & 0 \\ 0 & 1 & 0 & -z_n & 0 & x_n \\ 0 & 0 & 1 & 0 & -x_n & 0 \end{bmatrix} \begin{Bmatrix} u_0 \\ v_0 \\ w_0 \\ \varphi_x \\ \varphi_y \\ \varphi_z \end{Bmatrix} \tag{7.7}$$

Thus, the displacement and rotation of the massless foundation in terms of time away from the blast are given by

$$u(t) = \sum_{-n}^{n} u_{0n} e^{in\omega_0 t} \tag{7.8a}$$

$$v(t) = \sum_{-n}^{n} v_{0n} e^{in\omega_0 t} \tag{7.8b}$$

$$w(t) = \sum_{-n}^{n} w_{0n} e^{in\omega_0 t} \tag{7.8c}$$

$$\varphi_x = \sum_{-n}^{n} \varphi_{x_n} e^{in\omega_0 t} \tag{7.9a}$$

$$\varphi_y = \sum_{-n}^{n} \varphi_{y_n} e^{in\omega_0 t} \tag{7.9b}$$

$$\varphi_z = \sum_{-n}^{n} \varphi_{z_n} e^{in\omega_0 t} \tag{7.9c}$$

7.4 Results and Discussions

The numerical values of the nondimensional displacement functions of u_1, u_2, v_1, and v_2 for Poisson ratio of 0.25 have been computed. These results are compared with those of Holzlohner (1980) and are shown in Figs. 7.2 and 7.3, where the abscissa is the dimensionless frequency ($a_0 = \omega r/c_2$). The comparison shows a close agreement between these independent results. These results can be used to determine the response of the surface on an elastic half-space due to vertical harmonic point force. Results indicate that both of these displacement amplitudes are decreasing exponentially with distance from the source.

The provided results are used in Eqs. 7.8 and 7.9 to determine the harmonic response of the six displacements of the rigid foundation for each term of the Fourier expansion of the blast force. In this analysis, the duration of the blast force is 2 s with a constant magnitude of 1 N. The Fourier expansion for this blast force is conducted by considering 101 terms with a period of 5 s. This expansion is shown in Fig. 7.3. The computed responses for the three translational and three rotational motions of the rigid base due to the blast are provided in Figs. 7.4 and 7.5. Figure 7.4 demonstrates that vertical displacement of the rigid foundation is more significant than the other two sliding motions. Similarly, due to the position considered for the blast force with respect to position of the rigid foundation, the rotational motion

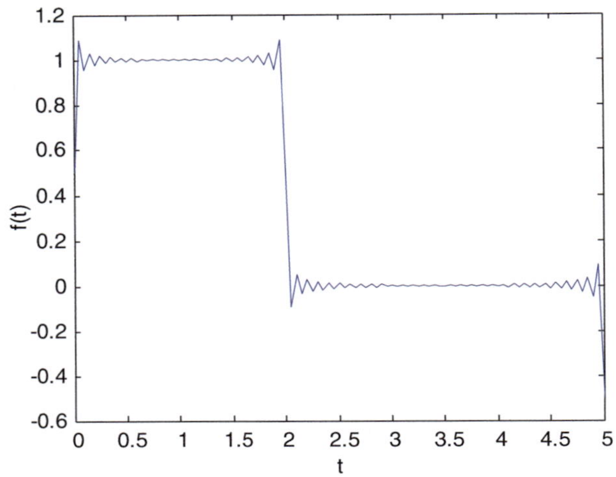

Fig. 7.3 Blast force at a distance r from the massless foundation with respect to time

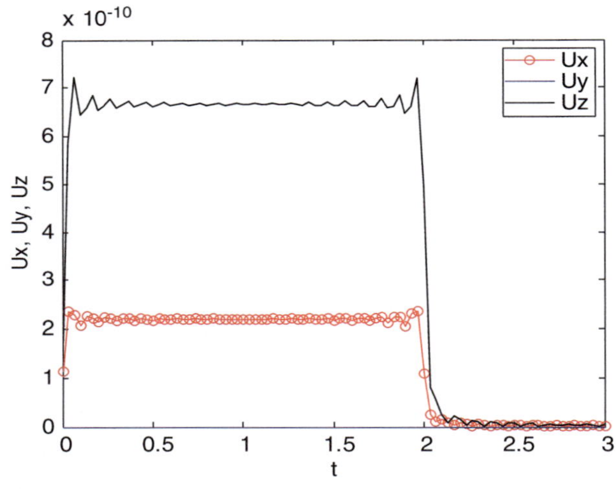

Fig. 7.4 x, y and z axes displacement of the massless foundation in terms of time away from the blast

about x-axis is higher than the other two rotational motions. It should be noted that blast force is along the x-axis at a distance of 30 m away from the center of the rigid square (10 m by 10 m) foundation. The analysis is performed for elastic half-space medium with the following properties: shear modulus of 5.41047E07 Pa, Poisson ratio of 0.25, and density of 1,760 kg/m^3.

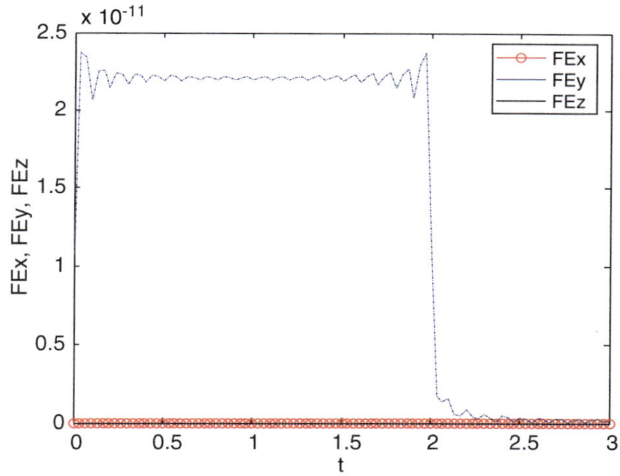

Fig. 7.5 *x*, *y* and *z* axes rotation of the massless foundation in terms of time away from the blast

7.5 Conclusion

An analytical solution for the response of a rigid foundation on the surface of an elastic half-space medium excited by a distant blast force is developed. The motions for the rigid foundation are considered to have six degrees of freedom. Time responses of these motions are computed for a square wave blast force with a magnitude of 1 N for a duration of 2 s by expanding this time response by its Fourier series with a period of 5 s. The medium used has a shear modulus of 6×10^6 Pa, density of 0.6667 kg/m^3, and Poisson ratio of 0.25.

Chapter 8
Identification of Vertical Exciting Force on the Surface of an Elastic Half-Space Using Sensor Fusion

Abstract An analytical technique for identification of the location of an unknown vertical exciting force on the surface of ground using sensor fusion is presented. The analysis is based on the dynamic responses of points on the surface of an elastic half-space medium subjected to a vertical, harmonic, and concentrated force on the surface. The medium is assumed to be an elastic, isotropic and homogeneous half-space. The problem is analytically formulated by employing double Fourier transforms, and the solution is obtained in the form of integral expressions in terms of Rayleigh functions. Numerical techniques are utilized for the computation of integrals presented by the inverse transforms. Nondimensional values for the in-phase and quadrature components of the displacements for any position on the surface of the unloaded half-space in terms of frequency and position of the exciting force are presented for a Poisson ratio of 0.25.

Keywords Force identification • Sensor fusion • Force location determination

An analytical technique for identification of the location of an unknown vertical exciting force on the surface of ground using sensor fusion is presented. The analysis is based on the dynamic responses of points on the surface of an elastic half-space medium subjected to a vertical, harmonic, and concentrated force on the surface. The medium is assumed to be an elastic, isotropic and homogeneous half-space. The problem is analytically formulated by employing double Fourier transforms, and the solution is obtained in the form of integral expressions in terms of Rayleigh functions. Numerical techniques are utilized for the computation of integrals presented by the inverse transforms. Nondimensional values for the in-phase and quadrature components of the displacements for any position on the surface of the unloaded half-space in terms of frequency and position of the exciting force are presented for a Poisson ratio of 0.25.

H.R. Hamidzadeh et al., *Wave Propagation in Solid and Porous Half-Space Media*, 153
DOI 10.1007/978-1-4614-9269-6_8, © Springer Science+Business Media New York 2014

8.1 Introduction

Object localization techniques have recently received great importance for identification of position and the transmitted wave of unexpected source on the surface of the ground. Application of accurate ground vibration sensors combined with analytical and numerical methods in solving ground wave propagation can robustly and accurately identify location of objects even in the presence of significant amounts of noise. The main reason for this ability is the fact that they rely on the integration of several different sensors instead of just a single sensor. Vibration sensor integration is effective because it allows the perception system to be applicable in a greater number of situations than would be possible with a single sensor. The ground object identification system described in this research work relates to detected ground motion by a number of sensors and requires the knowledge of dynamic response of the surface of the ground subjected to the forcing function of the object source.

Assuming that the ground can be modeled as an elastic half-space, the response of any point on the surface due to a harmonic vertical concentrated load can be determined analytically. Once this response is known, the solution may extend to obtain the response of the surface due to any time-dependent excitation on the surface by employing fast Fourier transformation (Hamidzadeh and Grootenhuis 1981). Although in the past considerable effort has been paid into finding the response of the foundation itself, little interest has been shown in the response of the surface of the half-space at distances away from the footing.

The first attempt at a mathematical formulation of the problem was by Lamb (1904). He gave integral representation for vertical and radial displacements on the surface of the medium for a vertical concentrated force. Nakano (1930) considered the same problem for normal and tangential forces and presented an integral solution. Barkan (1962) gave an approximate solution for vertical displacement when applying normal harmonic force; his solution is only valid for a small frequency range. Surface motions produced by a vertical point load varying with time like the Heaviside functions are considered by Pekeris (1955a, b). Elorduy et al. (1967) used Pekeris results, and by applying Duhamel's integral, he poorly estimated the vertical response due to a harmonic vertical point force. A few authors have given integral solutions to a class of three-dimensional pulse propagation problems such as Papadopulus (1963) and Aggarwal and Ablow (1967), who did not present any numerical results to this problem.

Numerical results for the vertical and horizontal displacements for a wide range of frequencies and the distances are given in Chap. 3, where, based on the above results, some problems on soil-foundation interaction were investigated (Hamidzadeh 1978). Holzlohner (1980) also presented a numerical result to this problem, which is based on his previous solution for a vertical load distributed over a square area, by limiting the size of the square to 0.

The objective of this work is to develop an analytical-numerical technique to determine the location of a periodic vertical force on the surface of an elastic half-space using sensors to estimate vertical surface displacements and the number of known positions. This requires the knowledge of the dynamic response of any location on the surface of a homogeneous elastic half-space subjected to a harmonic vertical force. The formal solution for response of the surface will be obtained by the integral transform method.

8.2 Numerical Techniques

For the isotropic homogeneous elastic half-space ($y \geq 0$), the governing equation of motion in terms of harmonic shown. The governing equation of motion in terms of harmonic displacements of u, v, and w was presented in Chaps. 2 and 3. As it was discussed, the integral in Eq. 3.26 cannot be correctly evaluated by means of the usual Gaussian quadrature method. Krylov (1962) has shown that the remainder error of the Gaussian method is proportional to the nth derivative of the integrand at the boundaries of $\eta = \gamma$ and $\eta = 1$. Since all the derivatives in this case are infinite at both boundaries, the modified Gaussian quadrature method given by Krylov (1962) and Stroud and Secrest (1966) is used. Similarly, the second integral in Eq. 3.27 has an infinite nth derivative in the lower boundary. To find a numerical result for this integral, another modified Gaussian quadrature method given in Krylov (1962) and Holzlohner (1980) was employed. In the case of the infinite integral in Eq. 3.27, since the denominator of the integrand has a real root in the range of the integration, the Longman technique (Longman 1958) for evaluating the Cauchy principal value of this integral was used. Also another Longman technique for the integration of oscillatory integrands with infinite upper limit was employed (Longman 1956).

The numerical values of the nondimensional displacement functions of α_1,α_2, β_1, and β_2 for Poisson ratio of 0.25 have been computed. These results are compared with those of Holzlohner (1980) and are shown in Figs. 8.1 and 8.2, where the abscissa is the dimensionless frequency ($k = \omega r/c_2$). The comparison shows a close agreement between these independent results. The magnitudes and phases of these dimensionless complex displacement functions for vertical and radial displacements are also illustrated in Fig. 8.3. It should be noted that the range of the dimensionless frequency covered is more than sufficient for practical purposes. These results can be used to determine response of the surface on an elastic half-space due to vertical harmonic point force.

8.3 Determination of the Source Location

Figure 8.4 depicts the location of displacement sensors relative to the unknown vertical periodic force on the surface of an elastic half-space. As shown the distance of R_m is the distance between the source of the period force $f(t)$ and the position of the sensor number *m*.

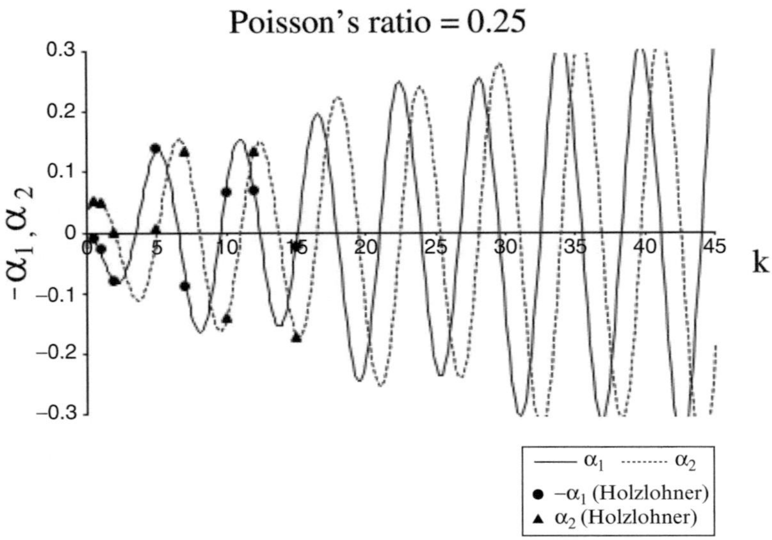

Fig. 8.1 Displacement functions β_1 and β_2 versus the frequency factor (k) for the vertical motion

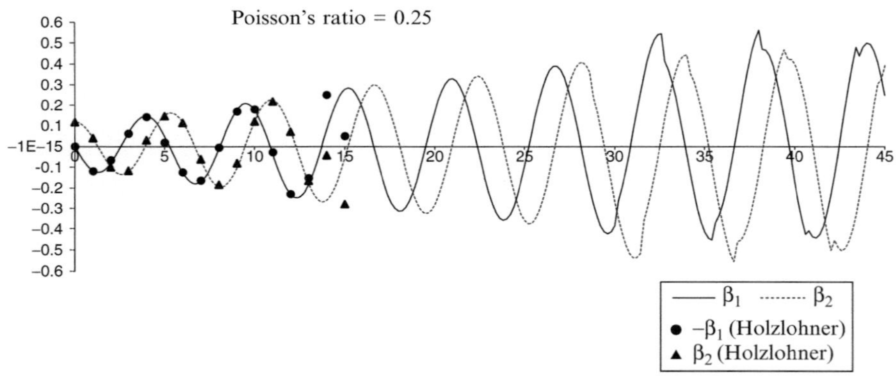

Fig. 8.2 Displacement functions $\alpha 1$ and $\alpha 2$ versus the frequency factor (k) for the radial motion

The periodic force at the source can be expanded in its Fourier terms as

$$f(t) = \sum_{n=1}^{\infty} F_n e^{j(n\omega_0 t + \varphi_n)} \tag{8.1}$$

Using equation (), then the vertical displacement component at location of the mth sensor can be presented by

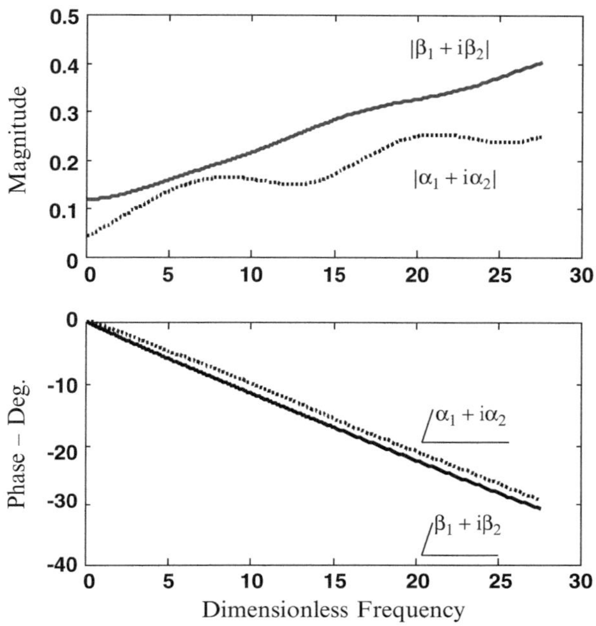

Fig. 8.3 Magnitudes and phases for displacement functions versus frequency factor (k)

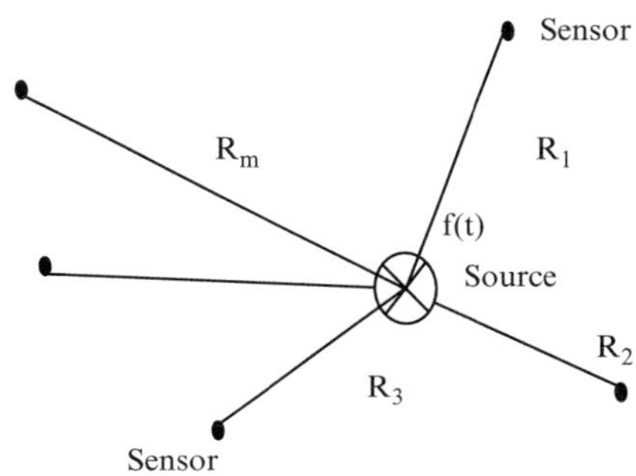

Fig. 8.4 Position of vertical period force excitation relative to the location of displacement sensors on the surface of an elastic half-space

$$v_m(t) = \sum_{n=1}^{\infty} \frac{F_n}{GR_m} |\alpha_1 + i\alpha_2|_m e^{i(n\omega_0 t + \varphi_n)} \tag{8.2}$$

Conducting analysis on the signals from the sensors, fundamental frequency of the periodic force at the source ω_0 and amplitudes of the displacements at all sensors for the fundamental frequency V_m $m = 1, \ldots, M$ can be determined.

$$V_m(t) = \frac{F_1}{GR_m} |\alpha_1 + i\alpha_2|_m$$

Consequently, the ratio of V_m/V_1 for different values of m can be presented as

$$\frac{V_m}{V_1} = \frac{R_1}{R_m} \frac{|\alpha_1 + i\alpha_2|_m}{|\alpha_1 + i\alpha_2|_1}$$

Having the ratio of V_m/V_1 for different number of sensors and the results provided in Fig. 8.3 for $|\alpha_1 + i\alpha_2|_m$, one can determine estimated values of R_m for $m = 1, \ldots,$ M by iteration method. It should be noted that the estimated values of R_m should be adequate for identifying the location of the source force on the surface of the medium. In fact this location is the common intersection of all circles drawn from location of the sensors with their corresponding radii.

8.4 Conclusions

The displacements of a point on the surface of an elastic half-space due to a vertical, harmonic, and concentrated force depend largely upon in-phase and quadrature components of the nondimensional functions. The agreement between the present results and the results given in Holzlohner (1980) provides a good basis for confidence in the computed values for the nondimensional displacement functions. Another important parameter that has a significant effect on the dynamic response is the Poisson ratio. The analytical solution for the determination of the displacement response of the surface of the medium was used to accurately determine the location of the vertical periodic exciting force.

Chapter 9
Surface Vibration of a Multilayered Elastic Medium Due to Harmonic Concentrated Force

Abstract Dynamic response of a multilayered elastic medium subjected to harmonic surface concentrated load is considered. In development of the analytical solution, the three-dimensional theory of elasto-dynamics is utilized for derivation of the governing partial differential equations for each layer. These equations are solved in the Fourier domain by employing the double complex Fourier transform technique. In the analysis, each layer of the medium is assumed to be extended infinitely in the horizontal x and z directions and has uniform depth in the y direction and is considered to be linearly elastic, homogeneous, and isotropic. Utilizing the integral Fourier transform, displacements and stresses at any point in each layer can be determined in terms of boundary stresses for each layer. Also, the presented solution provides the relation between stress and displacement vectors for the top and bottom of each layer in matrix notation. By satisfying the compatibility of displacements and stresses for each interface, a propagator matrix relating displacements and stresses at the top of the medium to the bottom interface will be obtained. This relates the displacement and stress vector on the top surface to the bottom interface by eliminating similar information for the other interfaces. In this study, the displacements on the surface of the layered medium are computed for the two cases where the surface of the medium is subjected to a concentrated harmonic vertical or horizontal harmonic force.

Keywords Surface vibration • Harmonic concentrated force

Nomenclature

σ_{ii} Normal stress in "i" direction
σ Constant normal stress
τ_{ij} Shear stress in plane i and in direction j
τ Constant shear stress
f_1 Real part of displacement function

H.R. Hamidzadeh et al., *Wave Propagation in Solid and Porous Half-Space Media*,
DOI 10.1007/978-1-4614-9269-6_9,

f_2	Imaginary part of displacement function
f_x	Force per unit volume in x direction
f_y	Force per unit volume in y direction
f_z	Force per unit volume in z direction
F	Fourier transform
ρ	Density
u	Displacement of medium in x direction
v	Displacement of medium in y direction
w	Displacement of medium in z direction
t	Time
Ω	Circular frequency
E	Young's modulus
υ	Poisson ratio
ϵ_{ii}	Linear strain in "i" direction
λ	Lamb's constant
G	Shear modulus
∇^2	Laplace operator for Cartesian coordinates
Ω_1	$= \Omega / c_1$
Ω_2	$= \Omega / c_1$
p	Variable in the Fourier domain
q	Variable in the Fourier domain
r_{ij}	Shear strain plane "i" and in direction "j"
ϵ	$= \epsilon_{xx} + \epsilon_{yy} + \epsilon_{zz}$ (volumetric strain)
ω_x	Elastic rotation about x-axis
ω_y	Elastic rotation about y-axis
ω_z	Elastic rotation about x-axis

9.1 Introduction

In the present investigation, the response of a three-dimensional layered medium to a dynamic input at the bounding surface is considered. The formulation is kept as general as possible, in order not to restrict the results of this study to geophysical applications alone.

Each layer considered is homogeneous, isotropic, of uniform height, and of infinite extent. The equations governing the motion in each layer are the equations of motion coupled with the constitutive relations. Neglecting the body forces and supplementing the constitutive relations, the solution of the equations of motion can be expressed in terms of nine stresses and displacements. For convenience, since the six stress functions involved are functions of one dilatation gradient and two rotation gradients only, three of these stress functions are eliminated to yield six functions of stresses and displacements.

The double complex Fourier transform is used to remove the variables in the direction of infinite extents. The system of partial differential equations is thus reduced to a system of first-order ordinary differential equations in which the independent variable is the depth of the layered medium and the dependent variable consists of a state vector whose six components are the transformed stresses and displacements.

The general solution of the transformed equations of motion yields the transformed stresses and displacements anywhere in the field in terms of the same quantity at the surface. For a half-space medium, since the depth of the medium is infinite, the independent variable did not exist anymore in the transformed stresses and displacements. The properties of the medium and the applied force on the surface will directly decide what the displacements will be on the surface of the medium.

For a layered medium, the method described is particularly advantageous because the transfer matrix approach results directly in the transformed stresses and displacements at any interface and the state vector produces all of the desired field quantities simultaneously. The continuity of the interface stresses and displacements is then automatically assured, as the intermediate state vector is eliminated upon multiplication by the transfer matrix associated with the next layer.

The continued multiplication of the initial state vector by a chain of layer transfer matrices produces the state vector anywhere in the field. The initial state vector is not usually completely known, and the missing components are found by converting the boundary value problem to an initial value problem. The integral transform inversion must in general be accomplished on a computer using a fast Fourier transform technique.

The procedure described retains a six-by-six matrix throughout. The number of constants to be determined does not increase with the number of layers.

9.2 Equation of Motion

By substituting stress–strain relations into the Newton's second law of motion equilibrium equations for cubic element in the elastic medium, we have

$$\frac{\partial}{\partial x}(\lambda\epsilon + 2G\epsilon_{xx}) + \frac{\partial}{\partial y}G\gamma_{yx} + \frac{\partial}{\partial z}G\gamma_{zx} + f_x = \rho\frac{\partial^2 u}{\partial t^2} \tag{9.1a}$$

$$\frac{\partial}{\partial y}\left(\lambda\epsilon + 2G\epsilon_{yy}\right) + \frac{\partial}{\partial x}G\gamma_{xy} + \frac{\partial}{\partial z}G\gamma_{zy} + f_y = \rho\frac{\partial^2 v}{\partial t^2} \tag{9.1b}$$

$$\frac{\partial}{\partial z}(\lambda\epsilon + 2G\epsilon_{zz}) + \frac{\partial}{\partial x}G\gamma_{xz} + \frac{\partial}{\partial y}G\gamma_{yz} + f_z = \rho\frac{\partial^2 w}{\partial t^2} \tag{9.1c}$$

Volumetric strain and displacements in strain equation become

$$(\lambda + G)\frac{\partial}{\partial x}\epsilon + G\nabla^2 u + f_x = \rho\frac{\partial^2 u}{\partial t^2} \tag{9.2a}$$

$$(\lambda + G)\frac{\partial}{\partial y}\epsilon + G\nabla^2 v + f_y = \rho\frac{\partial^2 v}{\partial t^2} \tag{9.2b}$$

$$(\lambda + G)\frac{\partial}{\partial z}\epsilon + G\nabla^2 w + f_z = \rho\frac{\partial^2 w}{\partial t^2} \tag{9.2c}$$

where $\nabla^2 = \partial^2/\partial x^2 + \partial^2/\partial y^2 + \partial^2/\partial x^2$ is the Laplacian operator. If the body force is neglected and variation of the displacements are harmonic,

$$u = U\,(x, y, z)\, e^{i\Omega t}, \quad v = V\,(x, y, z)\, e^{i\Omega t}, \quad \text{and } w = W\,(x, y, z)\, e^{i\Omega t}$$

then

$$\frac{\partial^2 u}{\partial t^2} = -\Omega^2 U\,(x, y, z)\, e^{i\Omega t} = -\Omega^2 u \tag{9.3a}$$

$$\frac{\partial^2 v}{\partial t^2} = -\Omega^2 v \tag{9.3b}$$

and

$$\frac{\partial^2 w}{\partial t^2} = -\Omega^2 w \tag{9.3c}$$

where Ω is circular frequency.

Substituting Eqs. 9.3a–9.3c into Eqs. 9.2a–9.2c,

$$(\lambda + G)\frac{\partial}{\partial x}\epsilon + G\nabla^2 u + f_x = -\Omega^2 u \tag{9.4a}$$

$$(\lambda + G)\frac{\partial}{\partial y}\epsilon + G\nabla^2 v + f_y = -\Omega^2 v \tag{9.4b}$$

and

$$(\lambda + G)\frac{\partial}{\partial z}\epsilon + G\nabla^2 w + f_z = -\Omega^2 w \tag{9.4c}$$

These equations can be written in terms of dilatation and rotations by differentiating equations (9.4a–9.4c) with respect to x, y, z and then adding them together:

$$(\lambda + G)\,\nabla^2\epsilon + \left(G\nabla^2 + \rho\Omega^2\right)\left(\frac{\partial u}{\partial x} + \frac{\partial v}{\partial y} + \frac{\partial w}{\partial z}\right) = 0 \tag{9.5}$$

Taking into account the definition of volumetric strain, the above equation becomes

$$(\lambda + 2G)\,\nabla^2\epsilon + \rho\Omega^2\epsilon = 0 \tag{9.6a}$$

or

$$\left(\nabla^2 + \Omega_1^2\right)\epsilon = 0 \tag{9.6b}$$

where

$$\Omega_1^2 = \frac{\Omega^2}{C_1^2} \text{ and } C_1 = \sqrt{\frac{\lambda + 2G}{\rho}} \tag{9.6c}$$

where C_1 is the velocity of dilatation in the media. (C_1 is also called "velocity of compression wave" or "P-wave.")

Differentiating Eq. 9.5 with respect to y

$$(\lambda + G)\,\frac{\partial^2}{\partial x \partial y}\epsilon + G\nabla^2\frac{\partial u}{\partial y} = -\Omega^2\frac{\partial u}{\partial y} \tag{9.7}$$

Subtracting the derivation equation (Eq. 9.5) with respect to x from the derivation of Eq. 9.5 with respect to y gives

$$(\lambda + G)\,\frac{\partial^2\epsilon}{\partial x \partial y} + G\nabla^2\frac{\partial u}{\partial y} - (\lambda + G)\,\frac{\partial^2\epsilon}{\partial x \partial y} - G\nabla^2\frac{\partial v}{\partial x} = -\rho\Omega^2\frac{\partial u}{\partial y} + \Omega^2\frac{\partial v}{\partial x} \tag{9.8}$$

This can be simplified to

$$\left(G\nabla^2 - \rho\Omega^2\right)\left(\frac{\partial u}{\partial y} - \frac{\partial v}{\partial x}\right) = 0 \tag{9.9}$$

Substituting elastic rotation ω_z into the above equation, it yields

$$\left(\nabla^2 + \Omega_2^2\right)w_z = 0 \tag{9.10a}$$

where

$$\Omega_2^2 = \frac{\Omega^2}{C_2^2} \text{ and } C_2 = \sqrt{\frac{G}{\rho}}$$

C_2 is the velocity of wave of distortion in the media. (C_2 is also called "velocity of shear wave" or "S-wave.")

Following a similar procedure as above, the third equation of motion will be given by the following equation:

$$\left(\nabla^2 + \Omega_2^2\right) w_x = 0 \tag{9.10b}$$

9.2.1 Displacement Equations

Since general solutions of equations of motion are given in terms of dilatation and rotation components, to determine displacements, they should be expressed in terms of dilatation and rotation components. The Laplacian of the displacement in the x direction is

$$\nabla^2 u = \frac{\partial^2 u}{\partial x^2} + \frac{\partial^2 v}{\partial y^2} + \frac{\partial^2 w}{\partial z^2} \tag{9.11}$$

Using the definition of strains and elastic relations in terms of displacements,

$$\nabla^2 u = \frac{\partial}{\partial x}\epsilon_{xx} + \frac{\partial}{\partial y}\left(\frac{\partial v}{\partial x} - 2w_z\right) + \frac{\partial}{\partial z}\left(2w_y + \frac{\partial w}{\partial x}\right) \tag{9.12}$$

After simplification,

$$\nabla^2 u = \frac{\partial}{\partial x}\epsilon + 2\frac{\partial w_y}{\partial z} - 2\frac{\partial w_z}{\partial y} \tag{9.13a}$$

Similarly the corresponding expressions for the Laplacians of v and w are

$$\nabla^2 v = \frac{\partial}{\partial y}\epsilon + 2\frac{\partial w_z}{\partial x} - 2\frac{\partial w_x}{\partial z} \tag{9.13b}$$

$$\nabla^2 w = \frac{\partial}{\partial z}\epsilon + 2\frac{\partial w_x}{\partial y} - 2\frac{\partial w_y}{\partial x} \tag{9.13b}$$

By substituting these equations into Eqs. 9.4a–9.4c, displacements will be found in terms of dilatation and components of rotation:

$$(\lambda + G)\frac{\partial \epsilon}{\partial x} + G\left(\frac{\partial \epsilon}{\partial x} + 2\frac{\partial w_y}{\partial z} - 2\frac{\partial w_z}{\partial y}\right) = -\rho\Omega^2 u \tag{9.14a}$$

$$(\lambda + G)\frac{\partial \epsilon}{\partial y} + G\left(\frac{\partial \epsilon}{\partial y} + 2\frac{\partial w_z}{\partial x} - 2\frac{\partial w_x}{\partial z}\right) = -\rho\Omega^2 v \tag{9.14b}$$

$$(\lambda + G)\frac{\partial \epsilon}{\partial z} + G\left(\frac{\partial \epsilon}{\partial z} + 2\frac{\partial w_x}{\partial y} - 2\frac{\partial w_y}{\partial x}\right) = -\rho\Omega^2 w \tag{9.14c}$$

After some reduction to the above equations, we get

$$u = -\frac{1}{\Omega_1^2}\frac{\partial \epsilon}{\partial x} - \frac{2}{\Omega_2^2}\left(\frac{\partial w_y}{\partial z} - \frac{\partial w_z}{\partial y}\right) \tag{9.15a}$$

$$v = -\frac{1}{\Omega_1^2}\frac{\partial \epsilon}{\partial y} - \frac{2}{\Omega_2^2}\left(\frac{\partial w_z}{\partial x} - \frac{\partial w_x}{\partial z}\right) \tag{9.15b}$$

$$w = -\frac{1}{\Omega_1^2}\frac{\partial \epsilon}{\partial z} - \frac{2}{\Omega_2^2}\left(\frac{\partial w_x}{\partial y} - \frac{\partial w_y}{\partial x}\right) \tag{9.15c}$$

9.2.2 Stress Equations

In order to get the stresses in terms of the dilatation and rotation components, Eq. 9.6b can be written as

$$\sigma_{yy} = \lambda\epsilon + 2G\epsilon_{yy} = \lambda\epsilon_{xx} + (\lambda + 2G)\,\epsilon_{yy} + \lambda\epsilon_{zz} \tag{9.16}$$

and, in terms of the displacement component,

$$\sigma_{yy} = \lambda\frac{\partial u}{\partial x} + (\lambda + 2G)\frac{\partial v}{\partial y} + \lambda\frac{\partial w}{\partial z} \tag{9.17}$$

$$\sigma_{yy} = -\left(\frac{\lambda+2G}{\Omega_1^2}\frac{\partial^2\epsilon}{\partial y^2}\right) + \frac{2(\lambda+2G)}{\Omega_2^2}\left(\frac{\partial^2 w_x}{\partial y\partial z} - \frac{\partial^2 w_z}{\partial x\partial z}\right) + \frac{2\lambda}{\Omega_2^2}\left(\frac{\partial^2 w_x}{\partial x\partial y} - \frac{\partial^2 w_y}{\partial x\partial z}\right)$$
$$-\frac{\lambda}{\Omega_1^2}\left(\frac{\partial^2\epsilon}{\partial x^2}\right) + \frac{2\lambda}{\Omega_2^2}\left(\frac{\partial^2 w_y}{\partial x\partial z} - \frac{\partial^2 w_x}{\partial y\partial z}\right) - \frac{\lambda}{\Omega_1^2}\left(\frac{\partial^2\epsilon}{\partial z^2}\right) \tag{9.18}$$

This equation, after simplification, becomes

$$\sigma_{yy} = -\frac{\lambda}{\Omega_1^2}\nabla^2\epsilon - \frac{2G}{\Omega_1^2}\left(\frac{\partial^2\epsilon}{\partial y^2}\right) - \frac{4G}{\Omega_2^2}\left(\frac{\partial^2 w_z}{\partial x\partial y} - \frac{\partial^2 w_x}{\partial y\partial z}\right) \tag{9.19a}$$

Similarly, the other direct stresses are

$$\sigma_{xx} = -\frac{\lambda}{\Omega_1^2}\nabla^2\epsilon - \frac{2G}{\Omega_1^2}\left(\frac{\partial^2\epsilon}{\partial x^2}\right) - \frac{4G}{\Omega_2^2}\left(\frac{\partial^2 w_y}{\partial x\partial z} - \frac{\partial^2 w_z}{\partial x\partial y}\right) \tag{9.19b}$$

$$\sigma_{zz} = -\frac{\lambda}{\Omega_1^2}\nabla^2\epsilon - \frac{2G}{\Omega_1^2}\left(\frac{\partial^2\epsilon}{\partial z^2}\right) - \frac{4G}{\Omega_2^2}\left(\frac{\partial^2 w_x}{\partial y\partial z} - \frac{\partial^2 w_y}{\partial x\partial z}\right) \tag{9.19c}$$

9.2.3 Shear Stress Equations

Shear stresses in dilatation and rotation components can be obtained as follows; for instance, the shear stress τ_{xy} in terms of displacement is

$$\tau_{xy} = G\gamma_{xy} = G\left(\frac{\partial u}{\partial y} + \frac{\partial v}{\partial x}\right) \tag{9.20}$$

Substituting for u and v from Eqs. 9.15a and 9.15b, the above equation yields

$$\tau_{xy} = -\frac{2G}{\Omega_1^2}\frac{\partial^2\epsilon}{\partial x\partial y} - \frac{2G}{\Omega_2^2}\left(\frac{\partial^2 w_y}{\partial z\partial y} - \frac{\partial^2 w_z}{\partial y^2} + \frac{\partial^2 w_z}{\partial x^2} - \frac{\partial^2 w_x}{\partial x\partial z}\right) \tag{9.21}$$

Differentiating three elastic rotations in terms of displacements with respect to z, x, and y, respectively, gives

$$\frac{\partial\omega_z}{\partial z} = \frac{1}{2}\left(\frac{\partial^2 v}{\partial z\partial x} - \frac{\partial^2 u}{\partial z\partial y}\right) \tag{9.22a}$$

$$\frac{\partial \omega_x}{\partial x} = \frac{1}{2}\left(\frac{\partial^2 w}{\partial x \partial y} - \frac{\partial^2 v}{\partial x \partial z}\right) \tag{9.22b}$$

$$\frac{\partial \omega_y}{\partial y} = \frac{1}{2}\left(\frac{\partial^2 u}{\partial y \partial z} - \frac{\partial^2 w}{\partial y \partial x}\right) \tag{9.22c}$$

Adding all three equations gives

$$\frac{\partial w_y}{\partial y} = -\frac{\partial w_x}{\partial x} - \frac{\partial w_z}{\partial z} \tag{9.23}$$

Substituting for $\frac{\partial w_y}{\partial y}$ from the above equation into Eq. 9.20 and simplifying,

$$\tau_{xy} = -\frac{2G}{\Omega_1^2}\frac{\partial^2 \epsilon}{\partial x \partial y} + \frac{4G}{\Omega_2^2}\frac{\partial^2 w_x}{\partial x \partial z} + \frac{2G}{\Omega_2^2}\left(\frac{\partial^2}{\partial z^2} + \frac{\partial^2}{\partial y^2} - \frac{\partial^2}{\partial x^2}\right) w_z \tag{9.24a}$$

Following the same procedure, the other shear stress becomes

$$\tau_{zy} = -\frac{2G}{\Omega_1^2}\frac{\partial^2 \epsilon}{\partial z \partial y} - \frac{4G}{\Omega_2^2}\frac{\partial^2 w_z}{\partial x \partial z} - \frac{2G}{\Omega_2^2}\left(\frac{\partial^2}{\partial x^2} + \frac{\partial^2}{\partial y^2} - \frac{\partial^2}{\partial z^2}\right) w_x \tag{9.24b}$$

9.2.4 Solutions of the Governing Equations

In order to reduce the system of partial differential equations to a system of first-order ordinary differential equations, double complex Fourier transformation is used. The boundary conditions must then be treated in the same way so that, instead of having relations in partial derivatives with respect to x, y, and z, relations are obtained in terms of the derivatives with respect to y only.

If $f(x,y,z)$ is a function of three independent variables, then the Fourier transform of this function with respect to x is

$$f_1(p, y, z) = \int_{-\infty}^{\infty} f(x, y, z)\, e^{-i2\pi px} \mathrm{dx} \tag{9.25}$$

and the Fourier transform of this function regarded as a function of z will be

$$\overline{f}(p, y, q) = \int_{-\infty}^{\infty} f_1(p, y, z)\, e^{-i2\pi qz} \mathrm{dz} \tag{9.26}$$

Therefore, the last equation in terms of $f(x,y,z)$ will be given by the combination of the last two transformations, as

$$\overline{f}\,(p,y,q) = \int_{-\infty}^{\infty}\int_{-\infty}^{\infty} f\,(x,y,z)\,e^{-i2\pi(px+qz)}\mathrm{dxdz} \tag{9.27}$$

This is the definition of a double Fourier transform of $f(x,y,z)$, and p and q are variables in the transformed domain. The double complex Fourier transform of the first two derivatives of $f(x,y,z)$ is given by

$$F\left[\frac{\partial f\,(x,y,z)}{\partial x}\right] = i2\pi p\overline{f}_x \tag{9.28a}$$

$$F\left[\frac{\partial f\,(x,y,z)}{\partial z}\right] = i2\pi q\overline{f}_z \tag{9.28b}$$

$$F\left[\frac{\partial^2 f\,(x,y,z)}{\partial x^2}\right] = -4\pi^2 p^2\overline{f}_x \tag{9.28c}$$

$$F\left[\frac{\partial^2 f\,(x,y,z)}{\partial z^2}\right] = -4\pi^2 q^2\overline{f}_z \tag{9.28d}$$

Here transforms with respect to x and z are

$$\overline{f}_x\,(p,y,z) = \int_{-\infty}^{\infty} f\,(p,y,z)\,e^{-i2\pi px}\mathrm{dx} \tag{9.29a}$$

$$\overline{f}_z\,(x,y,q) = \int_{-\infty}^{\infty} f\,(p,y,z)\,e^{-i2\pi qz}\mathrm{dz} \tag{9.29b}$$

The equations of motion (Eq. 9.7) contain three independent variables. Their solutions will be facilitated by the application of the double complex Fourier transforms, having parameters p and q which will eliminate two independent variables x and z with ranges from $-\infty$ to $+\infty$.

By applying the Eqs. 9.28a–9.28d to the equations of motion, the reduced equation becomes

$$\frac{\partial^2\overline{\epsilon}}{\partial \mathrm{y}^2} + \left(\Omega_1^2 - 4\pi^2\mathrm{p}^2 - 4\pi^2\mathrm{q}^2\right)\overline{\epsilon} = 0 \tag{9.30a}$$

Transformations of Eq. 9.9 are

$$\frac{\partial^2 \overline{w_x}}{\partial y^2} + \left(\Omega_1^2 - 4\pi^2 p^2 - 4\pi^2 q^2\right) w_x^2 = 0 \tag{9.30b}$$

$$\frac{\partial^2 \overline{w_z}}{\partial y^2} + \left(\Omega_1^2 - 4\pi^2 p^2 - 4\pi^2 q^2\right) w_z^2 = 0 \tag{9.30c}$$

where $\overline{\epsilon}$, $\overline{w_x}$, and $\overline{w_z}$ are the corresponding double complex Fourier transformations of ϵ, w_x and w_z. Solving the ordinary differential equations (Eqs. 9.30a–9.30c), the general solutions of the transformed equations of motion are

$$\overline{\epsilon} = A_{1\epsilon} e^{-r_1 y} + A_{2\epsilon} e^{r_1 y} \tag{9.31a}$$

$$\overline{w_x} = A_{1x} e^{-r_2 y} + A_{2x} e^{r_2 y} \tag{9.31b}$$

$$\overline{w_z} = A_{1z} e^{-r_3 y} + A_{2z} e^{r_3 y} \tag{9.31c}$$

where

$$r_1^2 = 4\pi^2 p^2 + 4\pi^2 q^2 - \Omega_1^2 \tag{9.32a}$$

$$r_2^2 = r_3^2 = 4\pi^2 p^2 + 4\pi^2 q^2 - \Omega_2^2 \tag{9.32b}$$

From the last two equations, it is obvious that $r_2^2 = r_3^2$ and $A_{1\epsilon}, A_{2\epsilon}, A_{1x}, A_{2x}, A_{1z}$, and A_{2z} are arbitrary functions of the transform parameters p and q to be determined from the mixed boundary conditions. It is seen from Eqs. 9.32a and 9.32b that r_1 and r_2 are associated with dilatational waves and shear waves, respectively.

It is necessary to find the double complex Fourier transform of the stresses. This is due to the fact that the yet unknown arbitrary functions in the general solutions (Eqs. 9.29a and 9.29b) are only related to the transforms of dilatation and two components of rotation.

Taking the double complex Fourier transforms of (9.14a), (9.14b), and (9.14c) by applying (9.28a–9.28d), first with respect to x and then with respect to z, will yield the following:

$$\overline{\sigma}_{yy} = \left(\lambda - \frac{2G}{\Omega_1^2}\frac{\partial^2}{\partial y^2}\right)\overline{\epsilon} + \frac{\mathrm{i}8\pi \mathrm{q}G}{\Omega_2^2}\frac{\partial \overline{w}_x}{\partial y} - \frac{\mathrm{i}8\pi \mathrm{p}G}{\Omega_2^2}\frac{\partial \overline{w}_z}{\partial y} \tag{9.33a}$$

$$\overline{\tau}_{xy} = -\frac{\mathrm{i}4\pi \mathrm{p}G}{\Omega_1^2}\frac{\partial \overline{\epsilon}}{\partial y} - \frac{16\pi^2 \mathrm{pq}G}{\Omega_2^2}\overline{w}_x + \frac{2G}{\Omega_2^2}\left(\frac{\partial^2}{\partial y^2} - 4\pi^2\mathrm{q}^2 + 4\pi^2\mathrm{p}^2\right)\overline{w}_z \tag{9.33b}$$

$$\overline{\tau}_{zy} = -\frac{\mathrm{i}4\pi \mathrm{q}G}{\Omega_1^2}\frac{\partial \overline{\epsilon}}{\partial y} - \frac{2G}{\Omega_2^2}\left(\frac{\partial^2}{\partial y^2} - 4\pi^2\mathrm{p}^2 + 4\pi^2\mathrm{q}^2\right)\overline{w}_x + \frac{16\pi^2 \mathrm{pq}G}{\Omega_2^2}\overline{w}_z \tag{9.33c}$$

Applying the Fourier transform to Eq. 9.11, the transformed surface motions become in terms of $\overline{\epsilon}$, $\overline{w_x}$, $\overline{w_y}$ and $\overline{w_z}$

$$\overline{u} = -\frac{\mathrm{i}2\pi \mathrm{p}}{\Omega_1^2}\overline{\epsilon} - \frac{2}{\Omega_2^2}\left(\mathrm{i}2\pi \mathrm{q}\overline{w}_y - \frac{\partial \overline{w}_z}{\partial y}\right) \tag{9.34a}$$

$$\overline{v} = -\frac{1}{\Omega_1^2}\frac{\partial \overline{\epsilon}}{\partial y} - \frac{2}{\Omega_2^2}\left(\mathrm{i}2\pi \mathrm{p}\overline{w}_z - \mathrm{i}2\pi \mathrm{q}\overline{w}_x\right) \tag{9.34b}$$

$$\overline{w} = -\frac{\mathrm{i}2\pi \mathrm{q}}{\Omega_1^2}\overline{\epsilon} - \frac{2}{\Omega_2^2}\left(\frac{\partial \overline{w}_x}{\partial y} - \mathrm{i}2\pi \mathrm{p}\overline{w}_y\right) \tag{9.34c}$$

where $\overline{w}_\mathrm{y} = \mathrm{i}\pi \mathrm{q}\overline{u} - \mathrm{i}\pi \mathrm{p}\overline{w}$

Substituting for $\overline{w}_\mathrm{y}$ from the above equation into Eqs. 9.34a and 9.34c and solving for $\overline{u}$, $\overline{w}$ results in terms of $\overline{\epsilon}$, $\overline{w_x}$, and $\overline{w_z}$, then the double complex Fourier transforms of the displacements will be

$$\overline{u} = -\frac{\mathrm{i}2\pi \mathrm{p}}{\Omega_1^2}\overline{\epsilon} + \frac{8\pi^2 \mathrm{pq}}{\mathrm{r}_2^2\Omega_2^2}\frac{\partial \overline{w}_x}{\partial y} - \frac{8\pi^2 p^2 - 2\Omega_2^2}{\mathrm{r}_2^2\Omega_2^2}\frac{\partial \overline{w}_z}{\partial y} \tag{9.35a}$$

$$\overline{v} = -\frac{1}{\Omega_1^2}\frac{\partial \overline{\epsilon}}{\partial y} - \frac{\mathrm{i}4\pi \mathrm{q}}{\Omega_2^2}\overline{w}_x - \frac{\mathrm{i}4\pi \mathrm{p}}{\Omega_2^2}\overline{w}_z \tag{9.35b}$$

$$\overline{w} = -\frac{\mathrm{i}2\pi \mathrm{q}}{\Omega_1^2}\overline{\epsilon} + \frac{2\Omega_2^2 - 8\pi^2 q^2}{\mathrm{r}_2^2\Omega_2^2}\frac{\partial \overline{w}_x}{\partial y} + \frac{8\pi^2 \mathrm{pq}}{\mathrm{r}_2^2\Omega_2^2}\frac{\partial \overline{w}_z}{\partial y} \tag{9.35c}$$

Substituting the general solutions of the transformed equations of motion in Eqs. 9.31a–9.31c into Eqs. 9.33a–9.33c, the Fourier transformed stress and shear stresses in terms of depth, y, are

$$\overline{\sigma}_{yy} = \left(\frac{\lambda\Omega_1^2 - 2Gr_1^2}{\Omega_1^2}\right)(A_{1\epsilon}e^{-r_1 y} + A_{2\epsilon}e^{r_1 y}) - \frac{i8\pi q G r_2}{\Omega_2^2}(-A_{1x}e^{-r_2 y} + A_{2x}e^{r_2 y}) - \frac{i8\pi p G r_2}{\Omega_2^2}(-A_{1z}e^{-r_2 y} + A_{2z}e^{r_2 y}) \tag{9.36a}$$

$$\overline{\tau}_{xy} = -\frac{i4\pi p G r_1}{\Omega_1^2}(-A_{1\epsilon}e^{-r_1 y} + A_{2\epsilon}e^{r_1 y}) - \frac{16\pi^2 pqG}{\Omega_2^2}(A_{1x}e^{-r_2 y} + A_{2x}e^{r_2 y}) + \frac{16G\pi^2 p^2 - 2G\Omega_2^2}{\Omega_2^2}(A_{1z}e^{-r_2 y} + A_{2z}e^{r_2 y}) \tag{9.36b}$$

$$\overline{\tau}_{zy} = -\frac{i4\pi q G r_1}{\Omega_1^2}(-A_{1\epsilon}e^{-r_1 y} + A_{2\epsilon}e^{r_1 y}) - \frac{16G\pi^2 q^2 - 2G\Omega_2^2}{\Omega_2^2}(A_{1x}e^{-r_2 y} + A_{2x}e^{r_2 y}) + \frac{16\pi^2 pqG}{\Omega_2^2}(A_{1z}e^{-r_2 y} + A_{2z}e^{r_2 y}) \tag{9.36c}$$

Substituting Eqs. 9.31a–9.31c into Eqs. 9.35a–9.35c, the transformed displacements are

$$\overline{u} = -\frac{i2\pi p}{\Omega_1^2}(A_{1\epsilon}e^{-r_1 y} + A_{2\epsilon}e^{r_1 y}) - \frac{8\pi^2 pq}{r_2\Omega_2^2}(-A_{1x}e^{-r_2 y} + A_{2x}e^{r_2 y}) + \frac{8\pi^2 p^2 - 2\Omega_2^2}{r_2\Omega_2^2}(-A_{1z}e^{-r_2 y} + A_{2z}e^{r_2 y}) \tag{9.37a}$$

$$\overline{v} = -\frac{r_1}{\Omega_1^2}(-A_{1\epsilon}e^{-r_1 y} + A_{2\epsilon}e^{r_1 y}) - \frac{i4\pi q}{\Omega_2^2}(A_{1x}e^{-r_2 y} + A_{2x}e^{r_2 y}) - \frac{i4\pi p}{\Omega_2^2}(A_{1z}e^{-r_2 y} + A_{2z}e^{r_2 y}) \tag{9.37b}$$

$$\overline{w} = -\frac{i2\pi q}{\Omega_1^2}(A_{1\epsilon}e^{-r_1 y} + A_{2\epsilon}e^{r_1 y}) + \frac{2\Omega_2^2 - 8\pi^2 q^2}{r_2\Omega_2^2}(-A_{1x}e^{-r_2 y} + A_{2x}e^{r_2 y}) + \frac{8\pi^2 pq}{r_2\Omega_2^2}(-A_{1z}e^{-r_2 y} + A_{2z}e^{r_2 y}) \tag{9.37c}$$

Modal stresses and displacements can be written in terms of p, q and the properties of the medium:

$$\overline{\sigma}_{yy} = \{S_{11}, S_{12}, S_{13}, S_{14}, S_{15}, S_{16}\}\{A\} \tag{9.38a}$$

$$\overline{\tau}_{xy} = \{S_{21}, S_{22}, S_{23}, S_{24}, S_{25}, S_{26}\}\{A\} \tag{9.38b}$$

$$\overline{\tau}_{zy} = \{S_{31}, S_{32}, S_{33}, S_{34}, S_{35}, S_{36}\}\{A\} \tag{9.38c}$$

$$\overline{u} = \{S_{41}, S_{42}, S_{43}, S_{44}, S_{45}, S_{46}\}\{A\} \tag{9.38d}$$

$$\overline{v} = \{S_{51}, S_{52}, S_{53}, S_{54}, S_{55}, S_{56}\}\{A\} \tag{9.38e}$$

$$\overline{w} = \{S_{61}, S_{62}, S_{63}, S_{64}, S_{65}, S_{66}\}\{A\} \tag{9.38f}$$

Here $\{A\} = \{A_{1\epsilon}, A_{2\epsilon}, A_{1x}, A_{2x}, A_{1z}, A_{2z}\}^T$, the elements of S_{ij}, $i = 1,2,3,4,5,6,$ and $j = 1,2,3,4,5,6$.. Consider a layered medium with perfect bonding along all interfaces. This implies the continuity of the transformed stresses and displacements across each layer. To enforce this condition, Eqs. 9.38a–9.38f can be written in a matrix form:

$$\begin{Bmatrix} \overline{\sigma}_{yy} \\ \overline{\tau}_{xy} \\ \overline{\tau}_{yz} \\ \overline{u} \\ \overline{v} \\ \overline{w} \end{Bmatrix} = \begin{bmatrix} S_{11} & S_{12} & S_{13} & S_{14} & S_{15} & S_{16} \\ S_{21} & S_{22} & S_{23} & S_{24} & S_{25} & S_{26} \\ S_{31} & S_{32} & S_{33} & S_{34} & S_{35} & S_{36} \\ S_{41} & S_{42} & S_{43} & S_{44} & S_{45} & S_{46} \\ S_{51} & S_{52} & S_{53} & S_{54} & S_{55} & S_{56} \\ S_{61} & S_{62} & S_{63} & S_{64} & S_{65} & S_{66} \end{bmatrix} \begin{Bmatrix} A_{1\epsilon} \\ A_{2\epsilon} \\ A_{1x} \\ A_{2x} \\ A_{1z} \\ A_{2z} \end{Bmatrix} \tag{9.39}$$

The matrix form is simplified to

$$\{\overline{R}\} = [S]\{A\} \tag{9.40}$$

where the vector $\{\overline{R}\} = \{\overline{\sigma}_{yy}, \overline{\tau}_{xy}, \overline{\tau}_{yz}, \overline{u}, \overline{v}, \overline{w}\}^T$.

The above vector is for any point of any layer. The composite medium is assumed to be formed by multiple different layers bounded by upper and lower interfaces. The thickness of each layer is h_1, h_2 … h_m … and h_n. The elastic layer is characterized by a Lamb's constant λ, shear modulus G, and density ρ. The mathematical formulation is used to develop a relationship of displacements and stresses at boundaries between consecutive layers. Consequently we obtain a propagator matrix relating displacements and stresses at boundaries of the lower layer to the upper layer. To build the propagator matrix for an n-layered medium, the relationships of stresses and displacements between any layers to the top layer are assumed to be the same at any location in the multilayered medium with infinite length and width.

As presented in Eqs. 9.38a–9.38f, the displacements and stresses vector for any point on the medium of each layer can be given as follows:

$$\begin{Bmatrix} \overline{\sigma}_{yy} \\ \overline{\tau}_{xy} \\ \overline{\tau}_{yz} \\ \overline{u} \\ \overline{v} \\ \overline{w} \end{Bmatrix}_i = \begin{bmatrix} & & \\ & S & \\ & & \end{bmatrix}_i \begin{Bmatrix} A_{1\epsilon} \\ A_{2\epsilon} \\ A_{1x} \\ A_{2x} \\ A_{1z} \\ A_{2z} \end{Bmatrix}_i \tag{9.41}$$

where

i indicates the ith layer (i = 1, 2, … n).

σ is the normal stress in the y direction.

τ_{xy} is the shear stress in the x direction.

τ_{zy} is the shear stress in the z direction.

u is the displacement in the x direction.

v is the displacement in the y direction.

w is the displacement in the z direction.

$$\{\overline{R}\}_i = \{\overline{\sigma}_{yy}, \overline{\tau}_{xy}, \overline{\tau}_{yz}, \overline{u}, \overline{v}, \overline{w}\}_i^T \tag{9.42}$$

$$\{A\}_i = \{A_{1\epsilon}, A_{2\epsilon}, A_{1x}, A_{2x}, A_{1z}, A_{2z}\}_i^T \tag{9.43}$$

We may simplify Eq. 9.41 into

$$\left\{\overline{R}(p, y, q)\right\}_i = [S(\lambda, G, \rho, y, p, q)]_i \{A\}_i \tag{9.44}$$

From Eq. 9.44, the following equations can be written for any point in the medium. Considering the first layer with depth $y = h_1$, for interfacing between the first layer and second layer, the transformed stresses and displacements are

$$\left\{\overline{R}(p, h_1, q)\right\}_1 = \left[S_{y=h_1}\right]_1 \{A\} \tag{9.45a}$$

The transformed stresses and displacements on the top surface of the first layer are

$$\left\{\overline{R}(p, 0, q)\right\}_1 = \left[S_{y=0}\right]_1 \{A\} \tag{9.45b}$$

The contracted form of Eqs. 9.45a and 9.45b is

$$\left\{\overline{R}(p, h_1, q)\right\}_1 = \left[S_{y=h_1}\right]_1 \left[S_{y=0}\right]_1^{-1} \left\{\overline{R}(p, 0, q)\right\}_1 \tag{9.46}$$

$$\text{or } \left\{\overline{R}(p, h_1, q)\right\}_1 = [P_1] \left\{\overline{R}(p, 0, q)\right\}_1 \tag{9.47}$$

where

$$[P_1] = \left[S_{y=h_1}\right]_1 \left[S_{y=0}\right]_1^{-1}$$

Considering the second layer, following the above procedure and assuming the depth of second layer to be $= h_2$, we have

$$\left\{\overline{R}(p, h_2, q)\right\}_2 = [P_2] \left\{\overline{R}(p, 0, q)\right\}_2 \tag{9.48}$$

where

$$[P_2] = \left[S_{y=h_2}\right]_2 \left[S_{y=0}\right]_2^{-1} \tag{9.49}$$

assuming that, at the interface between two layers, the displacements and stresses are the same. So we have

$$\left\{\overline{R}(p, 0, q)\right\}_2 = \left\{\overline{R}(p, h_2, q)\right\}_1 \tag{9.50}$$

Substituting Eqs. 9.49 and 9.47 into Eq. 9.48, the result yields

$$\left\{\overline{R}(p,h_2,q)\right\}_2 = [\mathrm{P}_2]\,[\mathrm{P}_1]\left\{\overline{R}(p,0,q)\right\}_1 \tag{9.51}$$

Writing Eq. 9.51 for all layers and eliminating vector {R} for all the points on the interfaces results in a relation between displacements and stresses at the top and bottom boundaries of the multilayered medium. We have

$$\left\{\overline{R}(p,h_2,q)\right\}_2 = [\mathrm{P}_n]\,[\mathrm{P}_{n-1}]\cdots[\mathrm{P}_1]\left\{\overline{R}(p,0,q)\right\}_1 \tag{9.52a}$$

or

$$\left\{\overline{R}(p,h_n,q)\right\}_{\mathrm{n}} = [\mathrm{P}]\left\{\overline{R}(p,0,q)\right\}_1 \tag{9.52b}$$

where the propagator matrix [P] is a 6 × 6 matrix. For notational simplicity $\mathrm{P_n}$ designates $\mathrm{P}(\lambda_n, G_n, \rho_n, h_n, p, q)$. In Eq. 9.52a the missing initial functions on the boundary surface of the medium can be obtained from the corresponding boundary conditions at the other end of the medium. Equation 9.52b then describes the overall response of the layered medium where all intermediate state vectors which occur in the same equation have been eliminated upon continued matrix multiplication.

Local information consisting of state vectors at interfaces can now be obtained by terminating the matrix multiplication at the appropriate interface. The state vector in an arbitrary layer can now be obtained by the relation

$$\left\{\overline{R}(p,y,q)\right\} = \left[\mathrm{P}(\mathrm{y})\prod_{\mathrm{i}=1}^{\mathrm{m}-1}\mathrm{P_i}\right]\left\{\overline{R}(p,0,q)\right\} \tag{9.53}$$

in which the y coordinate is the local depth within the layer m, ranging from zero to $\mathrm{h_m}$, and P(y) denotes $\mathrm{P}(\lambda_n, G_n, \rho_n, h_n, p, q)$. For a concentrated surface load applied on a multilayered medium with rigid bottom, the displacements at the interface between the bottom layer and the rigid bottom are zero; Eq. 9.53 can be written as

$$\begin{Bmatrix} \overline{\sigma}_{yy} \\ \overline{\tau}_{xy} \\ \overline{\tau}_{yz} \\ 0 \\ 0 \\ 0 \end{Bmatrix} = \begin{bmatrix} \mathrm{P}_{11} & \mathrm{P}_{12} & \mathrm{P}_{13} & \mathrm{P}_{14} & \mathrm{P}_{15} & \mathrm{P}_{16} \\ \mathrm{P}_{21} & \mathrm{P}_{22} & \mathrm{P}_{23} & \mathrm{P}_{24} & \mathrm{P}_{25} & \mathrm{P}_{26} \\ \mathrm{P}_{31} & \mathrm{P}_{32} & \mathrm{P}_{33} & \mathrm{P}_{34} & \mathrm{P}_{35} & \mathrm{P}_{36} \\ \mathrm{P}_{41} & \mathrm{P}_{42} & \mathrm{P}_{43} & \mathrm{P}_{44} & \mathrm{P}_{45} & \mathrm{P}_{46} \\ \mathrm{P}_{51} & \mathrm{P}_{52} & \mathrm{P}_{53} & \mathrm{P}_{54} & \mathrm{P}_{55} & \mathrm{P}_{56} \\ \mathrm{P}_{61} & \mathrm{P}_{62} & \mathrm{P}_{63} & \mathrm{P}_{64} & \mathrm{P}_{65} & \mathrm{P}_{66} \end{bmatrix} \begin{Bmatrix} \overline{\sigma}_{yy} \\ \overline{\tau}_{xy} \\ \overline{\tau}_{yz} \\ \overline{u} \\ \overline{v} \\ \overline{w} \end{Bmatrix} \tag{9.54}$$

For vertical concentrated surface load, Eq. 9.54 becomes

$$\begin{bmatrix} \mathrm{P}_{44} & \mathrm{P}_{45} & \mathrm{P}_{46} \\ \mathrm{P}_{54} & \mathrm{P}_{55} & \mathrm{P}_{56} \\ \mathrm{P}_{64} & \mathrm{P}_{65} & \mathrm{P}_{66} \end{bmatrix} \begin{Bmatrix} \overline{u} \\ \overline{v} \\ \overline{w} \end{Bmatrix} = \begin{Bmatrix} \mathrm{P}_{41}\overline{\sigma}_{yy} \\ \mathrm{P}_{51}\overline{\sigma}_{yy} \\ \mathrm{P}_{61}\overline{\sigma}_{yy} \end{Bmatrix} \tag{9.55a}$$

For a horizontal concentrated surface load, Eq. 9.54 becomes

$$\begin{bmatrix} P_{44} & P_{45} & P_{46} \\ P_{54} & P_{55} & P_{56} \\ P_{64} & P_{65} & P_{66} \end{bmatrix} \begin{Bmatrix} \overline{u} \\ \overline{v} \\ \overline{w} \end{Bmatrix} = \begin{Bmatrix} P_{42}\overline{\tau}_{xy} \\ P_{52}\overline{\tau}_{xy} \\ P_{62}\overline{\tau}_{xy} \end{Bmatrix} \tag{9.55b}$$

or

$$\begin{bmatrix} P_{44} & P_{45} & P_{46} \\ P_{54} & P_{55} & P_{56} \\ P_{64} & P_{65} & P_{66} \end{bmatrix} \begin{Bmatrix} \overline{u} \\ \overline{v} \\ \overline{w} \end{Bmatrix} = \begin{Bmatrix} P_{43}\overline{\tau}_{yz} \\ P_{53}\overline{\tau}_{yz} \\ P_{63}\overline{\tau}_{yz} \end{Bmatrix} \tag{9.55c}$$

Equations (9.55a–9.55c) provide the surface motional equations for the system. And it should indicate that elements of the transfer matrix are all related to the necessary properties, such as density, shear modulus, Lamb's constant, and the geometry of each layer.

9.2.5 General Solutions of Transformed Equations of Motion

Recall Eqs. 9.29a and 9.29b. When the depth of the medium is infinite, the value of the transformed functions will become infinitely large, which is unacceptable. In fact, values of any function at $y = \infty$ will be eliminated. Therefore, in Eqs. 9.29a and 9.29b, we will ignore the second term and the value of the transformed dilatation, and elastic rotations will become

$$\overline{\epsilon} = A_{\epsilon} e^{-r_1 y} \tag{9.56a}$$

$$\overline{w}_x = A_x e^{-r_2 y} \tag{9.56b}$$

$$\overline{w}_z = A_z e^{-r_3 y} \tag{9.56c}$$

Substituting the above in Eqs. 9.28a–9.28d results in

$$r_1^2 = 4\pi^2 p^2 + 4\pi^2 q^2 - \Omega_1^2 \tag{9.57a}$$

and

$$r_2^2 = r_3^2 = 4\pi^2 p^2 + 4\pi^2 q^2 - \Omega_2^2 \tag{9.57b}$$

Values of A_ϵ, A_x, and A_z are functions of p and q, and they depend on the boundary conditions of the problem. This means that they are dependent on the double complex Fourier transform of the stresses which act as external excitation on the surface of the half-space medium. In order to evaluate these arbitrary functions, Eqs. 9.33a–9.33c must be satisfied by the boundary conditions which requires that at $y = 0$ (surface of half-space) there exist three stresses components which are σ_{yy}, τ_{xy}, and τ_{yz}. Having put $y = 0$ in Eqs. 9.33a–9.33c and equating them to the double complex Fourier transform of the above stresses, they become

$$\overline{\sigma}_{yy} = \frac{\lambda\Omega_1^2 - 2G\mathrm{r}_1^2}{\Omega_1^2}A_\epsilon - \frac{\mathrm{i}8\pi\mathrm{q}G\mathrm{r}_2}{\Omega_2^2}A_x + \frac{\mathrm{i}8\pi\mathrm{p}G\mathrm{r}_2}{\Omega_2^2}A_z \tag{9.58a}$$

$$\overline{\tau}_{xy} = \frac{\mathrm{i}4\pi\mathrm{p}G\mathrm{r}_1}{\Omega_1^2}A_\epsilon - \frac{16\pi^2\mathrm{pq}G}{\Omega_2^2}A_x + \frac{16\pi^2\mathrm{p}^2G - 2G\Omega_2^2}{\Omega_2^2}A_z \tag{9.58b}$$

$$\overline{\tau}_{zy} = \frac{\mathrm{i}4\pi\mathrm{q}G\mathrm{r}_1}{\Omega_1^2}A_\epsilon - \frac{16\pi^2\mathrm{q}^2G - 2G\Omega_2^2}{\Omega_2^2}A_x + \frac{16\pi^2\mathrm{pq}G}{\Omega_2^2}A_z \tag{9.58c}$$

Similarly, setting $y = 0$ in Eqs. 9.33a–9.33c, we have

$$\overline{u} = -\frac{\mathrm{i}2\pi\mathrm{p}}{\Omega_1^2}A_{1\epsilon}e^{-r_1y} - \frac{8\pi^2\mathrm{pq}}{r_2\Omega_2^2}A_{1x}e^{-r_2y} + \frac{8\pi^2p^2 - 2\Omega_2^2}{r_2\Omega_2^2}A_{1z}e^{-r_2y} \tag{9.59a}$$

$$\overline{v} = \frac{r_1}{\Omega_1^2}A_{1\epsilon}e^{-r_1y} - \frac{\mathrm{i}4\pi\mathrm{q}}{\Omega_2^2}A_{1x}e^{-r_2y} - \frac{\mathrm{i}4\pi\mathrm{p}}{\Omega_2^2}A_{1z}e^{-r_2y} \tag{9.59b}$$

$$\overline{w} = -\frac{\mathrm{i}2\pi\mathrm{q}}{\Omega_1^2}A_{1\epsilon}e^{-r_1y} - \frac{2\Omega_2^2 - 8\pi^2q^2}{r_2\Omega_2^2}A_{1x}e^{-r_2y} - \frac{8\pi^2\mathrm{pq}}{r_2\Omega_2^2}A_{1z}e^{-r_2y} \tag{9.59c}$$

The solutions to the above set of equations are given by the values of A_ϵ, A_x, and A_z. These values can be expressed as

$$A_\epsilon = \frac{D_\epsilon}{D},\ A_x = \frac{D_x}{D},\ \text{and}\ A_z = \frac{D_z}{D} \tag{9.60}$$

where

$$D = \begin{vmatrix} \frac{\lambda\Omega_1^2 - 2G\mathrm{r}_1^2}{\Omega_1^2} & \frac{-\mathrm{i}8\pi\mathrm{q}G\mathrm{r}_2}{\Omega_2^2} & \frac{\mathrm{i}8\pi\mathrm{p}G\mathrm{r}_2}{\Omega_2^2} \\ \frac{\mathrm{i}4\pi\mathrm{p}G\mathrm{r}_1}{\Omega_1^2} & \frac{-16\pi^2\mathrm{pq}G}{\Omega_2^2} & \frac{16\pi^2\mathrm{p}^2G - 2G\Omega_2^2}{\Omega_2^2} \\ \frac{\mathrm{i}4\pi\mathrm{q}G\mathrm{r}_1}{\Omega_1^2} & \frac{-16\pi^2\mathrm{q}^2G + 2G\Omega_2^2}{\Omega_2^2} & \frac{16\pi^2\mathrm{pq}G}{\Omega_2^2} \end{vmatrix} \tag{9.61a}$$

After expansion this becomes

$$D = \frac{4G^3}{\Omega_1^2\Omega_2^2}\phi\,(p, q) \tag{9.61b}$$

and

$$\phi\,(p, q) = \left[\Omega_2^2 - 8\pi^2\left(\mathrm{p}^2 + \mathrm{q}^2\right)\right]^2 - 16\pi^2\mathrm{r}_1\mathrm{r}_2\left(\mathrm{p}^2 + \mathrm{q}^2\right) \tag{9.62}$$

This is the well-known function associated with Rayleigh surface waves. Also

$$D_\epsilon = \begin{vmatrix} -\overline{\sigma}_{yy} & \frac{\mathrm{i}8\pi\mathrm{q}G\mathrm{r}_2}{\Omega_2^2} & \frac{\mathrm{i}8\pi\mathrm{q}G\mathrm{r}_2}{\Omega_2^2} \\ -\overline{\tau}_{xy} & \frac{-16\pi^2\mathrm{pq}G}{\Omega_2^2} & \frac{16\pi^2\mathrm{p}^2G - 2G\Omega_2^2}{\Omega_2^2} \\ -\overline{\tau}_{zy} & \frac{-16\pi^2\mathrm{q}^2G + 2G\Omega_2^2}{\Omega_2^2} & \frac{16\pi^2\mathrm{pq}G}{\Omega_2^2} \end{vmatrix} \tag{9.63a}$$

$$D_x = \begin{vmatrix} \frac{\lambda\Omega_1^2 - 2G\mathrm{r}_1^2}{\Omega_1^2} & -\overline{\sigma}_{yy} & \frac{\mathrm{i}8\pi\mathrm{q}G\mathrm{r}_2}{\Omega_2^2} \\ \frac{\mathrm{i}4\pi\mathrm{p}G\mathrm{r}_1}{\Omega_1^2} & -\overline{\tau}_{xy} & \frac{16\pi^2\mathrm{p}^2G - 2G\Omega_2^2}{\Omega_2^2} \\ \frac{\mathrm{i}4\pi\mathrm{q}G\mathrm{r}_1}{\Omega_1^2} & -\overline{\tau}_{zy} & \frac{16\pi^2\mathrm{pq}G}{\Omega_2^2} \end{vmatrix} \tag{9.63b}$$

and

$$D_z = \begin{vmatrix} \frac{\lambda\Omega_1^2 - 2G\mathrm{r}_1^2}{\Omega_1^2} & \frac{\mathrm{i}8\pi\mathrm{q}G\mathrm{r}_2}{\Omega_2^2} & -\overline{\sigma}_{yy} \\ \frac{\mathrm{i}4\pi\mathrm{p}G\mathrm{r}_1}{\Omega_1^2} & \frac{-16\pi^2\mathrm{pq}G}{\Omega_2^2} & -\overline{\tau}_{xy} \\ \frac{\mathrm{i}4\pi\mathrm{q}G\mathrm{r}_1}{\Omega_1^2} & \frac{-16\pi^2\mathrm{q}^2G + 2G\Omega_2^2}{\Omega_2^2} & -\overline{\tau}_{zy} \end{vmatrix} \tag{9.63c}$$

9.2.6 Harmonic Response of the Surface Due to a Concentrated Vertical Load

The elastic medium is defined as infinitely deep and occupying an infinitely wide horizontal plane, which forms its only boundary $y=0$. It is assumed to be homogeneous and isotropic. This definition ensures that waves propagated from a source die out and are not reflected, thus eliminating any possible interference of waves. For vertical harmonic point force on the elastic medium, boundary stresses on the surface of the half-space are

$$\tau_{xy} = 0 \qquad 0 \leq |x| < \infty \wedge 0 \leq |z| < \infty \tag{9.64a}$$

$$\tau_{zy} = 0 \qquad 0 \leq |x| < \infty \wedge 0 \leq |z| < \infty \tag{9.64b}$$

Since the normal applied force is a harmonic point force, the corresponding stress can be expressed by Dirac delta function as

$$\sigma_{yy} = \sigma \cdot \delta\left(x, z\right) e^{i\Omega t} \tag{9.64c}$$

This means that σ_{yy} will be eliminated everywhere except that point at which the force is applied. It should be noted that o is a constant and it can be obtained by writing the relation between force and stress which is

$$F_y = \int_{-\infty}^{\infty}\int_{-\infty}^{\infty} \sigma \cdot \delta\left(x, z\right) dxdz = \sigma \tag{9.65}$$

Therefore,

$$\sigma = F_y \tag{9.66}$$

where F_y is the amplitude of the applied force.

Double complex Fourier transformations of these stresses with respect to x and then z are

$$\overline{\sigma}_{yy} = \int_{-\infty}^{\infty}\int_{-\infty}^{\infty} \sigma_{yy} e^{-i2\pi(px+qz)} dxdz \tag{9.67a}$$

$$\overline{\sigma}_{yy} = \int_{-\infty}^{\infty}\int_{-\infty}^{\infty} \sigma\delta\left(x, z\right) e^{-i2\pi(px+qz)} dxdz \tag{9.67b}$$

Delta functions for two variables can be separated as $\delta(x)\delta(z)$. Therefore,

$$\overline{\sigma}_{yy} = \sigma\int_{-\infty}^{\infty}\delta(x)e^{-i2\pi px}dx\int_{-\infty}^{\infty}\delta(z)e^{-i2\pi qz}dz = F_y \tag{9.68a}$$

$$\overline{\tau}_{xy} = 0 \tag{9.68b}$$

$$\overline{\tau}_{zy} = 0 \tag{9.68c}$$

Similarly, the horizontal harmonic applied point force finds that the boundary stresses on the surface of the half-space are

$$\overline{\sigma}_{yy} = 0\ ,\overline{\tau}_{xy} = F_y \quad \text{and}\ \overline{\tau}_{zy} = 0 \tag{9.69a}$$

$$\overline{\sigma}_{yy} = 0\ ,\overline{\tau}_{xy} = 0\ \text{and}\ \overline{\tau}_{xy} = F_y \tag{9.69b}$$

Having substituted for the transforms of the stresses from Eqs. 9.69a and 9.69b into Eqs. 9.63a–9.63c, values of D_ϵ, D_x, and D_z can be determined by

$$D_\epsilon = \frac{4\overline{\sigma}_{yy}G^2}{\Omega_2^2}\left[\Omega_2^2 - 8\pi^2\left(p^2 + q^2\right)\right] \tag{9.70a}$$

$$D_x = \frac{-4\overline{\sigma}_{yy}G^2}{\Omega_1^2}\left(i2\pi q r_1\right) \tag{9.70b}$$

$$D_x = \frac{4\overline{\sigma}_{yy}G^2}{\Omega_1^2}\left(i2\pi p r_1\right) \tag{9.70c}$$

Substituting for these values in Eqs. 9.60, arbitrary values of Eqs. 9.56a–9.56c become

$$A_\epsilon = \frac{\Omega_1^2\overline{\sigma}_{yy}}{G\phi\left(p,q\right)}\left[\Omega_2^2 - 8\pi^2\left(p^2 + q^2\right)\right] \tag{9.71a}$$

$$A_x = \frac{-\Omega_2^2\overline{\sigma}_{yy}}{G\phi\left(p,q\right)}\left(i2\pi q r_1\right) \tag{9.71b}$$

$$A_x = \frac{\Omega_2^2 \overline{\sigma}_{yy}}{G\phi\,(\mathrm{p},\mathrm{q})}\,(\mathrm{i}2\pi \mathrm{p}r_1) \tag{9.71c}$$

To obtain the solution for the displacements in the Fourier transformation domain, we may substitute the above values of Eqs. 9.71a–9.71c into general equations of displacement (Eqs. 9.59a–9.59c). After simplification the displacements equations in the Fourier domain for "y" equals zero become functions of p and q only. Similarly having substituted for the transformed stresses from Eqs. 9.69a and 9.69b into Eqs. 9.59a–9.59c, we may obtain the solutions for the transformed displacements for the horizontal harmonic applied point force on the elastic half-space medium.

9.3 Results and Discussions

To verify the validity of the developed procedure and to determine the frequency factors and their corresponding layered mediums, several case studies are considered. The computer results are discussed and presented in this chapter. Cases considered are one-layered elastic medium and two-layered elastic medium.

The displacements in terms of frequency factors were computed with different Poisson ratios for different directions of external concentrated force applied on the surfaces in all cases. In order to transfer the results from the Fourier domain to spatial domain, the method of fast Fourier transform (FFT) was utilized. All the case studies were solved and analyzed here by using MATLAB.

9.3.1 Vertical and Horizontal Surface Load on the One-Layered Mediums

As shown in Fig. 9.1, the vertical and horizontal harmonic point forces were applied with a magnitude of 0.1 N/m^2 on the surface of the mediums. For the computations, the following assumptions were used:

$\omega = 10\ rad/s$ (excitation frequency)

$\upsilon = 0.25$ (Poisson ratio of the medium)

$G = 1$ (shear modulus)

$p = 1\ Kg/m^3$ (density of the medium)

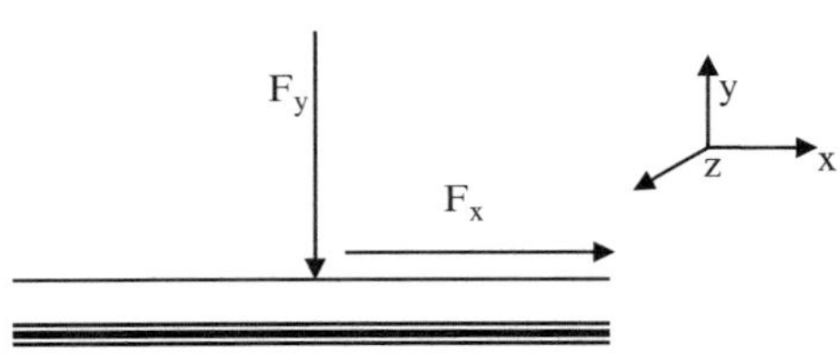

Fig. 9.1 Concentrated vertical and horizontal load on one-layered medium

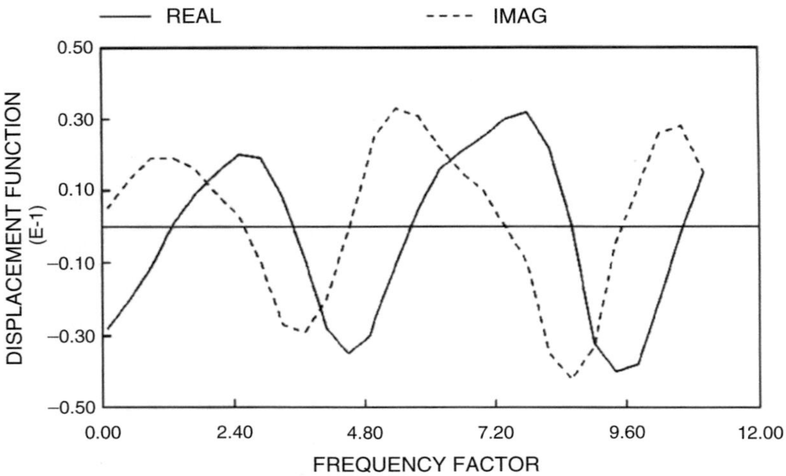

Fig. 9.2 Real and imaginary part of vertical displacement function versus frequency factor for vertical harmonic point force on one-layered medium

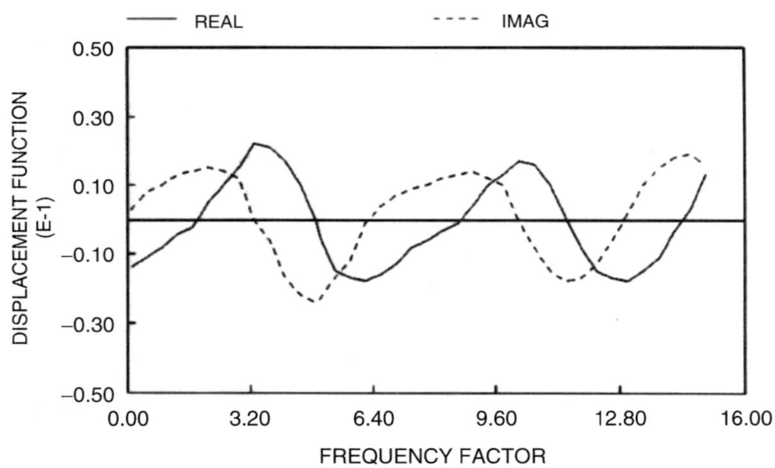

Fig. 9.3 Real and imaginary part of horizontal displacement function versus frequency factor for vertical harmonic point force on one-layered medium

$y = 1$m (depth of the medium)

The numerical values have been presented in Figs. 9.2 through 9.4. The present results can be compared with the reported values of elastic half-space medium. Theoretically, the one-layered medium with only 1 m depth should be stiffer than the half-space medium. That is, it is predictable that the present results should be smaller than the results of half-space medium with the same properties.

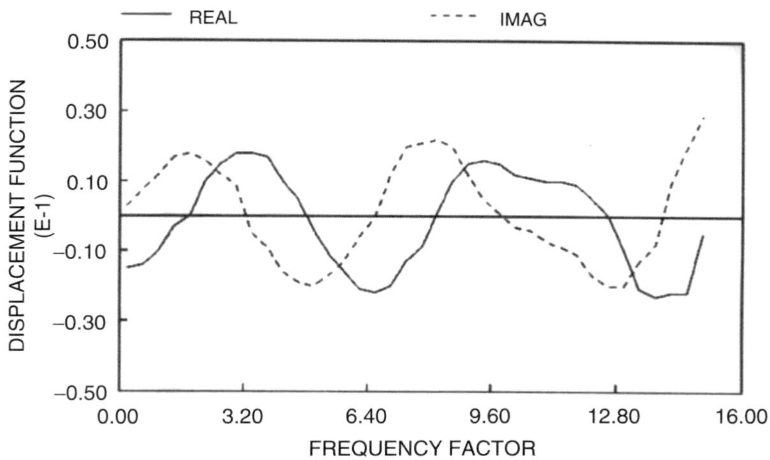

Fig. 9.4 Real and imaginary part of vertical displacement function versus frequency factor for harmonic point force on one-layered medium

Fig. 9.5 Concentrated vertical and horizontal load on two-layered medium

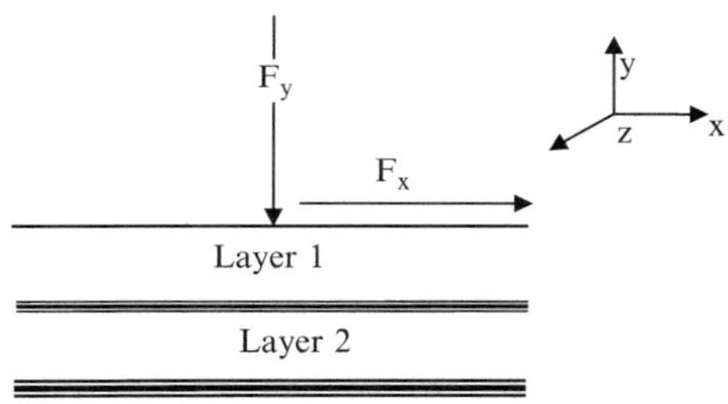

9.3.2 Vertical and Horizontal Surface Load on the Two-Layered Mediums

As shown in Fig. 9.5, the vertical and horizontal harmonic point forces were applied with a magnitude of 0.1 N/m^2 on the surface of the mediums. For the computations, the following assumptions were used:

$\omega = 10$ *rad/s* (excitation frequency)
$\upsilon = 0.25$ (Poisson ratio of the medium)
$G = 1$ (shear modulus)
$p = 1$ Kg/m^3 (density of the medium)
$y_1 = 1$m (depth of the first layer medium)
$y_2 = 1$m (depth of the second layer medium)

The numerical values have been presented in Figs. 9.6 through 9.8. The present results can be compared with the reported values of elastic half-space medium and one-layered medium. Theoretically, the stiffness of the two-layered medium with two 1 m depth layers should be larger than half-space medium but smaller than one-

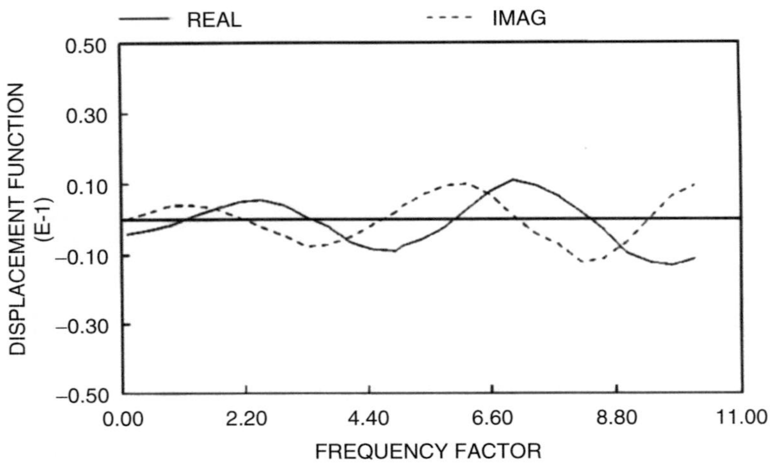

Fig. 9.6 Real and imaginary part of vertical displacement function versus frequency factor for vertical harmonic point force on two-layered medium

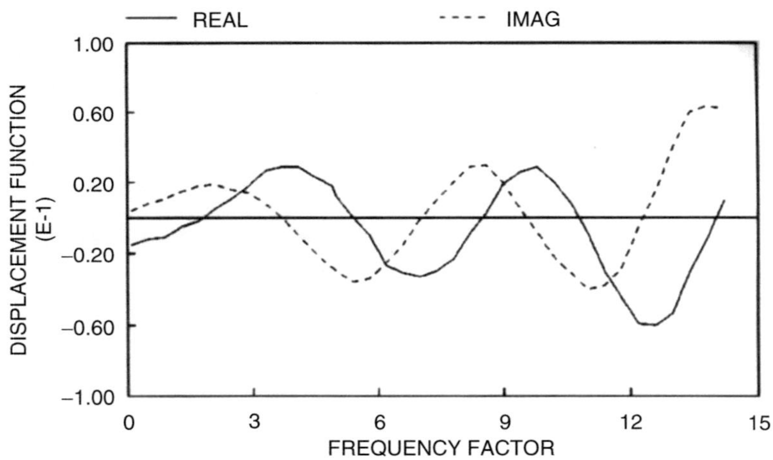

Fig. 9.7 Real and imaginary part of horizontal displacement function versus frequency factor for vertical harmonic point force on one-layered medium

layered medium with 1 m depth. That is, the present results should be between the results of half-space medium and the results of one-layered medium with the same properties.

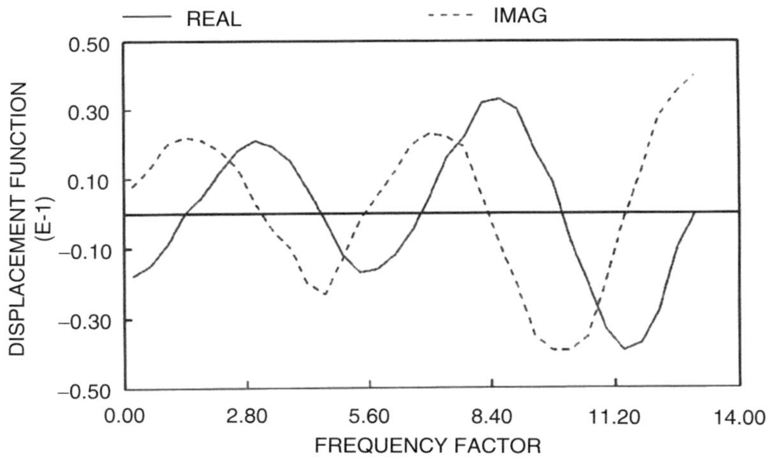

Fig. 9.8 Real and imaginary part of vertical displacement function versus frequency factor for horizontal harmonic point force on one-layered medium

9.4 Conclusion

The design and analysis of a vibration system are based on mathematical models of complex systems. The mathematical models, which follow from the physical laws of the process, are partial differential equations. The Fourier transform method is one way to compute the solution of the differential equations. The fast Fourier transform (FFT) is a particular method to rapidly compute the transformed solutions.

This paper presents an approximate solution for the dynamic response of the surface of multilayered elastic mediums subjected to harmonic concentrated surface loading. After comparison with available results, the numerical results presented are in close agreement over a short range of frequency factors for half-space medium. As to the results of multilayered medium, one way to check the present values is to compare them with the results obtained from a half-space model. Theoretically, these two kinds of results should be comparable if the depth of the layered stratum is large enough. However, there are two limitations in our computer program. One is that it cannot compute the displacements of the medium with Poisson ratio 0.5, because this will result in a zero value in the denominator of some elements in the propagator matrix. Another is that the depth in each layer of the multilayered medium cannot be over 10 m because of the round-off error. Except for the above limitations and on the basis of these results, it is felt that the analysis developed herein can be used to predict the surface response for engineering purposes.

Chapter 10
Three-Dimensional Wave Propagations in Porous Half-Space Subjected to Multiple Energy Excitations

Abstract This chapter deals with three-dimensional wave propagation in pores materials. Most materials containing pores in them are considered as porous media and porous media widely exist in the world. Basically, a porous medium consists of pore portion and solid matrix or skeletal portion. The porous media in nature are usually filled with liquid or gas. Strictly speaking, most of solid materials in nature are porous media. With the significant development of the research and applications on porous media in recent years, the concept of porous media can be seen in applying in many areas of applied science and engineering, such as petroleum engineering, acoustics, rock mechanics, geosciences, and biology. This chapter concentrates on the wave propagations, characteristics of the waves and the applications of the waves in porous media of half-space. The porous materials considered in this chapter are continuous in the half-space with respect to both the solid skeleton and liquid saturated in the pores of the media.

Keywords Porous media • 3D wave propagation • Liquid mobilization • Multi-source waves • EOR with vibratory stimulations • Relative displacements in porous media • Viscous fluid • Multiphase liquid mobilization

Nomenclature

a	Radius of the hill
a_{mn}	Binary function of m and n
A	Cross-sectional area
$A_m,\ B_m, C_m, D_m$	Unknown coefficients
c_s	Velocity of the shear wave
$H_m^{(1)}(\,)$	m-Order Hankel function of the first kind
$J_m(\,)$	m-Order Bessel function of the first kind
k	Wave number of the input wave
R	Radius of the hole

H.R. Hamidzadeh et al., *Wave Propagation in Solid and Porous Half-Space Media*,
DOI 10.1007/978-1-4614-9269-6_10,

t	Time
$\xi, \overline{\xi}$	Complex variables
P	Mass density
θ	Angle coordinate in the polar coordinate system
θ_1	Angle coordinate of the polar coordinate system with origin O_1
μ	Shear modulus
$\tau_{\theta z}, \quad \tau_{rz}, \quad \tau_{r_1 z}$	Shear stress on the surface
ω	Circular frequency of the waves
η	Introduced variable
C_1, C_2	Amplitudes of the waves propagating in solid and fluid
d_j	Distance from a source to the origin
e	Volume strains of solid
$Exp(\cdot)$	Exponential function
H	Introduced physical parameter
k	Permeability
K_b	Bulk modulus of the skeletal frame
K_f	Bulk modulus of the fluid
K_s	Bulk modulus of the solid grain
l	Wave number of compressional wave
p	Pressure of fluid
Q	Mass flow
r	Radius coordinate in polar system
r_j	Distance from a point P to the *j*th wave sources
s_{ij}	Stresses acting on the fluid of a porous medium
$\mathbf{u}$	Displacement vector of fluid
u_{0j}, U_{0j}	Displacements of solid and fluid of the *j*th source, respectively ($j = 1, 2, \ldots, n$)
$\mathbf{u}_{0j}, \mathbf{U}_{0j}$	Displacement vectors of solid and fluid excited by the *j*th source, respectively ($j = 1, 2, \ldots, n$)
$\mathbf{U}$	Displacement vector of solid
V_1	Dilatation wave velocity with respect to first compressible wave
V_2	Dilatation wave velocity with respect to second compressible wave
V_c	Ratio of H and ρ
V	Reference wave velocity
x, y	Coordinates of a Cartesian coordinate system
z_j	Introduced complex variable
α	Coefficient related to porosity
δ_{ij}	Kronecker symbol
ε	Volume strains of fluid
θ	Angular coordinate in polar system
μ_s	Shear modulus of the material
v_s	The Poisson ratio of the solid
ξ	Ratio between reference velocity and wave velocity
ξ_I, ξ_{II}	Roots

μ	Viscosity of the fluid
ρ	Density parameter
ρ_{11}, ρ_{12}, ρ_{22}	Density terms of the porous medium
ρ_f	Mass density of fluid
ρ_s	Mass density of solid
σ_{ij}	Total stresses of a porous medium
σ_{ij}^s	Stresses acting on the solid frame of a porous medium
ϕ	Porosity of the medium
φ_s	Scalar potential of solid
φ_f	Scalar potential of fluid
$\boldsymbol{\psi}_f$	Vector potential of solid
$\boldsymbol{\psi}_s$	Vector potential of fluid
ω	Frequency of wave
∇, ∇^2	Laplacians

10.1 Introduction

Most materials containing pores in them are considered as porous media and porous media widely exist in the world. Basically, a porous medium consists of pore portion and solid matrix or skeletal portion. The porous media in nature are usually filled with liquid or gas. Strictly speaking, most of solid materials in nature are porous media. With the significant development of the research and applications on porous media in recent years, the concept of porous media can be seen in applying in many areas of applied science and engineering, such as petroleum engineering, acoustics, rock mechanics, geosciences, and biology. This chapter concentrates on the wave propagations, the characteristics of the waves, and the applications of the waves in porous media of half-space. The porous materials considered in this chapter are continuous in the half-space with respect to both the solid skeleton and liquid saturated in the pores of the media.

Wave propagations in porous media have great application significances in the areas of geophysics, earthquake, civil engineering, and petroleum engineering. One of the major applications of the Earth wave propagation study, probably the most economically sound application in engineering, is on the enhanced oil recovery (EOR) with seismic wave and vibration stimulations for the waves propagating in half-space porous media (Baviere 2007; Cook and Sheppard 1989; Kouznetsova et al. 1998; Simkin and Surguchev 1991; Westermark and Brett 2001). According to the literature and available field test reports, the increase in oil production with vibration stimulation ranges from 10 to 65 %. With the increasing prices and oil and gas, various methods have been searched and some of them have been employed to enhance oil recovery practices, such as water and gas flooding, hydraulic and explosive fracturing, and layer burning (Beresnev and Johnson 1994). However, based on the available knowledge and techniques, more than 60 % of the residual oil is still left in the reservoirs and cannot be produced with the existing technologies.

Therefore, new methods are being demanded by the industries to increase the crude oil recovery from the reservoirs. To increase the effectiveness of current oil production, the elastic-wave stimulation technique is becoming an appealing alternative method for oil extraction. Investigations involving vibration and seismic stimulation tests have also identified the effects of wave motion and vibration on oil production. This occurs as vibrations propagate into a reservoir, creating elastic waves that positively impact the fluids' flow in porous rocks in various ways. Numerous investigations on EOR with vibration and seismic stimulation have been carried out, and a considerably large number of theoretical research results have been reported (Huh 2006; Igor and Beresnev 1994; Pujol 2003; Roberts et al. 2003; Serdyukov and Kurlenya 2007; Steven et al. 2008).

The research in wave propagation in half porous media is also important in seismic analyses and other related areas. The research results of wave propagation provide theoretical foundations and practical guidance to the researchers, engineers, and industrial practitioners in many areas such as petroleum engineering, geology, geophysics, and geotechnical engineering. In investigating wave motions such as wave propagations in porous media, the main concerns are both the waves regarding the media in which the waves propagate and the energy sources by which the waves are generated. In fact, the wave motions are affected by a variety of factors such as topologic and material properties of the media, characteristics of wave energy sources, medium porosity and permeability, properties of the fluids saturated in the media, as well as combinations of the waves propagating in the domain considered. The behaviors of the wave motions are therefore very complex.

Wave propagation in half-space porous media saturated with fluids is very complex compared to wave propagation in solids due to the presence of fluids in the pores of the media. The study of wave propagation in a fluid-saturated porous media has been, and is still, a compelling area of engineering and science. The wave behavior in a fluid-saturated medium is affected not only by the separate motions of the solid and fluid but also by the relative motion between the solid and fluid. It is generally accepted that the waves attenuate while they propagate in the medium due to the material properties and the presence of the pore fluid in the porous media. In fact, wave velocity and wave attenuation are two key aspects of wave propagation in porous media. These aspects are important in analyzing the dynamic response of the media with respect to the properties of both the media and the wave sources, such as viscosity, porosity, and frequency (Biot 1956a, b; Berryman 1985; Bardet and Sayed 1993; Kelder and Smeulders 1997; Carcione et al. 2000; Pridea and Garambois 2002).

A number of methods and theories have been published to contribute to the comprehension of wave propagation in porous half-space media, whose properties influence the behaviors of the wave motion in a semi-infinite domain. To improve the extraction of fluids from compressible geological medium, a solution method is proposed for the consolidation of a saturated, porous elastic half-space due to pumping of a pore fluid at a constant rate from a point sink embedded beneath the surface (Carter and Booker 1987). It is demonstrated that compressibility can have a significant influence on the rate of consolidation around the sink. To withdraw fluids from a single soil layer, analytical solutions are presented for the steady state of displacement and stresses in a half-space subjected to a point sink (Chen 2005). From the analysis of a two-layered soil and an inhomogeneous half-space, it is concluded that the numerical evaluation of the solution can be easily achieved with very high accuracy and efficiency. A model is established to provide a representation of a semi-infinite porous media domain with materially incompressible constituents (Heider et al. 2010). An example is given to show the numerical applicability of the proposed model.

For a half-space porous media under isothermal conditions, Stokes' first problem for a Rivlin–Ericksen fluid of second grade is considered (Jordan and Puri 2003). Laplace transform techniques are used to determine the exact solution, temporal limits, small-time expressions, and displacement thickness. In the near field of a layer over a half-space for a point source, to investigate the effects of source type, source depth, range, and structure on the waveforms, a set of seismograms based on the impulse to the generalized ray source are displayed for P-wave (pressure or compressional wave), SV wave (shear vertical wave), and SH wave (shear horizontal wave) sources using the Cagniard–de Hoop method (Nautiyal 1972). For a point source in linear elastic diffusive half-space governed by Biot's theory and interior point forces (both vertical and horizontal), the fundamental solutions are derived and numerical verification of the solutions obtained is also provided (Chau 1996). It is concluded that, inside a linear elastic diffusive half-space, the point force solutions can be used to synthesize force doublet (with or without moment), then center of dilation, center of shear, concentrated couple, double couple, etc.

In a liquid-saturated porous solid layer under a uniform layer of homogeneous liquid, dispersion of Rayleigh-type surface waves is investigated (Shama et al. 1991) lying over a transversely isotropic elastic half-space. To observe the effects of the depths on the phase velocity, dispersion curves for the phase velocity have been plotted for different values of the ratio of the depths of the two layers. It is found that if dissipation is taken into account, waves will become dispersive and dissipative. The vertical transient motion of an elastic plate, on a fluid-saturated poroelastic half-space, is studied using integral equations (Jin and Liu 2000). The numerical examples indicate that the drainage boundary conditions, at the surface of a saturated poroelastic half-space, can influence the transient displacement of the elastic plate. At a plane interface between a uniform elastic half-space and a fractured porous half-space containing two immiscible fluids, reflection and transmission phenomena are analyzed due to incidence of plane longitudinal/transverse wave from uniform elastic half-space (Arora and Tomar 2010). The presence of fractures in the porous half-space was found to affect the reflection of waves. This is found responsible for raising the reflection and lowering the transmission coefficients.

From a free boundary of the half-space, the reflection of inclined incident plane wave is derived, and both the cases of the transverse and longitudinal incident wave are considered (Ciarletta and Sumbatyan 2003). It is shown that only the transverse wave can propagate in the solid without attenuation after having reflected from the free boundary surface. From a free boundary surface of a porous half-space medium consisting the mixture of a micropolar elastic solid and Newtonian liquid, a problem of reflection of coupled longitudinal waves is investigated (Singh and Tomar 2006). Of the ten waves found, the two coupled longitudinal waves, involved in the problem of reflection from a boundary surface of a half-space, are investigated, and the results indicate that Erigen's theory differs from the classical theory of elasticity.

To obtain the dynamic response of poroelastic half-space to tangential surface loading, solutions for asymmetric conditions are presented (Jin and Liu 2001). It is out that the solution obtained can be employed in a variety of asymmetric wave propagation problems of poroelastic medium. Propagation of inhomogeneous waves in anisotropic porous layered medium, of which the interface between porous layer and elastic half-space is considered as imperfect, is studied (Vashishth and Khurana 2002). With the theory of transfer matrix, the analytical expression for the surface impedance is derived, of which the role in seismological studies and in the study of composites is discussed. To study generalized surface waves in an anisotropic poroelastic solid half-space, a new method is introduced (Sharma 2004). To verify the numerical applicability of the proposed method, numerical work is carried out for the model of a crustal rock, and the propagation of surface waves is studied numerically for three anisotropies: triclinic, monolithic, and orthorhombic. During reflection-of-plane waves in a fluid-saturated poroelastic half-space, surface displacements, surface strain, rocking, and energy partitioning are discussed (Lin et al. 2005). From the results, it may be concluded that the effects of a second P-wave become noticeable in poroelastic media with low dry-frame stiffness.

Along the boundary of a poroelastic solid half-space saturated by two immiscible fluids, the dispersion equation of Rayleigh-type surface waves is obtained (Arora and Tomar 2007). It is found that the density of fluids affects the Rayleigh mode. In a transversely isotropic fluid-saturated porous layered half-space, the complex dispersion equation for Love waves is derived, based on Biot's theory for transversely isotropic fluid-saturated porous media (Ke et al. 2005). Besides dispersion and attenuation, another feature of the Love waves in a loss system is that they become leaky due to the energy transfer through the interface. In a fluid-saturated porous layer under a rigid boundary and lying over an elastic half-space, mathematical modeling of the propagation of Love waves has been considered (Ghorai et al. 2009). It is shown that increase in porosity will lead to increase in phase velocity of Love wave.

For a completely saturated poroelastic half-space and full space domains, subjected to axisymmetric ring loads and ring flow sources, explicit solutions for Laplace transformations of displacements, tractions, pore pressure, and pore fluid are derived (Puswewala and Rajapakse 1988). It is shown that solutions corresponding to buried circular path loads and flow sources with intensity specified by polynomial of arbitrary degree could be deduced from solutions corresponding to ring loads and source configurations through appropriate substations. In a medium comprising of an impervious elastic layer resting on a water-saturated porous half-space, stress and pore pressure within are investigated, for the condition that the upper surface of the layer is acted upon by a normal stress field varying harmonically (Chander 1997). It is found that the stress is constant along one horizontal space direction while it varies in a prescribed manner along the direction perpendicular to the horizontal space, and that the phase shifts in the induced pore pressure represent the effect of volume changes of solid material, pore spaces, and water in the half-space under the induced stress, as well as diffusion of water under these changes. The stresses and pore pressure due to finite reservoir load in a porous elastic half-space are obtained with Green's function and the calculation for the actual water level changes of the reservoir load (Chander 2000). The results help to understand the pore pressure effect in a porous elastic medium. The non-axisymmetric Biot consolidation problem for multilayered porous half-space media is studied, taking stresses, pore pressure, and displacement at layer interface as basic unknown functions (Wang and Fang 2003). By solving a linear equation system for discrete values of Laplace-Hankel transformation, the time histories of stresses, pore pressure, and displacement are obtained.

For a layered poroelastic half-space subjected to moving loads, the transmission and reflection matrices (TRM) method is developed (Xu et al. 2008). Numerical results show that the occurrence of a softer middle layer in the layered half-space will enhance the vertical displacement and the pore pressure of the layered half-space. To analyze the isolation of moving-load induced vibration using pile rows embedded in a layered poroelastic half-space, a numerical model is developed (Lu et al. 2009). It is concluded that the speed of moving loads has an important influence on the isolator of vibration by pile rows. In addition, in petroleum engineering, evaluation of the reservoir compaction and the induced land subsidence has also been an active research area. A DDM-FEM coupling method is applied to reservoir simulation to evaluate the reservoir behavior over a compacting oil reservoir in half-space (Yin et al. 2006). Displacement discontinuity method (DDM) is employed to account for the semi-infinite surroundings, while finite element method (FEM) is used in the fully coupled simulation of the reservoir itself. The advantages and the feasibility of the coupling method are shown in the numerical simulation.

Although numerous investigations have been performed on wave propagations and motions of solid matrix and liquid in the pores, however, there is still a lack of comprehensive understanding on the mechanism of the technique and the mobilization of the fluids in the porous media under the excitations of the wave motions generated with seismic and/or vibration stimulations. This has become a serious hurdle for developing and applying the vibration technique in EOR practice. A thorough and systemic study on wave propagations in fluid-saturated porous half-space media is therefore needed. The progress in such a study will certainly contribute to the comprehensive understanding of the mechanism of the technique and enhance its application. Development of a methodology and associated mathematical model for evaluating the dynamics of wave propagation in a porous half-space medium saturated with a single fluid and/or two immiscible fluids will be displayed in this chapter. The relative displacement between the solid and fluid is the major contributor to the mobilization of the fluid, and relative displacement is critical to help understand the mechanism of seismic vibration technique used in the oil production industry.

Development of a new methodology for analyzing the behavior of wave motion in a fluid-saturated, elastic, isotropic, and homogeneous porous medium that is excited by multiple circular-cylindrical energy sources is to be presented in this chapter. Establishment of a model considering the porous medium subjected to multiple energy sources will also be demonstrated. This model allows quantitative description for the vibration of any point in the domain considered under the wave propagations of the multiple energy sources. Under the multiple energy sources, a desired controllable wave field can be obtained by selecting proper frequencies, amplitudes, and locations of the wave sources. Implementation of the model established simulates and analyzes the dynamic behavior of the domain of the porous medium under the excitations of multiple circular-cylindrical wave sources.

The methodology developed in this chapter helps to understand the mechanism of the seismic vibration technology and provides a scientific foundation for the implementation of the multi-energy model in EOR practice. A direct advantage of is that the energy can be saved in EOR with multiple energy sources. It also makes the availability of the investigation of the wave propagations in porous media under multiple energy sources, via analytical and numerical approaches. The wave equations are transferred in the form of displacements of solid and fluid. This makes the availability of the evaluation of the relative displacements between the fluid and the solid of the porous media subjected to the excitations of multiple energy sources. Availability of the behavior of any specified particle or point in the considered semi-infinite domain is to be quantified, and the relative displacement between the fluid and solid of the medium is to be conveniently determined with the established model, under the stimulation of the waves generated by multiple energy sources. The development of wave field with desired frequency, amplitudes, and durability will also be presented with controllable stimulation parameters of the multiple circular-cylindrical energy sources.

10.2 Porous Materials and Porous Media in Petroleum Industry

10.2.1 Porous Materials

The phenomenon of fluid flowing through medium with complex structures both exists and plays important roles in natural environment and industry and human life. There exist various kinds of porous materials which can be generally categorized as natural porous media and man-made porous media according to their formation process and constitute (Bear 1972). Natural porous media are widely seen in oil field. A core sample, a porous medium in natural form, is shown in Fig. 10.1a. This core is taken from an oil field with crude oil in the pores as shown in the figure (Carstens 2011). Man-made porous materials are also common in industry. Figure 10.1b illustrates a sample of a newly developed road pavement material

a Core sample from oil field

b ARC pavement material

Fig. 10.1 Examples of natural and man-made porous media

Asphalt Rubber Concrete (ARC) which is used by the author in examining the acoustic properties of the porous pavement material. ARC is manufactured with mixture of rubber and asphalt which has shown good acoustic properties (Dai and Lou 2008). The natural porous materials or media can be further distinguished as underground porous materials and biological porous materials. The underground porous media consist of rocks and soil, such as sandstone, limestone, and coal. The biological porous media refer to the one constructed by plants and animals, such as the root and stem of a plant as well as the lungs and livers of animals which contain many microtubules. The man-made porous materials refer to the porous materials that are synthesized or manufactured by human power.

A porous medium can be described as a solid containing pores or voids in it. Obviously, a hollow metal cylinder cannot be categorized as a porous media. Similarly, a solid with only isolated eyelets or voids cannot be treated as a porous medium either. From the perspective of fluid passing through the media, in general, a porous medium is considered of at least containing a few continuous paths from one side to another. That means the voids are considered as interconnected. However, this is still a rather coarse description. To further approach a detailed definition, the following major contents can be presented (Bear et al. 1968):

1. Multiphase materials together occupy a certain space. In the multiphase, there is at least one phase that is not composed of solid skeleton. They could be air or liquid. The part without any solid skeleton is called void space.
2. In the domain occupied by a porous medium, the solid skeleton should spread over the whole porous medium. One basic characteristic of porous media is that the surface ratio of solid skeleton is quite large, which determines the properties of fluid flow within the porous medium in many ways. Another characteristic of porous media is that the volume of void space is relatively small.
3. At least some voids or holes are interconnected. The interconnected void space is called effective pore space. The disconnected space then can sometimes be treated as solid skeleton.

These contents are only the major points in describing a porous medium. What mentioned above cannot be considered as a real and precise definition because the nomenclatures used are all for qualitative descriptions. However, it gave us a picture about the main characteristics of a porous media. It has been pointed out that it is extremely difficult to give a universally precise definition. For fluid flows through porous media, the solid surface plays a key role. Yet determination of the geometrical shapes of solid surface is of great difficulty.

10.2.1.1 Basic Concepts

Before approaching the theories investigations on porous media and most importantly the motion of the porous media under external excitations, it is useful to introduce some basic concepts that give quantitative indexes for describing porous media, though the majority of the archived documents on investigating the waves propagating in porous media do not consider the geometry of the pores. Porous media are commonly referred to as solid bodies that contain "pores." However, it is impossible to give an exact geometrical definition of pores because the real structures of porous media can be extremely complex.

Pores can be categorized from different perspectives. Depending on the pore's size, it can be distinguished as molecular interstices for extremely small voids and caverns for very large ones. Pores are deemed void spaces intermediate in size between caverns and molecular interstices (Scheidegger 1972).

Fluid flow through porous media is possible only if at least part of the pore space is interconnected. Therefore, from this point of view, the pores can be interconnected and non-interconnected. The interconnected part of pore space that allows fluid flow is also known as effective pore space of the porous media.

The porous media may include solid skeleton and at least one phase fluid. The surface of the solid skeleton and the size of the pores play a key role in the motion of the fluid. It is thus desirable and reasonable to classify porous media according to the types of pore spaces which they contain. The pores can be distinguished as voids, capillaries, and force spaces according to their sizes and effects. Voids usually refer to the pores whose sizes are large that the walls have an insignificant effect upon macro hydrodynamic phenomena in their interior. On the other hand, capillaries represent the pores whose sizes are small, and the surface of the pores plays an important role upon hydrodynamic phenomena in their interior, whereas the force spaces designate the pores with extremely small characteristic length in which the molecular structure of the fluid has to be taken into consideration. It should be noted that it is not possible for any porous media to contain only one type of pore spaces.

10.2.1.2 Continuum Hypothesis and Fractal Theories

Due to the fact that real structures of porous media are very complex, it is impossible to give an exact picture about the phenomenon involving the interaction of fluid

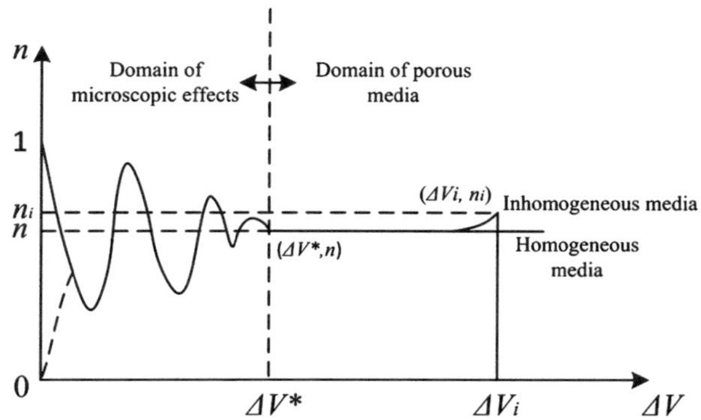

Fig. 10.2 Definition of representative volume element (RVE)

and solid surface. To overcome this difficulty, one can treat the porous media as a continuum consisting of a series of particles, which is very similar to the definition of fluid density.

Consider that ∇V_i is a small volume including a random point in porous media *P*(*x*,*y*,*z*) whose pore volume is denoted as $(\Delta V_p)_i$. The point *P* is the centroid of the small volume. As thus, the average porosity n_i may be defined as

$$n_i = \frac{(\Delta V_p)_i}{\Delta V_i} \tag{10.1}$$

At a certain time, selecting a series of such small volumes around point *P* and ensuring $\Delta V_1 > \Delta V_2 > \Delta V_3 > \ldots$, which also means $(\Delta V_p)_1 > (\Delta V_p)_2 > (\Delta V_p)_3 > \ldots$, we can obtain a series of average porosity noted as $n_1, n_2, n_3, \ldots$. Plotting these points $(\Delta V_i, n_i)$ under the coordinates $\Delta V \quad n$, a curve shown below can be obtained.

It can be seen from Fig. 10.2 that if the selected small volume ∇V_i reduces from a sufficient large value, for homogeneous media, the curve obtained becomes horizontal between ΔV_i and a certain value ΔV^*; for nonhomogeneous media, the curve varies a bit. If the volume continuously reduces, smaller than ΔV^*, the porosity will begin to oscillate, indicating that there are only a few pores contained. The oscillation becomes violent as the volume is further reduced. In the end, when ΔV^* is close to zero, namely, a geometrical point, it will be either in the pores or in the solid skeleton, which means the porosity will be 1 or 0.

The volume ΔV^* mentioned above is therefore called the representative volume element of porous media. The size of this volume is required to be much larger than that of a single pore, which indicates that a porous medium contains many pores, which are much less than the characteristic length of the whole flow field. The porous media can then be treated as a continuum consisting of a series of such elements. The porosity of point *P* is defined:

$$n(P) = \lim\nolimits_{\nabla V_i \to \Delta V^*} \frac{(\Delta V_p)_i}{\Delta V_i} \tag{10.2}$$

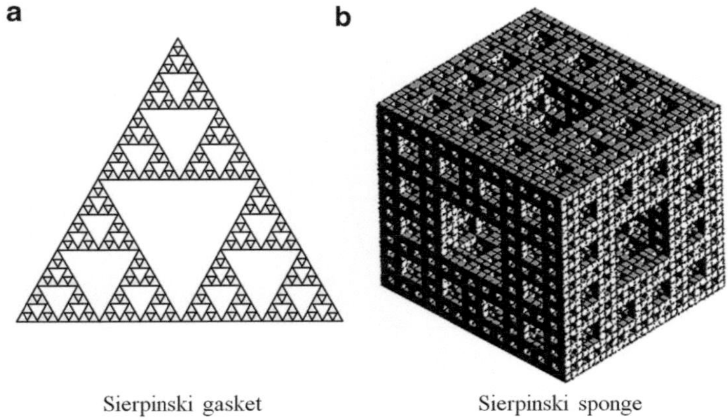

Fig. 10.3 Examples of fractal structures

For certain media, if the porosity is irrelevant to the spatial position, the media are called homogenous media. Otherwise, it is called inhomogeneous media.

The word fractal was first introduced by a mathematician, Mandelbrot, meaning fractional, irregular, and fragmentary (Mandelbrot 1982). The parameter for quantitative description of the property of fractal is called fractal dimension. Contrary to the classical Euclid geometry, whose dimension can only be assigned as an integer because it deals with regular geometrical figures, the dimension of fractal structure is a fraction. The fractal theory is applied to potential regulations from the superficial irregularity and has been widely used among many subjects.

The main characteristics of fractal geometry are self-similar and scale invariant which means part of a geometrical shape can be exactly or very close to a reduced size copy of the whole. Actually, it has been reported that the microstructure and the distribution of pore size possess the characteristics of fractal. Sierpinski gasket and Sierpinski sponge, as shown in Fig. 10.3, are typical models usually used in mathematically representing the geometries of porous media.

Since the invention of fractal theory, it has been applied to investigate the transportation problem in porous media. For the oil reservoirs with fractal properties, the porosity and permeability of porous media are related to fractal index.

10.2.1.3 Darcy's Equation

To quantitatively represent the behavior of fluid flowing through porous media, as one may expect, the governing equations may fundamentally be the same as those governing the motion of viscous fluids in ordinary free vessels, namely, Navier–Stokes equations. Navier–Stokes equations focus on the dynamical equilibrium in flow system subjected to inertial and viscous forces and those generated by extended body force and the inertial distribution of fluid pressures. However, due to the

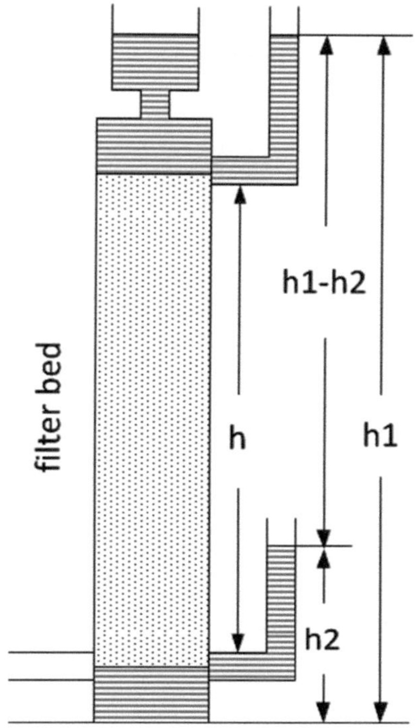

Fig. 10.4 Darcy's experiment

complexity of the real porous media, the difficulty of precise mathematical modeling applying these equations is entirely insurmountable.

Fluid flowing through porous media has been studied for years by many researchers and engineers. The earliest and of the greatest contribution was made by Darcy with the fame experiment on the flow of water through beds of sand, illustrated in Fig. 10.4.

Based on Darcy's experiment, Darcy's law was established. Darcy's law is deemed the real foundation of the quantitative theory of the flow through porous media. According to Darcy's law, the rate of flow q of fluid through the filter bed is proportional to the cross area A of the filter bed and the difference pressure $(p_1 - p_2)$ between the fluid heads at the inlet and outlet faces of bed, but inversely proportional to the thickness of the bed. Darcy's law is a mathematical statement which neatly summarizes the properties of laminar fluid flow through homogeneous porous media. The relation can be expressed analytically as

$$Q = -\frac{kA}{\mu}\frac{(p_1 - p_2 + \rho g h)}{h} \tag{10.3}$$

where A is the cross-sectional area, μ is the viscosity of the fluid, Q is the mass flow, ρ is the density of fluid, h is the height as shown in the figure, and k is the permeability.

However, Darcy's law is not applicable for all the fluid flow. Although Darcy's law is valid for a wide domain of flows and arbitrary small pressure differentials, it is invalid for liquids at high velocities and for gases at very low and very high velocities (Ahmed and Sunada 1969).

To extend its region of application and improve the accuracy of Darcy's law, the formula needs to be expressed in a more general form. To get one step further, isotropic porous media will be considered. The first thing is to exchange the scale parameters with vector parameters denoting the directions of pressure. We have

$$\boldsymbol{q} = -\frac{k}{\mu}(\nabla p - \rho \boldsymbol{g}) \tag{10.4}$$

where $\boldsymbol{g}$ is a vector in the direction of gravity, vertically downward, and of the magnitude of gravity. This expression is still rather coarse because it is unable to show what will happen if the permeability and viscosity are not constants, which is closer to real conditions. Therefore, to equally treat these parameters as variables, (10.2) can be rewritten as

$$\boldsymbol{q} = -\nabla\left(\frac{pk}{\mu}\right) + \frac{k\rho}{\mu}\boldsymbol{g} \tag{10.5}$$

Introduce a velocity-potential function ψ defined as

$$\psi = \frac{kp}{\mu} + \int_{z_0}^{z} k\rho g dz/\mu \tag{10.6}$$

where z represents the vertically upward direction opposite of gravity. Then (10.3) can be expressed in an equivalent and neat form by the gradient of the velocity-potential function

$$\boldsymbol{q} = -\nabla\psi \tag{10.7}$$

To make it clear, one may state that it is never possible to give an exact description for real porous media in one manner. For anisotropic porous media, the form of partial differential equations will be much more complicated, and more hypotheses and parameters need to be introduced even when the compressibility of the porous media is ignored. To save spaces, the content about these matters will be omitted. Interested readers are encouraged to refer to relevant materials. Tremendous work has been achieved by former scholars.

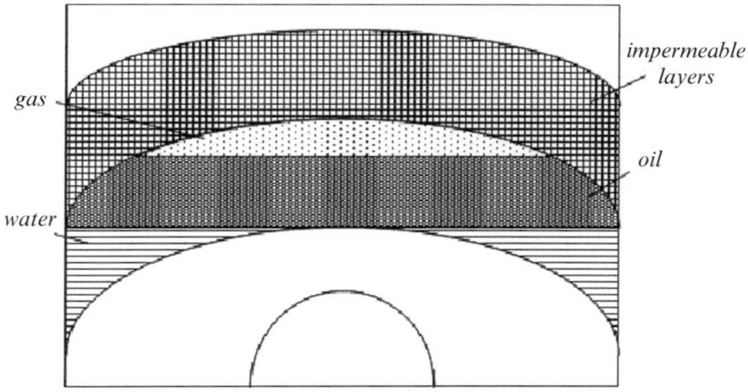

Fig. 10.5 Configuration of reservoir

10.2.2 Porous Media and Enhanced Oil Recovery in Petroleum Industry

It is generally accepted that petroleum is formed by the ancient organic matter. There used to be a great wealth of marine organism during the long geological age, whose bodies sank to the ocean floor after they died. Then, they were buried by sand and sediment and turned into sedimentary layers. For millions of years' isolation from external air, these bodies were gradually decomposed and turned into petroleum under the high pressure of the rocks and heat within the Earth.

Due to the lighter density than that of water, petroleum would intend to travel upward through one layer to another until it is trapped within porous rocks covered by impermeable layers. Thus, the petroleum became stationary and accumulated like being trapped in a box where it can neither evaporate nor flow away. As time passed by, an oil reservoir was formed gradually. Figure 10.5 demonstrates one typical reservoir.

Petroleum has been used since ancient times and now plays a key role in the society due to the invention of internal combustion engine. Nowadays a large percentage of the world's energy consumption relies on oil, ranging from 32 % for Europe and Asia, which is a quite low percentage, up to as high as 53 % for the MiddleEast. Oil accounts for over 90 % energy for transport. Besides, petroleum has been a vital element to many industries. It is of great significance to maintain the industrial civilization in its current configuration and thus becomes a critical concern for many nations. Therefore, the price of oil is closely related to the macroeconomy of the world.

The process of forming petroleum requires millions of years that it cannot be created in a short time. From this perspective, petroleum is a kind of nonrenewable energy. It is predicted that the oil reserves today are only available for 50 years. Since similar predictions were made before and none of them were realized ever, it is believed that the petroleum crisis is not going to happen. This is mainly because of the development of new technology developed and more and more petroleum reservoirs are discovered. However, it cannot be concluded that petroleum is inexhaustible within the limited space of the Earth.

- Enhanced Oil Recovery

Although nonrenewable, petroleum has been indispensable in our society. To meet the increasing requirement of oil, people need to explore more petroleum reservoirs. However, exploring new petroleum reservoirs is of great cost and of kind of risk that you may end up with exploring nothing after millions of dollars are spent. One relatively more feasible way for improving oil production is to increase the recovery rate of the oil reservoir. The recovery rate is defined as the oil recovered over the geologic reserves. Depending on the natural energy, such as elastic potential energy of rocks and fluid, the recovery rate is only 15–30 %; over 60 % still remained within the porous rocks beneath the ground. Therefore, a series of technologies are invented to increase the recovery rate, which are called EOR (Baviere 2007).

EOR, also called improved oil recovery, is often achieved through gas and water injection, chemical injection, thermal recovery, and external vibration. Since the oil is trapped in porous media, where the surface tension plays an important role, the oil will stick to the pores surface when natural energy dies out.

- Water and Gas Injection

Water injection is currently applied in most petroleum reservoirs in EOR. Through water injection, the pressure inside the reservoirs is maintained, and the remained oil will be forced to move upward toward the production well, thus increasing the production of the reservoirs. Usually the water is injected to the aquifer through several injection wells adjacent to the production well.

Gas injection works almost the same as water injection causing high pressure that makes the residual oil movable. What is more, the injected gas will mix with the crude oil whose viscosity will be reduced. The natural gas was first applied in gas injection for it can be arbitrarily mixed with oil. Nowadays carbon dioxide injection has been a hot subject since it can be easily obtained from natural environment and industrial carbon emissions, which could definitely minimize the operating cost.

- Chemical Injection

Chemical injection refers to injecting dilute solutions, including polymer, surfactant, microemulsion, etc. Injection of polymer solution into reservoirs will increase the viscosity of injected solution and thus decrease the viscous fingering. As a result, the amount of oil can be increased since the sweep area is extended. Through injecting solutions of surfactants, the interface tension of oil and water that prevents the residual oil moving through reservoirs will be lowered and more oil will be recovered. Microemulsion, which is made of hydrocarbon, water, and surfactant mixed with trace electrolyte, can increase viscosity of the injected solution and even eliminate the oil–water interface. Therefore, the recovery rate can be increased.

- Thermal Recovery

Thermal recovery has been proven an effective method to increase recovery rate of reservoirs, especially heavy oil reservoirs. Thermal recovery is conducted through various methods, such as steam injection, in situ combustion, and electrical heating. All these methods aim to heat the crude oil to reduce its viscosity and improve the sweep efficiency and the displacement efficiency. Thermal recovery is now the largest EOR projection in the world, occupying 44 % of the production, due to the rising demand for heavy oil in a lot of industries like manufacture of plastics.

- External Vibration

It was first observed in Russian fields when an earthquake occurred and the seismic wave propagated through the reservoir; the oil production was increased. Since then, vibration has been applied in oil recovery and proposed as a low-cost EOR method from successful field results. Vibration waves can easily propagate through high and low permeability zones in reservoirs with high heterogeneity, where injection of fluids becomes quite difficult, and squeeze the residual oil. However, the mechanism of vibration for oil recovery is very complicated and still not well known. As a result, process performance prediction and reliable projection design are hard to realize. In this chapter, waves generated by multiple energy sources are to be analyzed theoretically, and tools developed for the analysis are also to be presented. The waves propagating in the porous media containing oil reservoir may help oil recovery in a controlled manner, if the excitations of the energy sources are properly designed.

10.3 Development of General Governing Equations in Relative Displacements for Wave Propagations in Porous Media

Among the pioneers in developing the governing equations for the waves in porous media saturated with liquid, M.A. Biot (1905–1985) is probably the greatest one. Biot is the researcher who systematically analyzed the three-dimensional propagation of wave propagations in porous media saturated with liquid, and he firstly developed the corresponding governing wave equations on the basis of generalized elasticity theory. His general theoretical works allow the quantitative description of the waves in porous media with liquid filled in the pore spaces. A general theory of three-dimensional propagation of elastic waves in a fluid-saturated porous solid was presented by Biot (1956a). Biot is also the researcher who first theoretically proved the existence of the first and second compressible waves and one shear wave in porous media.

However, Biot's theory is not for quantifying explicitly and respectively the displacements of fluid and solid of the porous medium. For understanding the relative motion between the fluid and solid, a theoretical and systematic investigation on the relative motion between the fluid and solid of a porous medium is necessary. In recent years, three-dimensional governing equations describing displacements of the solid and liquid of porous media were developed on the basis of Biot's theory (Dai and Wang 2009; Wang et al. 2007a, b, 2009). In this chapter, based on these works, the governing wave equations in terms of displacements of the solid and liquid of a porous medium and the relative displacements between the solid and liquid will be developed, for describing the wave propagations in porous media.

However, Biot's theory is not for quantifying explicitly and respectively the displacements of fluid and solid of the porous medium. For understanding the relative motion between the fluid and solid, a theoretical and systematic investigation on the relative motion between the fluid and solid of a porous medium is necessary. In recent years, three-dimensional governing equations describing displacements of the solid and liquid of porous media were developed on the basis of Biot's theory (Dai and Wang 2009; Wang et al. 2007a, b, 2009; Wang 2008). In this chapter, based on these works, the governing wave equations in terms of displacements of the solid and liquid of a porous medium and the relative displacements between the solid and liquid will be developed, for describing the wave propagations in porous media.

10.3.1 Biot's Theory

Biot's theory (Biot 1956a) is of the propagation of the elastic waves in a system composed of a porous elastic solid saturated by a fluid. The relative motion of the fluid in the pores is of the Poiseuille type. Only the materials such that the walls of the main pores are impervious will be included, for which the pore size concentrated its average value. The development of the theory firstly introduced the stress and strain relations in the aggregate including the fluid pressure and dilatation. Relations were then established between these quantities for static deformation in analogy with a procedure followed in the theory of elasticity for porous materials developed.

Consider a volume of the solid–fluid system represented by a cube of unit size. The force S acting on the fluid part of the cube is proportional to the fluid pressure p with $-S = \beta p$, where β is the fraction of the fluid area per unit cross section.

The force component acting on the solid and fluid part of each face of the cube is respectively denoted as the tensor

$$\begin{Bmatrix} \sigma_x & \tau_z & \tau_y \\ \tau_z & \sigma_y & \tau_x \\ \tau_y & \tau_x & \sigma_z \end{Bmatrix} \text{ and } \begin{Bmatrix} s & 0 & 0 \\ 0 & s & 0 \\ 0 & 0 & s \end{Bmatrix} \tag{10.8}$$

The strain tensor in the solid is

$$\begin{Bmatrix} e_x & \frac{1}{2}\gamma_z & \frac{1}{2}\gamma_y \\ \frac{1}{2}\gamma_z & e_y & \frac{1}{2}\gamma_x \\ \frac{1}{2}\gamma_y & \frac{1}{2}\gamma_x & e_z \end{Bmatrix} \tag{10.9}$$

where $e_x = \frac{\partial u_x}{\partial x} \ldots, \gamma_x = \frac{\partial u_x}{\partial y} + \frac{\partial u_y}{\partial x} \ldots$

The strain in the fluid is defined by the dilatation

$$\varepsilon = \frac{\partial U_x}{\partial x} + \frac{\partial U_y}{\partial y} + \frac{\partial U_z}{\partial z} \tag{10.10}$$

where U_x, U_y, and U_z are the average fluid displacement vectors.

The potential energy W per unit volume of aggregate is given by

$$2W = \sigma_x e_x + \sigma_y e_y + \sigma_z e_z + \tau_x \gamma_x + \tau_y \gamma_y + \tau_z \gamma_z \tag{10.11}$$

With the assumption that the solid–fluid system is statistically isotropic, the directions of principal stress and principal strain coincide. Referring the stresses and strains to these directions and the symmetry properties of the material are taken into account, the stress–strain relations reduce to

$$\begin{aligned} \sigma_x &= 2Ne_x + Ae + Q\varepsilon \\ \sigma_y &= 2Ne_y + Ae + Q\varepsilon \\ \sigma_z &= 2Ne_z + Ae + Q\varepsilon \\ \tau_x &= N\gamma_x, \tau_y = N\gamma_y, \tau_z = N\gamma_z \\ S &= Qe + R\varepsilon \end{aligned} \tag{10.12}$$

where $e = e_x + e_y + e_z$ and N and A correspond to the Lame coefficients in the theory of the elasticity and are of positive sign. N represents the shear modulus of the medium.

With the Lagrangian viewpoint, consider the unit cube of the aggregate as an element, which is supposed to be small relative to the wavelength of the elastic waves and in turn the size of the element.

Now the consequence of this assumption is the fact that the microscopic velocity pattern is the same as if the fluid were incompressible. Hence, the microscopic flow pattern of the fluid relative to the solid depends only on the direction of the relative flow and not on its magnitude. When the viscosity is neglected, the microscopic velocity field will be a linear function of the six average velocity components of the solid and the fluid. The Lagrangian coordinates are chosen as the six average displacement components of the solid and the fluid, i.e., $u_x, u_y, u_z, U_x, U_y, U_z$.

The kinetic energy T of the system per unit volume may be expressed as

$$
\begin{aligned}
& \rho_{11}\left[\left(\frac{\partial u_x}{\partial t}\right)^2+\left(\frac{\partial u_y}{\partial t}\right)^2+\left(\frac{\partial u_z}{\partial t}\right)^2\right] \\
2T = & +2\rho_{12}\left[\left(\frac{\partial u_x}{\partial t}\right)\left(\frac{\partial U_x}{\partial t}\right)+\left(\frac{\partial u_y}{\partial t}\right)\left(\frac{\partial U_y}{\partial t}\right)+\left(\frac{\partial u_y}{\partial t}\right)\left(\frac{\partial U_y}{\partial t}\right)\right] \\
& +\rho_{22}\left[\left(\frac{\partial U_x}{\partial t}\right)^2+\left(\frac{\partial U_y}{\partial t}\right)^2+\left(\frac{\partial U_z}{\partial t}\right)^2\right]
\end{aligned}
\tag{10.13}
$$

where ρ_{11}, ρ_{12}, ρ_{22} in the equation are density terms, which can be expressed as

$$
\begin{cases}
\rho_{11} = (1-\phi)\,\rho_s, \\
\rho_{22} = \phi\rho_f, \\
\rho_{12} = -(\alpha-1)\,\phi\rho_f
\end{cases}
\tag{10.14}
$$

where ρ_s is the mass density of the solid grains and ρ_f is the mass density of the fluid in pores, $\alpha = (1/2)[\phi^{-1}+1]$, and ϕ is the porosity of the medium.

Consider a motion restricted to the x direction and denote q_x as the total force acting on the solid per unit volume and Q_x as the total force acting on the fluid per unit volume, derived from the Lagrangian equations; the following can be obtained:

$$
\begin{aligned}
\frac{\partial}{\partial t}\left(\frac{\partial T}{\partial \dot{u}_x}\right) &= \frac{\partial^2}{\partial t^2}\left(\rho_{11}u_x+\rho_{12}U_x\right) = q_x \\
\frac{\partial}{\partial t}\left(\frac{\partial T}{\partial \dot{U}_x}\right) &= \frac{\partial^2}{\partial t^2}\left(\rho_{12}u_x+\rho_{22}U_x\right) = Q_x
\end{aligned}
\tag{10.15a, b}
$$

In terms of stresses, the force components are expressed as stress gradients as

$$
q_x = \frac{\partial \sigma_x}{\partial x}+\frac{\partial \tau_z}{\partial y}+\frac{\partial \tau_y}{\partial z},\ Q_x = \frac{\partial s}{\partial x}
\tag{10.16}
$$

Therefore, the dynamic equations are

$$
\begin{aligned}
\frac{\partial \sigma_x}{\partial x}+\frac{\partial \tau_z}{\partial y}+\frac{\partial \tau_y}{\partial z} &= \frac{\partial^2}{\partial t^2}\left(\rho_{11}u_x+\rho_{12}U_x\right) \\
\frac{\partial s}{\partial x} &= \frac{\partial^2}{\partial t^2}\left(\rho_{12}u_x+\rho_{22}U_x\right)
\end{aligned}
\tag{10.17}
$$

By substituting (10.10), (10.11), (10.12), (10.13), (10.14) and (10.15), for the x direction,

$$
\begin{cases}
N\nabla^2 u_x+(A+N)\frac{\partial e}{\partial x}+Q\frac{\partial \varepsilon}{\partial x}\Big] = \frac{\partial^2}{\partial t^2}\left(\rho_{11}u_x+\rho_{12}U_x\right) \\
Q\frac{\partial e}{\partial x}+R\frac{\partial \varepsilon}{\partial x} = \frac{\partial^2}{\partial t^2}\left(\rho_{12}u_x+\rho_{22}U_x\right)
\end{cases}
\tag{10.18a, b}
$$

For the other two directions, with the vector notation

$$
\mathbf{u} = \left(u_x u_y u_z\right),\mathbf{U} = \left(U_x U_y U_z\right)
\tag{10.19}
$$

The wave equations in low frequency range in the absence of friction were developed as the following (Biot 1956a, b):

$$\begin{cases} N\nabla^2\mathbf{u} + \nabla\left[(A+N)e + Q\varepsilon\right] = \frac{\partial^2}{\partial t^2}\left(\rho_{11}\mathbf{u} + \rho_{12}\mathbf{U}\right) \\ \nabla\left[Qe + R\varepsilon\right] = \frac{\partial^2}{\partial t^2}\left(\rho_{12}\mathbf{u} + \rho_{22}\mathbf{U}\right) \end{cases} \quad (10.20\text{ a, b})$$

R is a measurement of the pressure on the fluid required to drive a unit volume of fluid into the porous medium, and *Q* describes the coupling between the volume change of the solid and that of fluid. The expressions for *A*, *N*, *Q*, and *R* will be further described in the following section.

It should be noticed that Biot's governing wave equations (Biot 1956a, b) for the porous media saturated with liquid are valid under the following assumptions and conditions:

- The solid and liquid waves governed by Biot's wave equations of simultaneous differential equations are different but the waves are coupled.
- The governing wave equations are valid in the low frequency range.
- The fluid in porous media is assumed to be compressible and may flow relative to the solid.
- In developing the governing equations of Biot, the effect of temperature variations due to dissipation of energy is assumed negligible.
- The relative motion of the fluid in pores is a laminar flow which follows Darcy's law.
- Friction in wave propagation is not considered.
- The porous medium considered is formed by unit solid–fluid elements, and the size of the unit solid–fluid element is geometrically larger in comparison with that of the pores.
- The wavelength of the elastic wave traveling in the porous media is much larger than that of the unit element.

10.3.1.1 Wave Governing Equations Derived in the Form of Displacements

As stated in Biot's studies (Biot 1956a), some other basic assumptions in elastic mechanics are also employed, such as homogeneity and isotropy of the porous media material and the impervious nature of the pore wall.

With this consideration, Helmholtz decomposition (Amrouche et al. 1998; Arfken et al. 2005) is applied to the displacement vectors corresponding to the solid and fluid of a porous medium. The general form of the governing equations for both shear waves and compressive waves needs to be derived. Specifically, applying, respectively, Helmholtz decomposition to the displacement vectors of the solid and fluid of a porous medium, the displacement vectors of the solid and fluid can be expressed as the following (Dai and Wang 2009; Wang et al. 2007a, b, 2009; Wang 2008):

$$\begin{cases} \mathbf{u} = \text{grad}\,(\varphi_s) + \text{curl}\,(\boldsymbol{\psi}_s) \\ \mathbf{U} = \text{grad}\,(\varphi_f) + \text{curl}\,(\boldsymbol{\psi}_f) \end{cases} \tag{10.21a, b}$$

where φ_s and φ_f are scalar potentials of the solid and fluid, respectively. Whereas $\boldsymbol{\psi}_s$ and $\boldsymbol{\psi}_f$ are vector potentials for the displacements of the solid and fluid. Also, $\boldsymbol{\psi}_s$ and $\boldsymbol{\psi}_f$ in (10.21) satisfy the following conditions:

$$\nabla \cdot \boldsymbol{\psi}_s = 0 \quad \text{and} \quad \nabla \cdot \boldsymbol{\psi}_f = 0 \tag{10.22}$$

For a compressible wave, also known as a P-wave or dilatational wave, the displacement is caused by the scalar potentials without rotation. This implies that

$$\nabla \times (\nabla \varphi_s) = 0 \quad \text{and} \quad \nabla \times (\nabla \varphi_f) = 0 \tag{10.23}$$

For a shear wave, also known as an S-wave or rotational wave, the displacement is due to the vector potentials, such that

$$\nabla \cdot (\nabla \times \boldsymbol{\psi}_\mathbf{s}) = 0 \quad \text{and} \quad \nabla \cdot (\nabla \times \boldsymbol{\psi}_\mathbf{f}) = 0 \tag{10.24}$$

With these expressions, substitute (10.21a) and (10.21b) into (10.20a) and (10.20b) and rearrange the terms according to the scalar and vector potentials; two sets of equations can be obtained corresponding to the scalar and vector potentials of the fluid and solid. The governing equations for the P- and S-waves can thus be established.

With the development above, pressure governing wave equations can be expressed as

$$\begin{cases} \nabla^2 \left(P\varphi_s + Q\varphi_f\right) = \frac{\partial^2}{\partial t^2} \left(\rho_{11}\varphi_s + \rho_{12}\varphi_f\right) \\ \nabla^2 \left[Q\varphi_s + R\varphi_f\right] = \frac{\partial^2}{\partial t^2} \left(\rho_{12}\varphi_s + \rho_{22}\varphi_f\right) \end{cases} \tag{10.25a, b}$$

In this equation, $P = A + 2N$ is defined, the subscript s represents the displacement of solid, and f designates the displacement of the fluid.

For S-wave,

$$\begin{cases} N\nabla^2 \boldsymbol{\psi}_s = \frac{\partial^2}{\partial t^2} \left(\rho_{11}\boldsymbol{\psi}_s + \rho_{12}\boldsymbol{\psi}_f\right) \\ 0 = \frac{\partial^2}{\partial t^2} \left(\rho_{12}\boldsymbol{\psi}_s + \rho_{22}\boldsymbol{\psi}_f\right) \end{cases} \tag{10.26a, b}$$

Equations (10.25) and (10.26) such expressed are the governing equations for wave propagations in terms of displacement potentials. On this basis, the governing wave equations in the forms of displacements of the solid and liquid of porous media can be derived.

In order to derive the wave equations in the form of displacements for a P-wave, take the divergence operator to (10.26), such that

$$\begin{cases} \nabla\left[\nabla^2\left(P\varphi_s + Q\varphi_f\right)\right] = \nabla\left[\frac{\partial^2}{\partial t^2}\left(\rho_{11}\varphi_s + \rho_{12}\varphi_f\right)\right] \\ \nabla\left[\nabla^2\left(Q\varphi_s + R\varphi_f\right)\right] = \nabla\left[\frac{\partial^2}{\partial t^2}\left(\rho_{12}\varphi_s + \rho_{22}\varphi_f\right)\right] \end{cases} \tag{10.27a, b}$$

Let φ be a general displacement scalar potential and $\mathbf{u}$ a general displacement vector; for the P-wave, the displacement vector $\mathbf{u}$ is defined merely in relating to the scalar potential φ as

$$\mathbf{u} = \nabla\varphi \tag{10.28}$$

As shown in (10.23), $\nabla \times (\nabla\varphi) = 0$, the scalar potential φ must satisfy the following property:

$$\begin{aligned} &\left(\nabla^2\left(\nabla\varphi\right)\right) \\ &= \nabla\left(\nabla\cdot\left(\nabla\varphi\right)\right) - \nabla\times\nabla\times\left(\nabla\varphi\right) \\ &= \nabla(\nabla\left(\nabla\cdot\varphi\right) \end{aligned} \tag{10.29}$$

At the same time,

$$\begin{aligned} &\nabla\left(\nabla^2\varphi\right) \\ &= \nabla\left(\nabla\left(\nabla\cdot\varphi\right) - \nabla\times\nabla\times\varphi\right) \\ &= \nabla\left(\nabla\left(\nabla\cdot\varphi\right) - \nabla(\nabla\times\nabla\times\varphi\right) \\ &= \nabla\left(\nabla\left(\nabla\cdot\varphi\right) - \nabla\times\nabla\times(\nabla\varphi\right) \\ &= \nabla(\nabla\left(\nabla\cdot\varphi\right) \end{aligned} \tag{10.30}$$

Therefore, the following expressions must be true:

$$\nabla\left(\nabla^2\varphi\right) = \nabla^2\left(\nabla\varphi\right) = \nabla^2\mathbf{u} \tag{10.31}$$

With these expressions, the displacement equation for P-wave can be given in the following general Laplacian operator form:

$$\begin{aligned} \nabla^2\left(P\mathbf{u} + Q\mathbf{U}\right) &= \frac{\partial^2}{\partial t^2}\left(\rho_{11}\mathbf{u} + \rho_{12}\mathbf{U}\right) \\ \nabla^2\left(Q\mathbf{u} + R\mathbf{U}\right) &= \frac{\partial^2}{\partial t^2}\left(\rho_{12}\mathbf{u} + \rho_{22}\mathbf{U}\right) \end{aligned} \tag{10.32a, b}$$

10.3.1.2 Shear Governing Wave Equations

For shear wave, the magnitude of rotation denoted as $\boldsymbol{v}$ is only related to the vector potential as $\mathbf{v} = \nabla \times \boldsymbol{\psi}$. To develop its governing equations in the form of displacement for shear wave, take the curl operation to (10.26a) and (10.26b) to obtain the shear wave equations as

$$\begin{aligned} \nabla \times \left[N \nabla^2 \boldsymbol{\psi}_{\mathbf{s}}) \right] &= \nabla \times \left[\frac{\partial^2}{\partial t^2} (\rho_{11} \boldsymbol{\psi}_{\mathbf{s}} + \rho_{12} \boldsymbol{\psi}_{\mathbf{f}}) \right] \\ 0 &= \nabla \times \left[\frac{\partial^2}{\partial t^2} (\rho_{12} \boldsymbol{\psi}_{\mathbf{s}} + \rho_{22} \boldsymbol{\psi}_{\mathbf{f}}) \right] \end{aligned} \tag{10.33a, b}$$

Based on (10.23), $(\nabla \cdot \boldsymbol{\psi}) = 0$, the vector potential has the following property:

$$\begin{aligned} &\nabla \times \left(\nabla^2 \boldsymbol{\psi} \right) \\ &= \nabla \times (\nabla (\nabla \cdot \boldsymbol{\psi}) - \nabla \times \nabla \times \boldsymbol{\psi}) \\ &= \nabla \times (\nabla (\nabla \cdot \boldsymbol{\psi})) - \nabla \times \nabla \times \nabla \times \boldsymbol{\psi} \\ &= -\nabla \times \nabla \times \nabla \times \boldsymbol{\psi} \end{aligned} \tag{10.34}$$

And since $\nabla \cdot (\nabla \times \boldsymbol{\psi}) = 0$,

$$\begin{aligned} &\nabla^2 (\nabla \times \boldsymbol{\psi}) \\ &= \nabla (\nabla \cdot (\nabla \times \boldsymbol{\psi})) - \nabla \times \nabla \times (\nabla \times \boldsymbol{\psi})) \\ &= -\nabla \times \nabla \times \nabla \times \boldsymbol{\psi} \end{aligned} \tag{10.35}$$

Therefore,

$$\nabla \times \left(\nabla^2 \boldsymbol{\psi} \right) = \nabla^2 (\nabla \times \boldsymbol{\psi}) = \nabla^2 \mathbf{v} \tag{10.36}$$

The governing equation in the form of displacement can then be given in the following form for shear wave:

$$\begin{aligned} N \nabla^2 \mathbf{v} &= \frac{\partial^2}{\partial t^2} (\rho_{11} \mathbf{v} + \rho_{12} \mathbf{V}) \\ 0 &= \frac{\partial^2}{\partial t^2} (\rho_{12} \mathbf{v} + \rho_{22} \mathbf{V}) \end{aligned} \tag{10.37a, b}$$

In (10.32) and (10.37), the parameters of the medium, P, Q, R, can be expressed as (Biot and Willis 1957; Lin et al. 2001; Plona and Johnson 1984)

$$P = \frac{(1 - \phi) \left[1 - \phi - \frac{K_b}{K_s} \right] K_s + \phi \frac{K_s}{K_f} K_b}{1 - \phi - \frac{K_b}{K_s} + \phi \frac{K_s}{K_f}} + \frac{4}{3} N \tag{10.38}$$

$$Q = \frac{\left[1 - \phi - \frac{K_b}{K_s}\right]\phi K_s}{1 - \phi - \frac{K_b}{K_s} + \phi\frac{K_s}{K_f}} \tag{10.39}$$

$$R = \frac{\phi^2 K_s}{1 - \phi - \frac{K_b}{K_s} + \phi\frac{K_s}{K_f}} \tag{10.40}$$

in which ϕ is the porosity. K_f, K_s, K_b, and N are property parameters of the material, K_f is the bulk modulus of the fluid, K_s is the bulk modulus of the solid, K_b designates the bulk modulus of the skeletal frame, and N is the shear modulus of the skeletal frame of the porous medium considered. The governing wave equations are thus all described in terms of displacements of solid and fluid in comparison with the volume strains used in Biot's paper (Biot 1956a).

10.4 Fractal Dimension Development of 3D Wave Model for Wave Propagations in Half-Space Porous Media

In real world, many waves propagate in half-space. A typical example of such wave propagations is the waves generated on the surface of the Earth which is considered as a porous medium. Studies on the wave motions in porous half-space are significant. The seismic and vibration stimulations on EOR (Dai and Wang 2009; Wang et al. 2007a, b, 2009; Wang 2008) are typical examples of the applications of waves in half-space porous media. This chapter focuses on the establishment of a three-dimensional model in quantifying the waves propagating in porous half-space.

The wave propagation described in the governing equations developed in the previous section completely relies on the characteristics of the energy source by which the wave is generated. Therefore, a model describing the wave field with the displacements and relative displacements between the solid and fluid for each point in the porous medium domain considered is necessary.

Consider a porous half-space subjected to a cylindrical energy source as shown in Fig. 10.1. As can be seen from the figure, a circular-cylindrical coordinate system is implemented to describe the propagation of the wave generated by the source which is also described in the circular-cylindrical coordinate system.

To emphasize on the relative displacements, focus is placed on a geometrical point in the porous half-space medium considered, the point P in Fig. 10.6, for its relative displacement between the fluid and solid. In developing the model, the following conditions and assumptions are made, in addition to the conditions made for Biot's theory as indicated in the previous section:

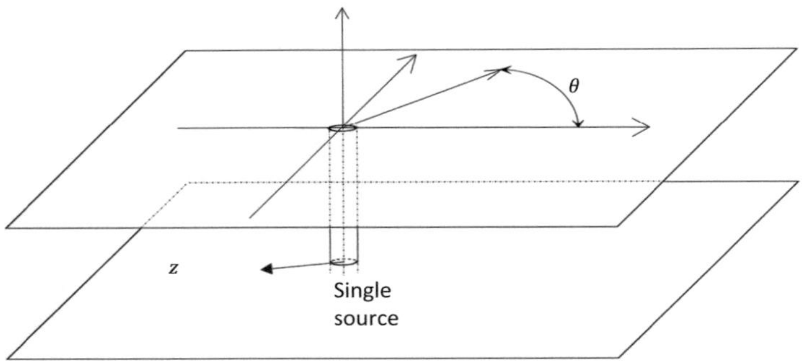

Fig. 10.6 The model of porous half-space media subjected to a single circular-cylindrical energy source

- The domain of porous media of half-space starts from the surface of the domain, the horizontal X-O-Y plane as shown in Fig. 10.6, and extends to infinity along the Z direction.
- The material of the porous medium is assumed to be isotropic and homogeneous the same as that of Biot.
- The energy source at the middle of the domain is uniformly distributed along the vertical axis of passing through the center of the source.
- The wave source generates a continuous harmonic wave with time.
- The wave generated by the energy source propagates as a compressible wave in the three-dimensional domain.
- The vertical circular-cylindrical source is perpendicular to the wave propagation direction of the P-waves generated by the energy source.
- The wave considered is in the low frequency range and propagates in the semi-infinite domain affected by the energy source.
- The dimensions of the energy source are negligible in comparison with the wavelength of the wave.

Under these conditions and assumptions, the waveform on any plane parallel to the horizontal plane of X-O-Y is the same for any given instance, and the wave propagating in the semi-infinite half-space is independent of depth along the Z direction. As such, the governing equations for such wave can be two dimensional. This significantly simplifies the development of the governing equations for describing and simulating the waves propagating in the field domain under consideration. It also provides the convenience to express the governing equations and corresponding solutions in cylindrical coordinate with the circular-cylindrical energy source.

10.4.1 Governing Equation Development

In the cylindrical coordinates, with r indicating the distance from the center of the energy source as shown in Fig. 10.6, the operators ∇ and ∇^2 can be given as

$$\begin{cases} \nabla = \frac{\partial}{\partial r} + \frac{1}{r} \\ \nabla^2 = \frac{\partial^2}{\partial r^2} + \frac{1}{r}\frac{\partial}{\partial r} \end{cases} \tag{10.41a, b}$$

For consistency of the research in this field, the following parameters are employed as those introduced by Biot (1956a):

$$V_c^2 = H/\rho \tag{10.42}$$

$$\begin{cases} \sigma_{11} = \frac{P}{H}, & \sigma_{12} = \frac{Q}{H}, & \sigma_{22} = \frac{R}{H} \\ \gamma_{11} = \frac{\rho_{11}}{\rho}, & \gamma_{12} = \frac{\rho_{12}}{\rho}, & \gamma_{22} = \frac{\rho_{22}}{\rho} \end{cases} \tag{10.43}$$

in which

$$H = P + R + 2Q, \qquad \rho = \rho_{11} + \rho_{22} + 2\rho_{12} \tag{10.44}$$

Applying the operators shown in (10.41) on (10.32), the governing equations for dilatational waves in a specific horizontal x-y plane at $z = z_{depth}$, of which $z_{depth} \in [0, \ +\infty)$, can be derived into the following form, with the cylindrical coordinates:

$$\begin{cases} \left(\frac{d^2}{dr^2} + \frac{1}{r}\frac{d}{dr}\right)(\sigma_{11}\mathbf{u} + \sigma_{12}\mathbf{U}) = \frac{1}{V_c^2}\frac{\partial^2}{\partial t^2}(\gamma_{11}\mathbf{u} + \gamma_{12}\mathbf{U}) \\ \left(\frac{d^2}{dr^2} + \frac{1}{r}\frac{d}{dr}\right)(\sigma_{12}\mathbf{u} + \sigma_{22}\mathbf{U}) = \frac{1}{V_c^2}\frac{\partial^2}{\partial t^2}(\gamma_{12}\mathbf{u} + \gamma_{22}\mathbf{U}) \end{cases} \tag{10.45a, b}$$

To describe propagating cylindrical wave solutions of the cylindrical wave equation, the Hankel functions are usually employed. Making use of the Sommerfeld Radiation Condition (Pao and Mow 1971), the wave propagating from a single cylindrical energy source can be described in terms of displacements of the solid and liquid of the porous medium, such that

$$\begin{cases} \mathbf{u} = C_1 H_0^{(1)}(lr)\ \exp(-i\omega t) \\ \mathbf{U} = C_2 H_0^{(1)}(lr)\ \exp(-i\omega t) \end{cases} \tag{10.46a, b}$$

where C_1 and C_2 are the amplitudes of the waves propagating in the solid and fluid, respectively. In the equations, l is the wave number and r is the distance from the considered point in the domain to the center of the source, the Z axis. The function $H_0^{(1)}(\cdot)$ is the zero-order Hankel function of the first kind. The subscript "0" represents the order of the function, $\exp(-i\omega t)$ in the equation designates the time factor of the harmonic wave generated by the energy source, $i = \sqrt{-1}$ is the complex unit, and ω represents the frequency of the wave.

In the above equations, the Hankel function $H_0^{(1)}(lr)$ is the first kind Hankel function which has the following form:

$$H_0^{(1)}(lr) = J_0(lr) + iY_0(lr) \tag{10.47}$$

in which $J_0(lr)$ is the Bessel function of the first kind and $Y_0(lr)$ is Bessel function of the second kind. The Bessel function of the first kind takes the form of

$$J_0(lr) = \sum_{k=0}^{\infty} \frac{(-1)^k}{(k!)^2} \left(\frac{1}{4}(lr)^2 \right)^k \tag{10.48}$$

and the Bessel function of the second kind

$$Y_0(lr) = \frac{2}{\pi} \left\{ \left[\ln\left(\frac{1}{2} lr \right) + \lim_{k \to \infty} \left(\sum_{i=1}^{k} \frac{1}{i} - \ln(k) \right) \right] \times J_0(lr) + \sum_{k=0}^{\infty} (-1)^{k+1} \left(\sum_{i=1}^{k} \frac{1}{i} \right) \frac{\left(\frac{1}{4}(lr)^2 \right)^k}{(k!)^2} \right\} \tag{10.49}$$

Using the superscript "(1)" to indicate the Hankel function of the first kind and employing the following equations of zero- and first-order Hankel functions (Andrews et al. 2001),

$$\begin{cases} \frac{d}{dx} H_0^{(1)}(x) = -H_1^{(1)}(x) \\ \frac{d}{dx} H_1^{(1)}(x) = \frac{1}{2} \left[H_0^{(1)}(x) - H_2^{(1)}(x) \right] \end{cases} \tag{10.50a, b}$$

For the displacements, taking the partial differentiation, one may obtain

$$\begin{cases} \frac{\partial^2}{\partial r^2} H_0^{(1)}(lr) = -\frac{l^2}{2} \left[H_0^{(1)}(lr) - H_2^{(1)}(lr) \right] \\ \frac{1}{r} \frac{\partial}{\partial r} H_0^{(1)}(lr) = -\frac{l^2}{2} \left[H_0^{(1)}(lr) + H_2^{(1)}(lr) \right] \end{cases} \tag{10.51a, b}$$

which may also be expressed in the following form:

$$\nabla^2 H_0^{(1)}(lr) = \left(\frac{\partial^2}{\partial r^2} + \frac{1}{r} \frac{\partial}{\partial r} \right) H_0^{(1)}(lr) = -l^2 H_0^{(1)}(lr) \tag{10.52}$$

By substituting the expressions of (10.52) into (10.45), the following equations can be obtained:

$$\begin{cases} -l^2\left(\sigma_{11}C_1+\sigma_{12}C_2\right)=-\frac{1}{V_c^2}\omega^2\left(\gamma_{11}C_1+\gamma_{12}C_2\right)\\ -l^2\left(\sigma_{12}C_1+\sigma_{22}C_2\right)=-\frac{1}{V_c^2}\omega^2\left(\gamma_{12}C_1+\gamma_{22}C_2\right)\end{cases} \quad (10.53a, b)$$

where V_c is as described in (10.42). The general equation for the velocities of the waves can be given as

$$V=\omega/l, \quad (10.54)$$

For the sake of clarification in a mathematical form, introduce a parameter:

$$\xi=V_c^2/V^2 \quad (10.55)$$

in which V is used as a reference velocity. Equation (10.53) may thus be rewritten as

$$\begin{cases} \xi\left(\sigma_{11}C_1+\sigma_{12}C_2\right)=\gamma_{11}C_1+\gamma_{12}C_2\\ \xi\left(\sigma_{12}C_1+\sigma_{22}C_2\right)=\gamma_{12}C_1+\gamma_{22}C_2\end{cases} \quad (10.56a, b)$$

To solve for ξ, eliminating C_1 and C_2 from (10.56), a quadratic equation for ξ can be obtained from the two equations in (10.56) as the following:

$$\left(\sigma_{11}\sigma_{22}-\sigma_{12}^2\right)\xi^2-\left(\sigma_{11}\gamma_{22}+\sigma_{22}\gamma_{11}-2\sigma_{12}\gamma_{12}\right)\xi+\left(\gamma_{11}\gamma_{22}-\gamma_{12}^2\right)=0 \quad (10.57)$$

In (10.57), σ_{ij} and γ_{ij} are physical parameters. Based on these physical parameters, it can be proven that (10.57) has two positive roots ξ_1 and ξ_2, which correspond to two velocities of dilatational or compressive wave propagations, $V1$ and $V2$, with respect to the first and second compressible waves. The two velocities can be expressed in the following form:

$$\begin{aligned} V_1^2&=V_c^2/\xi_1\\ V_2^2&=V_c^2/\xi_2\end{aligned} \quad (10.58a, b)$$

From (10.56), it can also be proven that the amplitudes C_1 and C_2 are identical to the following relation reported by Biot (1956a):

$$\xi=\frac{\gamma_{11}C_1^2+2\gamma_{12}C_1C_2+\gamma_{22}C_2^2}{\sigma_{11}C_1^2+2\sigma_{12}C_1C_2+\sigma_{22}C_2^2} \quad (10.59)$$

The first dilatational wave is analogous to the compressional wave in elastic media, while the second dilatational wave is much slower and highly attenuated, therefore difficult to observe. However, it should be noted that neither of the two dilatational waves propagates as a wave in fluid or in solid separately. In fact, both of the waves travel jointly in the solid and pore fluid. Once a porous medium is specified, the wave propagation velocities can be determined by (10.58), and the wave number can therefore be determined through (10.54). The frequency ω can also be determined once the properties of the wave are given. As such, all the parameters in (10.46) are determined, and the displacements of the solid and fluid can therefore be quantified by employing (10.46).

In practice of nonlinear characteristics diagnosis, the capacity dimension of a limit cycle considered as one, that of a two-period quasiperiodic orbit is two, and that of an m-period quasiperiodic orbit is m. Strange attractor, D_0, always takes fraction or non-integer values. For example, the Lorenz attractor is a good example of a strange attractor whose capacity dimension is 2.06 $\pm$0.01 and therefore taken as chaotic system. Due to the definition of the capacity dimension, in practice however, it is not always easy to determine a fractal dimension for a system that has strange attractor.

10.4.2 Establishment of Wave Propagation Model with Multiple Energy Sources

A great majority of research work found from the current literatures in the field of waves in porous media concentrates on studies where there is merely a single energy source in the domain and no wave superposition was considered, though superposition of waves and wave propagations with multiple wave sources are commonly seen in the real world. Waves of multiple energy sources are therefore more general though more complex. The advantages of multiple energy sources over that of the single energy source have been recognized recently in studying wave propagations in porous media (Han and Dai 2011a). As found in the recent investigations in the field, the joint effects of waves generated by multiple energy sources may contribute to creating vibrations at the particles with desirable relative displacements between the solid and liquid of the porous media and with high efficiency (Han and Dai 2011a; Wang et al. 2009). It is also found in the recent research (Han and Dai 2011b), with the same relative displacements in a domain considered, energy required with multiple energy sources is less than that of single energy source. More significantly, vibrations in the domain considered can be controlled with the waves generated by multiple energy sources in terms of amplitude, frequency, and duration of the vibrations. Multiple energy source vibration stimulation is therefore practically sound. It is especially significant in practicing seismic and vibration stimulated EOR with multiple energy sources.

It is therefore significant and practically meaningful to study the dynamic response of porous media and the relative displacement between solid and fluid when the domain is excited by multiple energy sources. In the model described in the previous section, a cylindrical wave is generated from circular-cylindrical energy sources, and the wave propagates in the porous half-space medium. In this section,

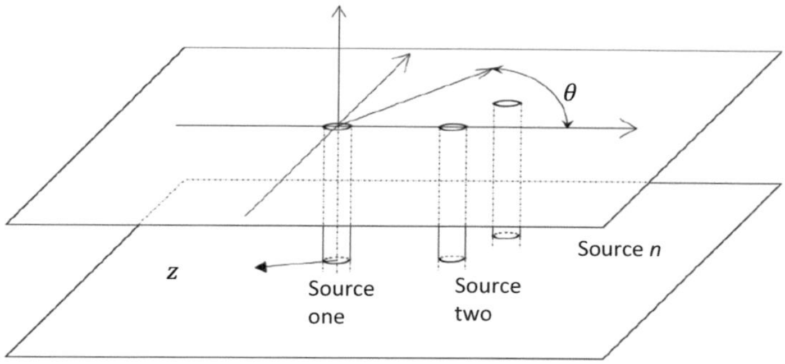

Fig. 10.7 3D model of multiple energy sources

a model that describes the wave propagations of multiple energy sources is to be established. Graphically, the model is illustrated in Fig. 10.7.

In order to develop a model for describing the wave propagation in porous half-space subjected to multiple energy sources, some additional conditions and assumptions need to be considered:

- All the conditions and assumptions made for the governing equations in the previous sections are valid. Following additional conditions and assumptions are needed for the wave equations made for the wave propagations of multiple energy sources in this section.
- All the energy sources are parallel to each other and perpendicular to the surface of the domain considered.
- The energy sources considered continuously generate harmonic waves and are uniformly distributed along the vertical axes passing through each of the centers of the sources.
- The vertical circular-cylindrical sources are all perpendicular to the wave propagation direction of the P-waves generated by the energy sources.
- The dimensions of the energy sources themselves are negligible in comparison with the wavelength of the waves.

With these conditions and assumptions, the superposition of the waves on any plane parallel to the horizontal plane of X-O-Y is the same at any given time, and the waves propagating in the half-space are independent of depth. Therefore, the waves and the superposition of the waves can be described in a two-dimensional fashion as that described in the previous section, and one may concentrate on a two-dimensional plain for analyzing the waves. This may significantly simplify the development of the model in quantifying the waves generated by multiple energy sources.

In the previous section, the propagating waves are expressed with the Hankel function in the form of displacement. In this section, for demonstration using numerical simulation, it is assumed that the n cylindrical compressible waves are

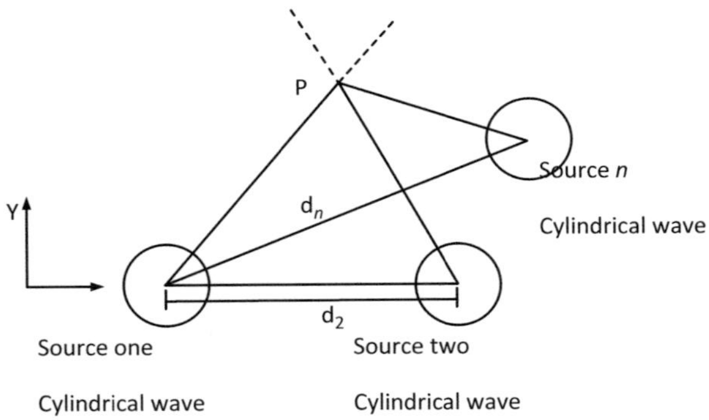

Fig. 10.8 Multisource model in an x-y plane

generated by multiple cylindrical sources, as shown in Fig. 10.3, in which the point P is specified on an x-y plane at $z = z_{depth}$ (Fig. 10.8).

The waves are also assumed to be continuous and harmonic and the waves are in steady state. Under all the conditions described above, the wave from each of the n multi-energy sources can be expressed in the following form in terms of displacements:

$$\begin{cases} \mathbf{u}_1 = u_{01}\mathrm{Re}\left[H_0^{(1)}\left(l_1 r_1\right)\exp\left(-i\omega_1 t\right)\right]\left[\frac{\vec{r}_1}{\left|\vec{r}_1\right|}\right] \\ \mathbf{U}_1 = U_{01}\mathrm{Re}\left[H_0^{(1)}\left(l_1 r_1\right)\exp\left(-i\omega_1 t\right)\right]\left[\frac{\vec{r}_1}{\left|\vec{r}_1\right|}\right] \end{cases} \tag{10.60a, b}$$

and

$$\begin{cases} \mathbf{u}_2 = u_{02}\mathrm{Re}\left[H_0^{(1)}\left(l_2 r_2\right)\exp\left(-i\omega_2 t\right)\right]\left[\frac{\vec{r}_2}{\left|\vec{r}_2\right|}\right] \\ \mathbf{U}_2 = U_{02}\mathrm{Re}\left[H_0^{(1)}\left(l_2 r_2\right)\exp\left(-i\omega_2 t\right)\right]\left[\frac{\vec{r}_2}{\left|\vec{r}_2\right|}\right] \end{cases} \tag{10.61a, b}$$

......

$$\begin{cases} \mathbf{u}_n = u_{0n}\mathrm{Re}\left[H_0^{(1)}\left(l_n r_n\right)\exp\left(-i\omega_n t\right)\right]\left[\frac{\vec{r}_n}{\left|\vec{r}_n\right|}\right] \\ \mathbf{U}_n = U_{0n}\mathrm{Re}\left[H_0^{(1)}\left(l_n r_n\right)\exp\left(-i\omega_n t\right)\right]\left[\frac{\vec{r}_n}{\left|\vec{r}_n\right|}\right] \end{cases} \tag{10.62a, b}$$

In the equations, u_{0j} and U_{0j} (j = 1, 2, ..., n) are the displacement amplitudes of the solid and fluid of the *j*th source. $\mathbf{u}_{0j}$ and $\mathbf{U}_{0j}$ (j = 1, 2, ..., n) are the corresponding displacement vectors of the solid and fluid excited at the *j*th source. In the equation, $\vec{r}_j$ is the position vector indicating the direction and distance from point P in the domain to the jth wave source. The unit vector $\left[\vec{r}_j / \left|\vec{r}_j\right|\right]$ is introduced to describe the direction of the displacements.

In order to investigate the superposed action of multiple waves efficiently, the expression for each wave is to be expressed in a common coordinate system.

To describe the jth wave in xyz-coordinates, let $r_j = r_p - d_j$. The displacements of a point P considered in the domain can be derived as

$$\begin{cases} u_j = u_{0j}\mathrm{Re}\left[H_0^{(1)}\left(l_j r_j\right)\exp\left(-i\omega_j t\right)\right]\left[\frac{\vec{r}_j}{\left|\vec{r}_j\right|}\right] \\ \quad = u_{0j}\mathrm{Re}\left[H_0^{(1)}\left(l_j\left|\vec{r}_p - d_j\right|\right)\exp\left(-i\omega_j t\right)\right]\left[\frac{\vec{r}_{p-d_j}}{\left|\vec{r}_{p-d_j}\right|}\right] \\ U_j = U_{0j}\mathrm{Re}\left[H_0^{(1)}\left(l_j r_j\right)\exp\left(-i\omega_j t\right)\right]\left[\frac{\vec{r}_j}{\left|\vec{r}_j\right|}\right] \\ \quad = U_{0j}\mathrm{Re}\left[H_0^{(1)}\left(l_j\left|\vec{r}_p - d_j\right|\right)\exp\left(-i\omega_j t\right)\right]\left[\frac{\vec{r}_{p-d_j}}{\left|\vec{r}_{p-d_j}\right|}\right] \end{cases} \tag{10.63a, b}$$

where d_j is the coordinate vector of the jth wave source in the common coordinate system and $\vec{r}_p$ denotes the common coordinate vector of the specified point P in the x-y plane at $z = z_{depth}$.

The total displacements of point P, jointly affected by the n energy sources, in the domain considered can thus be described in a common coordinate system. When the xyz-coordinate system is chosen as the common coordinate system, $d_1 = 0$ and $d_j \neq 0$ where $j = 2, 3, \cdots, n$. The superposed displacements of the point P in the domain can be given by

$$\begin{cases} \mathbf{u} = \sum_{j=1}^{n}\mathbf{u}_j = u_{01}\mathrm{Re}\left[H_0^{(1)}\left(l_1\left|\vec{r}_p\right|\right)\exp\left(-i\omega_1 t\right)\right]\left[\frac{\vec{r}_p}{\left|\vec{r}_p\right|}\right] + \cdots \\ \qquad + u_{0n}\mathrm{Re}\left[H_0^{(1)}\left(l_n\left|\vec{r}_p - d_n\right|\right)\exp\left(-i\omega_n t\right)\right]\left[\frac{\vec{r}_{p-d_n}}{\left|\vec{r}_{p-d_n}\right|}\right] \\ \mathbf{U} = \sum_{j=1}^{n}\mathbf{U}_j = U_{01}\mathrm{Re}\left[H_0^{(1)}\left(l_1\left|\vec{r}_p\right|\right)\exp\left(-i\omega_1 t\right)\right]\left[\frac{\vec{r}_p}{\left|\vec{r}_p\right|}\right] + \cdots \\ \qquad + U_{0n}\mathrm{Re}\left[H_0^{(1)}\left(l_n\left|\vec{r}_p - d_n\right|\right)\exp\left(-i\omega_n t\right)\right]\left[\frac{\vec{r}_{p-d_n}}{\left|\vec{r}_{p-d_n}\right|}\right] \end{cases} \tag{10.64a, b}$$

In implementing the model, as can be seen from the derivation of the equations, there are actually no restrictions on the location of the point considered and no restrictions on the number of energy sources or material constants. The expression shown in the above equations can therefore be conveniently used to analyze the behavior of the wave field for a variety of porous media under various excitation conditions.

As can be seen from the procedure in establishing the model, the equations and solutions developed allow separate descriptions of the displacements of the solid

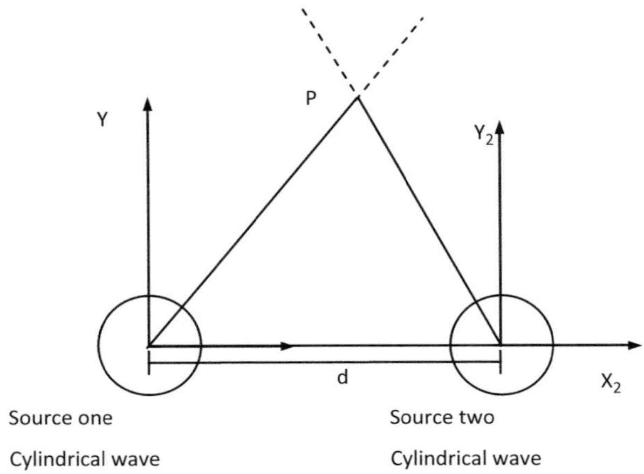

Fig. 10.9 Two-source model for numerical simulation

and fluid of the porous medium. Therefore, the relative displacement ($\mathbf{U}-\mathbf{u}$) of the fluid and solid can be quantified theoretically and numerically. This is significant in determining for the mobilization of liquid in the pore spaces of a porous medium. For the solid and fluid, respectively, let u_0 and U_0 designate fixed amplitudes of the sources. It should be noticed that setting the amplitudes of the sources as the constants u_0 and U_0 is merely for the sake of clarification. In actual numerical simulation, however, the amplitudes of the sources can be set differently if so desired.

Practically, this qualification of the displacements field is significant for theoretical analysis of the mobility of a fluid in an oil reservoir subjected to excitations of multiple elastic waves. Such a development has not been seen to date in the literature.

10.4.3 Numerical Study

With the methodology and the model developed as demonstrated in the previous section, the characteristics of the waves and displacement field in porous half-space subjected to multiple energy sources can be studied analytically and numerically. Specifically, the behavior of any specified point in the domain can be quantified, and the relative displacement between the fluid and solid of the medium can be conveniently determined by making use of the established model. This study may also demonstrate the application of the model developed.

Consider a simple case for the waves generated by two energy sources, as graphically shown in Fig. 10.9.

For the case as shown in Fig. 10.9, the distance between the two sources is noted as d, the position of point P in the field is expressed as $\mathrm{r}_p = x + iy$, and the frequencies of the two source waves are assumed as ω_1 and ω_2, respectively. For

Table 10.1 Parameters of the porous medium

Ø	v_s	μ_s	K_f	ρ_s	ρ_f	μ_s/K_f
0.3	0.4	2.25 GPa	3.0 GPa	2,700 kg/m^3	1,000 kg/m^3	0.75

Table 10.2 Parameters of the waves

V_{fast}	V_{slow}	C_1/C_2
3,277 m/s	177 m/s	1.038

C_1/C_2: The ratio of amplitudes of solid to fluid

the sake of simplification, as it is often assumed, the solid skeleton system is formed by spherical solid particles whose compressibility can be neglected. The parameters A, P, Q, and R in the previous equations may take the following forms as per Lin et al. (2001):

$$\begin{aligned} A &= \tfrac{2v_s}{1-2v_s}\mu_s + \tfrac{(1-\phi)^2}{\phi}K_f \\ P &= A + 2\mu_s \\ Q &= (1-\phi)\,K_f \\ R &= \phi K_f \end{aligned} \qquad (10.65\text{a, b, c, d})$$

where μ_s is the shear modulus of the material and v_s is the Poisson ratio of the solid.

Assume that the two sources are located at $x = 0$, $y = 0$ and $x = 1500$, $y = 0$, respectively. In general, the wave amplitudes decrease with increasing distance from the circular-cylindrical energy source. Moreover, the amplitude of the combined wave in a steady state is not simply the summation of the amplitudes of two waves.

With implementation of the wave model established, the wave motion in the domain considered relies on availability of the material properties of the porous medium and the characteristics of the waves initiated from the energy sources. The material properties of the porous medium used in the numerical simulations are shown in Table 10.1, and the wave characteristics such as wave velocities, amplitudes, and their calculated ratios are tabulated in Table 10.2.

It should be noticed that the mechanical and physical parameters of the porous half-space medium are not restrictive; thus, the established model can be applied to different porous half-space media as desired.

For the sake of clarification, U/u_0 and u/u_0 are introduced as nondimensional displacements of the fluid and solid, respectively. As described previously, it is the relative displacement $(U-u)$ that affects most of the mobilization of the liquid in the porous medium. A nondimensional relative displacement is defined as $(U-u)/u_0$. The utilization of the nondimensional displacements brings advantage of not considering the units during the calculation and, more importantly, the advantage of describing the displacements without considering the magnitude of the source amplitude. Therefore, the relative displacement such defined is the difference between the nondimensional displacements of the fluid and solid.

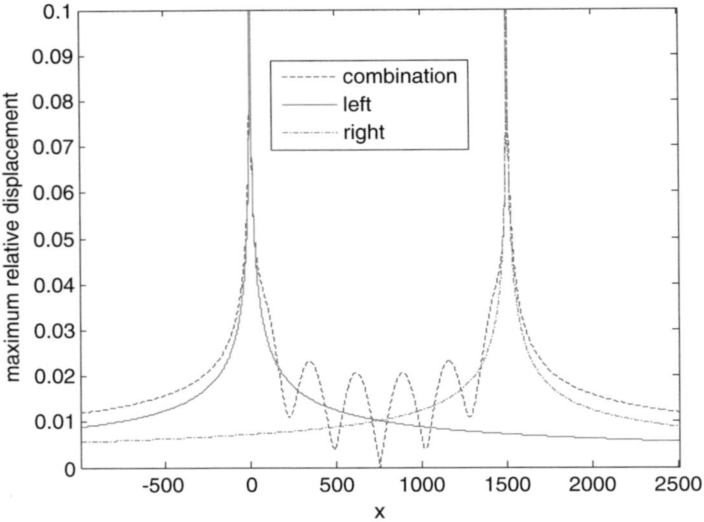

Fig. 10.10 The maximum relative displacements along the line connecting the two sources with identical frequencies

For the case considered, with the parametric values given, numerical results of the distributed maximum relative displacement along the distance between the two energy sources are plotted as shown in Fig. 10.10. In calculating for the relative displacements, the energy sources at the two ends of the connection line are identical in frequency and amplitude.

In Fig. 10.10, the center line represents the nondimensional maximum relative displacement amplitude of a single wave generated from the left source, and the dashed line is the maximum relative displacement of a single wave generated from the right source. For the purpose of comparison, the solid line illustrates the maximum relative displacement (plus and minus) jointly generated by the two identical sources together. One observes that the combined effect can be smaller as well as larger than the effect of merely one source.

Similarly, Fig. 10.11 illustrates the maximum relative displacement generated by two sources of different frequencies. Although the maximum relative displacements are more important in studying the mobilization of liquid in porous media, it is significant to visualize the actual relative displacements at a specified time. Figure 10.12 shows the profile of relative displacements at a given time. For any selected point in the semi-infinite domain, therefore, the relative displacement history of the point can be determined with the derived solutions.

When the porous half-space medium is excited by two circular-cylindrical energy sources, the wave response (amplitudes of the displacements) is totally different from that of the single source (represented by the curves of "left" and "right," respectively). In Fig. 10.10, with the maximum relative displacements generated

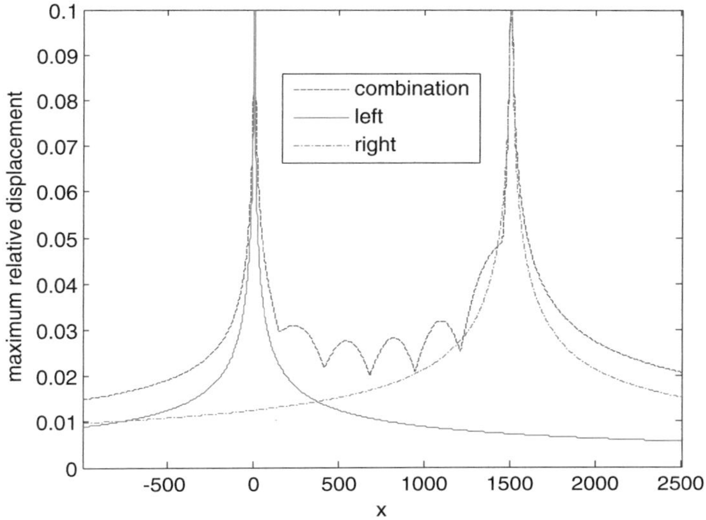

Fig. 10.11 The maximum relative displacements along the line connecting the two sources with different frequencies

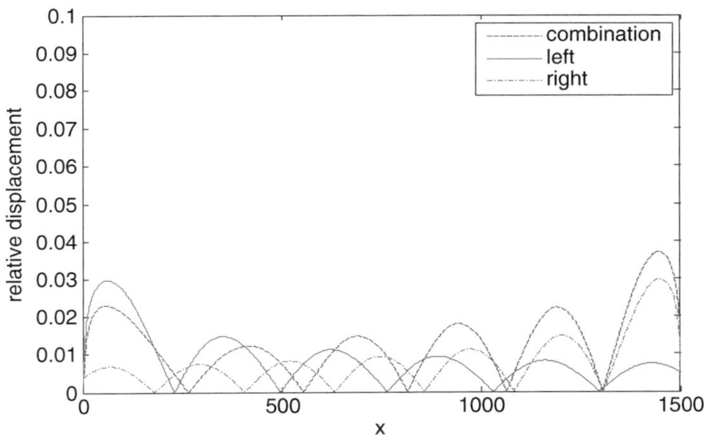

Fig. 10.12 The relative displacements along the line connecting the two sources at a specified time

by the source waves of the same frequency of 15, in some areas, the amplitude of the combined wave is smaller than that of single source, while for other areas, the amplitude is larger than that of the single source. The superposed wave generated by the two circular-cylindrical sources with identical frequency appears as a symmetrical motion, but the frequency of the superposed wave is different from that of the two source waves. One may also find out that the amplitude of the wave

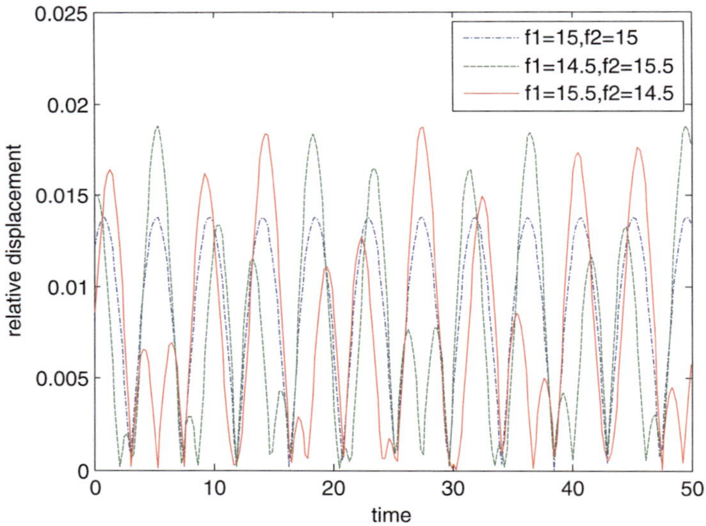

Fig. 10.13 The relative displacements in a time span

can be zero at a certain location between the two sources. This implies that this point can be permanently at rest under the given conditions. However, the location of such point depends on the frequencies of the two sources. With the comparison between Figs. 10.10 and 10.11, one may observe that the location of the points having peak relative displacement moves as the frequencies of the sources vary from both 15 (Fig. 10.10) to one 5 and another 25 (Fig. 10.11).

As previously described, the relative displacements can be quantified at any specified time for any given point in the considered domain by using the methodology established in this chapter. In general, the relative displacements of the porous half-space medium along the line connecting the two resources also form a wave at any specified time, as shown in Fig. 10.12.

For any selected point in the semi-infinite domain, the relative displacement history of the point can be determined with the solutions from the model developed. Figure 10.13 shows the relative displacement of a selected point located at $x = 750$, $y = 0$ in meters. The distance between the two sources, located at $x = 0$, $y = 0$ and $x = 1500$, $y = 0$, is 1,500 m. It illustrates the effects of the source frequencies on the relative displacements. For this specific case, the resulting displacement generated by the two circular-cylindrical sources with identical frequency appears as a periodic motion. One may also find that the displacement generated by the superposed two waves is evenly distributed when the frequencies of the two sources are the same.

To compare the displacements of single and two energy sources, a point along middle line in between the two sources is considered as shown in Fig. 10.14, in which the point M is taken at the middle point of the line connecting the two

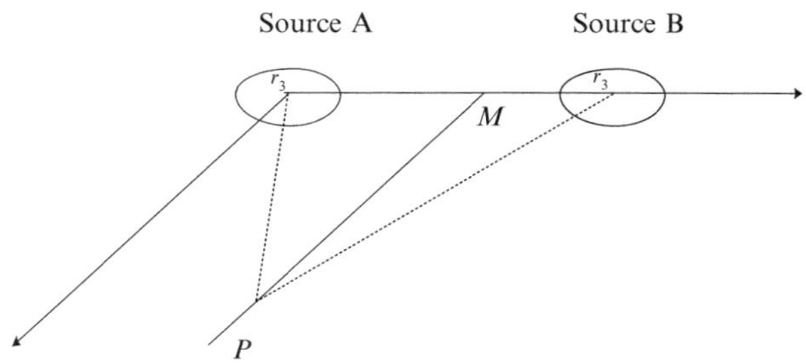

Fig. 10.14 The M-P line in space for comparing the effects of single and two sources

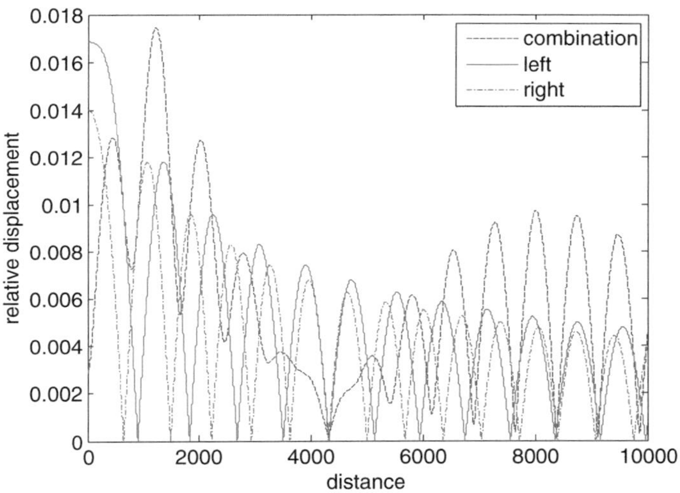

Fig. 10.15 Comparison between the single and multiple energy sources with identical frequencies in terms of relative displacement at a given time, $\omega_1 = \omega_2 = 15$

sources. As shown in Fig. 10.14, a source is located at (0, 0) and the other is located at (1,500, 0). Corresponding to a specified time, the relative displacement of the point P along the M-P line in Fig. 10.14 is plotted in Fig. 10.15, which compares the superposed actions of the two sources with that of a single source, and the frequencies are $\omega_1 = \omega_2 = 15$ with identical amplitude. The horizontal axis of Fig. 10.15 indicates the distance from point M along the M-P line.

Figure 10.15 shows that the relative displacement forms a sinusoidal wave as expected and reduces its magnitudes with the increase of the distance from point M. Generally, the relative displacements generated by the two sources are larger when compared with that of the single source especially at the location with distance from the energy sources. Significantly, the relative displacement of superposed waves of two sources with identical frequency (frequencies $\omega_1 = \omega_2 = 15$) is stabilized beyond a certain location along the line M-P. This implies that a stable relative

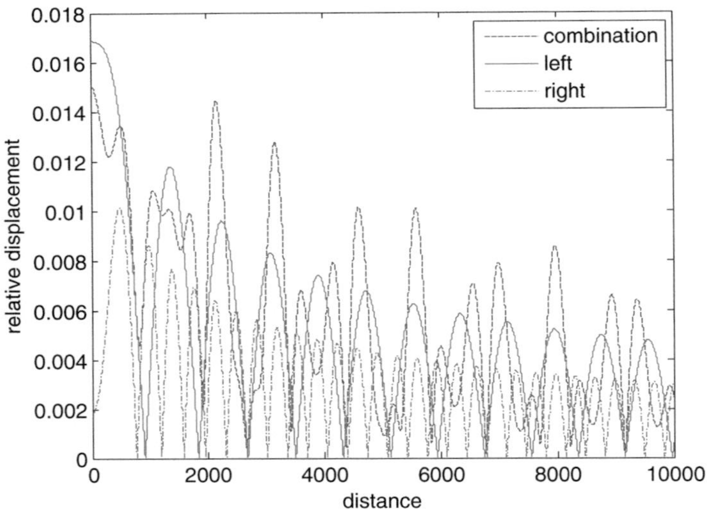

Fig. 10.16 Comparison between the single and multiple energy sources with two different frequencies in terms of relative displacement at a given time, $\omega_1 = 5,\ \omega_2 = 25$

displacement field, with larger amplitude when compared with that of single source, can be established with multiple energy sources of identical frequency.

Figure 10.16 also illustrates the instant distribution of relative displacements of single and two energy sources along the line M-P at a given time, whereas the frequencies in this case are $\omega_1 = 5$ and $\omega_2 = 25$, respectively. As can be observed from the figure, the relative displacements generated by the two sources are larger when compared with that of the single source.

Maximum relative displacements in the wave field are also important in evaluating the mobilization of liquid in porous media. The maximum relative displacements can be determined by employing the model established over a time range long enough after motion of the particle considered becomes stable. For most cases in the simulations, the time range used is from 0 to 50 s. The maximum relative displacements corresponding to the previous two cases of Figs. 10.11 and 10.12 are plotted in Figs. 10.13 and 10.14. As can be seen from Figs. 10.13 and 10.14, the maximum relative displacements produced by both the single and double sources decrease rapidly as the distance between points M and P increases. This time it decreases all the time without periodic fluctuation. However, excited by the waves of different frequencies, the maximum relative displacement of the double sources is generally larger than that of the single source, in the time range considered. In some areas, the maximum relative displacements of the combined waves are smaller than that of single course when the frequency of the source wave is the same.

As can be anticipated, the difference between the two maximum relative displacements will vanish at the position infinite from point M. In practice, however, the larger maximum relative displacement of double sources is significant as the

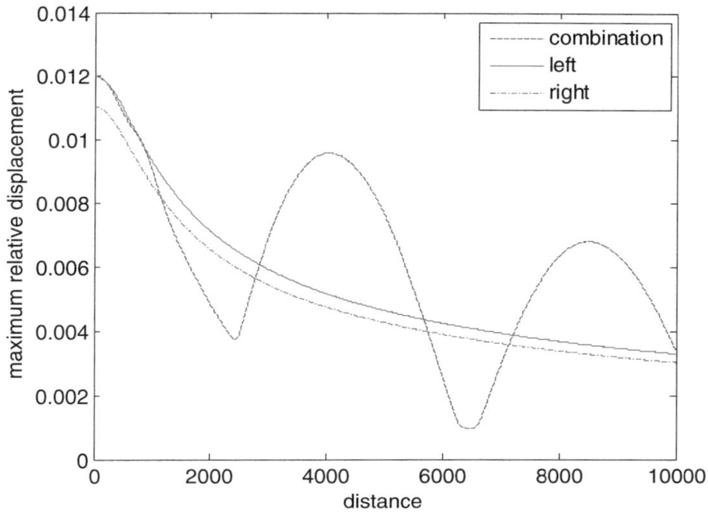

Fig. 10.17 Comparison between the single and multiple energy sources with identical frequency in terms of maximum relative displacement in the time range, $\omega_1 = \omega_2 = 15$

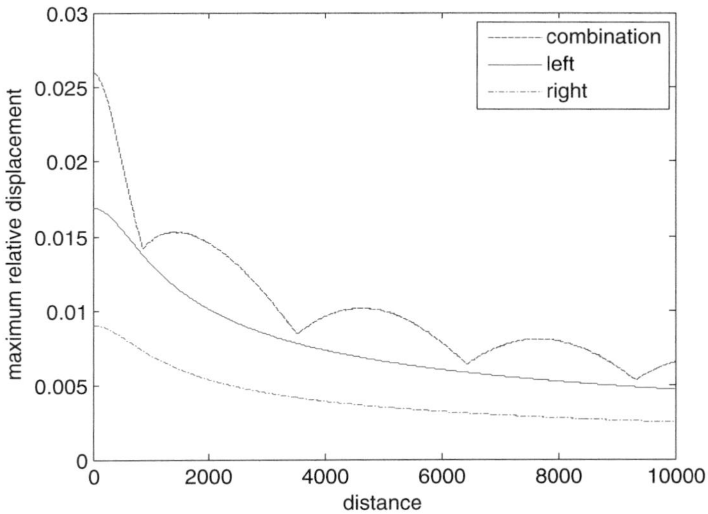

Fig. 10.18 Comparison between the single and multiple energy sources with different frequencies in terms of maximum relative displacement in the time range, $\omega_1 = 5$, $\omega_2 = 25$

field considered can only be finite. One may also learn from Fig. 10.17 that if improper locations, amplitudes, and frequencies of the sources are selected, the opposite results may appear; it is to say the maximum relative displacement of combined waves can be smaller than the case of single source (Fig. 10.18).

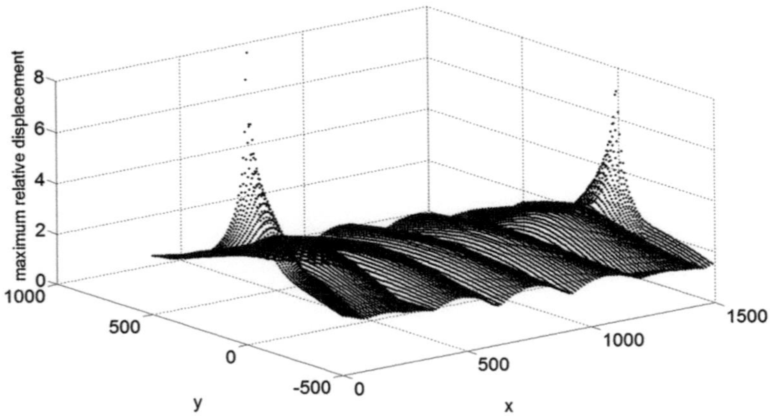

Fig. 10.19 Maximum relative displacement excited by two sources with different frequencies in a space domain

As can be concluded from the descriptions above, the relative displacements and maximum relative displacements of any point in the domain of porous half-space medium considered can be quantified by the model developed in the precious section, for at any time or time range desired. This implies that a displacement field or maximum relative displacement field in the domain subjected to multiple circular-cylindrical energy sources can be determined for a specified time or a time span desired. As such, the wave propagations and superposed action of the waves in the porous half-space medium consisting of fluid and solid can be quantified. Figures 10.19 and 10.16 illustrate the 3D wave shape of the relative displacement field in an x-y plane for a desired time interval. The frequencies of the two sources in Fig. 10.19 are $\omega_1 = 5$ and $\omega_2 = 25$, respectively. The frequencies of the two sources in Fig. 10.20 are $\omega_1 = \omega_2 = 15$. For the two cases, one wave source is located at $x = 0,\ y = 0$, while the other one is located at $x = 1500,\ y = 0$. The vertical axis is the maximum values of the nondimensional relative displacement with respect to different source frequencies. The diagrams in the two figures provide an excellent visualization of the 3D wave shapes of the maximum relative displacements and a graphical identification of the maximum relative displacement distribution over the domain considered. The shape of the waveform and the values of peak and valley maximum relative displacements are highly relying on the characteristics of the energy sources. This has application significance in mobilizing liquid in porous media under vibration or seismic stimulations.

The relative displacement response over a range of frequencies of different sources may also be examined with the model established. Figure 10.21 shows the relative displacement responding to the variations of the frequencies of the two sources at a specified time ($t = 35\,s$). Larger relative displacement occurs at the boundaries of the diagram. Over a time range from 0 to 50 s, Fig. 10.22 exhibits the maximum relative displacement corresponding to the frequency variations of the two energy sources. For the maximum relative displacement, the large values

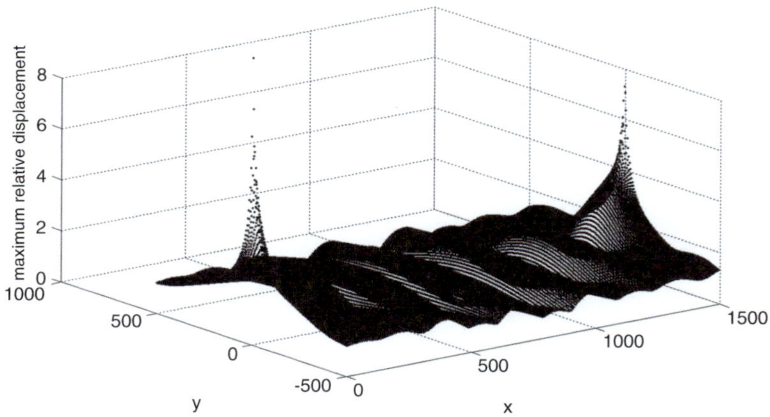

Fig. 10.20 Maximum relative displacement excited by two sources with different frequencies in a space domain

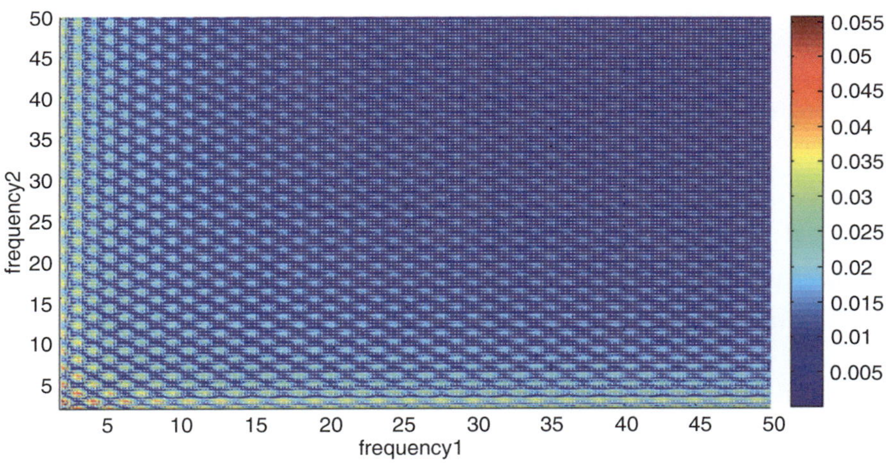

Fig. 10.21 Maximum relative displacement detected with different frequency range at a specified time

also happen at the boundary areas. One may therefore conclude, for obtaining larger relative displacement or maximum relative displacement, both or one of the frequencies should be small.

From the discussions above, one may conclude that the methodology and the model established make the availability of the evaluation of the dynamic response of the porous medium subjected to the excitations of multi-energy sources. The behavior of any specified point in the domain can be quantified, and the relative displacement and maximum relative displacement between the fluid and solid of the medium can be conveniently determined by making use of the established model.

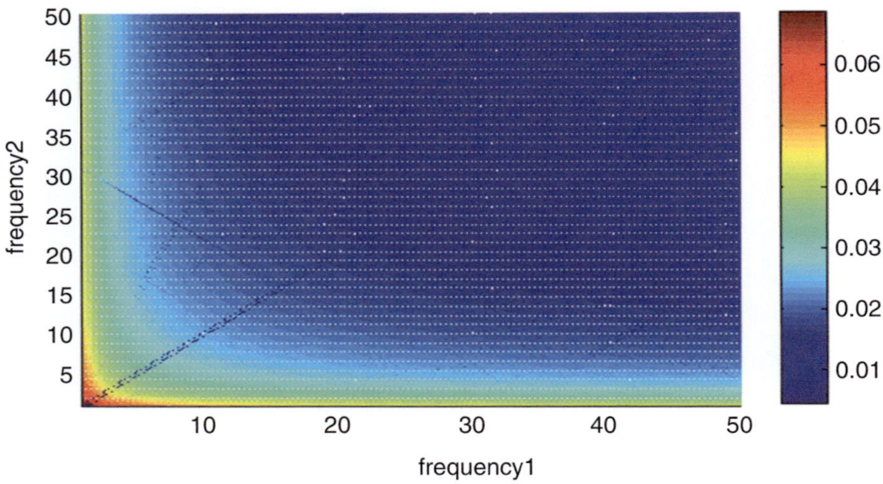

Fig. 10.22 Maximum relative displacements detected with different frequency range within a time span

It should be noticed, however, the mechanical and physical parameters of the porous half-space medium are not restrictive, and thus, the established model can be applied to different porous half-space media as desired.

10.5 Wave Field in Porous Half-Space Media Saturated with Newtonian Viscous Fluid

As indicated previously, the waves considered in the previous sections are propagating in the porous media saturated with zero viscous fluid. However, in the real world, viscosity exists in the fluid filled in the pores of the porous media and cannot be ignored in most cases in analyzing the waves propagating in the media. Considering the viscosity of the fluid in the porous media, energy dissipation and attenuation of the waves must be taken into consideration.

10.5.1 Development of Governing Wave Equations

In Biot's most representative papers (Biot 1956a, b), the equations of wave propagation in the low frequency range were presented (Smeulders 1992) with the assumption that the fluid in the porous media is compressible and may flow relative to the solid skeleton, which causes friction due to the viscosity of the fluid. Therefore, with the existence of solid and viscous fluid, the stresses in the porous media must be taken into consideration. Generally, the stresses acting on a porous medium can be separated into two parts: One is on the solid frame which can be written as σ_{ij}^{s}, and the other is on the fluid represented by $s_{ij} = -\phi p \delta_{ij}$, where ϕ is the porosity of the medium, p is the fluid pressure, δ_{ij} is Kronecker symbol, and the negative sign existing in the expression is for the association of directions between the fluid pressure and stress. Thus, the total stresses can be expressed by $\sigma_{ij} = \sigma_{ij}^{s} + s_{ij}$. With the stresses such expressed, viscosity of the fluid is involved. By utilizing the stress expressions for a porous medium and employing the force equilibrium relation, the governing wave equations of a porous medium saturated with viscous fluid can be derived in the following form:

$$\begin{cases} N\nabla^2 \mathbf{u} + \nabla\left[(A+N)\,e + Q\varepsilon\right] = \frac{\partial^2}{\partial t^2}\left(\rho_{11}\mathbf{u} + \rho_{12}\mathbf{U}\right) + b\frac{\partial}{\partial t}\left(\mathbf{u} - \mathbf{U}\right) \\ \nabla\left[Qe + R\varepsilon\right] = \frac{\partial^2}{\partial t^2}\left(\rho_{12}\mathbf{u} + \rho_{22}\mathbf{U}\right) - b\frac{\partial}{\partial t}\left(\mathbf{u} - \mathbf{U}\right) \end{cases} \tag{10.66a, b}$$

The coefficient b in the above equation is related to Darcy's coefficient of permeability k by

$$b = \frac{\beta\phi^2}{k} \tag{10.67}$$

where β in the equation is the fluid viscosity and ϕ is the porosity of the medium.

In (10.66), **u** and **U** are the displacement vectors of the fluid and solid, respectively. They consist of the quantities and directions of the displacements. e and ε in the equation are the volume strains of the solid and fluid, respectively, and can be expressed by

$$e = \nabla \cdot \mathbf{u}, \quad \varepsilon = \nabla \cdot \mathbf{U} \tag{10.68}$$

In (10.66), ρ_{11}, ρ_{12}, and ρ_{22} are density terms which can be expressed as

$$\rho_{11} = (1-\phi)\,\rho_s, \quad \rho_{22} = \phi\rho_f, \rho_{12} = -(\alpha-1)\,\phi\rho_f \tag{10.69}$$

in which ρ_s is the mass density of the solid grains, ρ_f is the mass density of the fluid in pores, $\alpha = (1/2)[\phi^{-1} + 1]$, and ϕ is the porosity of the medium. A, N, Q, and R in (10.66) are the physical parameters of the medium as defined in Sect. 10.4.

Biot proved mathematically that there would possibly be three waves existing in a porous medium saturated with fluid and presented the expressions of the waves in the form of the volume strain (Biot 1956a, b). However, it is not convenient to quantify the displacements from the volume strains of the fluid and the solid, especially when a two- or three-dimensional domain is considered. As indicated in Sect. 10.2, by applying Helmholtz decomposition to the displacement vectors for the solid and fluid, two sets of equations can be obtained corresponding to scalar potentials and vector potentials of the fluid and solid. Also, by substituting (10.21) into (10.66) and rearranging the terms according to the scalar and vector potentials, as Lin et al. (2001) conducted in their paper, the expressions for P- and S-waves in this case can be given as follows:

For a P-wave:

$$\begin{cases} \nabla^2\left(P\varphi_s + Q\varphi_f\right) = \frac{\partial^2}{\partial t^2}\left(\rho_{11}\varphi_s + \rho_{12}\varphi_f\right) + b\frac{\partial}{\partial t}\left(\varphi_s - \varphi_f\right) \\ \nabla^2\left[Q\varphi_s + R\varphi_f\right] = \frac{\partial^2}{\partial t^2}\left(\rho_{12}\varphi_s + \rho_{22}\varphi_f\right) - b\frac{\partial}{\partial t}\left(\varphi_s - \varphi_f\right) \end{cases} \tag{10.70a, b}$$

For an S-wave:

$$\begin{cases} N\nabla^2\boldsymbol{\psi}_s = \frac{\partial^2}{\partial t^2}\left(\rho_{11}\boldsymbol{\psi}_s + \rho_{12}\boldsymbol{\psi}_f\right) + b\frac{\partial}{\partial t}\left(\boldsymbol{\psi}_s - \boldsymbol{\psi}_f\right) \\ 0 = \frac{\partial^2}{\partial t^2}\left(\rho_{12}\boldsymbol{\psi}_s + \rho_{22}\boldsymbol{\psi}_f\right) - b\frac{\partial}{\partial t}\left(\boldsymbol{\psi}_s - \boldsymbol{\psi}_f\right) \end{cases} \tag{10.71a, b}$$

in which $P = A + 2N$ is introduced as a variable. Equations (10.70) and (10.71) are the governing equations of the waves propagating in porous media in terms of displacement potentials. These make it available to study the compression and shear waves separately or jointly for analyzing waves propagating in porous media.

As in the case of purely elastic waves, the body waves can be separated into uncoupled rotational and dilatational waves. The governing equations for a P-wave can be expressed in the form of displacements by applying the divergence operation to (10.70), and the equations for dilatational waves can be obtained in the following form:

$$\begin{cases} \nabla\left[\nabla^2\left(P\varphi_s + Q\varphi_f\right)\right] = \nabla\left[\frac{\partial^2}{\partial t^2}\left(\rho_{11}\varphi_s + \rho_{12}\varphi_f\right)\right] + \nabla\left[b\frac{\partial}{\partial t}\left(\varphi_s - \varphi_f\right)\right] \\ \nabla\left[\nabla^2\left(Q\varphi_s + R\varphi_f\right)\right] = \nabla\left[\frac{\partial^2}{\partial t^2}\left(\rho_{12}\varphi_s + \rho_{22}\varphi_f\right)\right] - \nabla\left[b\frac{\partial}{\partial t}\left(\varphi_s - \varphi_f\right)\right] \end{cases} \tag{10.72a, b}$$

By using the properties of scalar and vector, the governing equations of (10.57) for the dilatational waves can be written in the form of displacements as

$$\begin{cases} \nabla^2\left(P\mathbf{u}_{sp} + Q\mathbf{U}_{fp}\right) = \frac{\partial^2}{\partial t^2}\left(\rho_{11}\mathbf{u}_{sp} + \rho_{12}\mathbf{U}_{fp}\right) + b\frac{\partial}{\partial t}\left(\mathbf{u}_{sp} - \mathbf{U}_{fp}\right) \\ \nabla^2\left[Q\mathbf{u}_{sp} + R\mathbf{U}_{fp}\right] = \frac{\partial^2}{\partial t^2}\left(\rho_{12}\mathbf{u}_{sp} + \rho_{22}\mathbf{U}_{fp}\right) - b\frac{\partial}{\partial t}\left(\mathbf{u}_{sp} - \mathbf{U}_{fp}\right) \end{cases} \tag{10.73a, b}$$

in which the subscript s represents the displacement of the solid, f represents the displacement of the fluid, and p represents the displacement due to the P-wave. The parameters of the material in (10.73), P, Q, and R, are the same parameters as described in Sect. 10.2. Equation (10.73) is the governing equations for a P-wave propagating in the porous medium. It should be noted that the wave equations are all written in terms of displacements of the solid and fluid. The governing equations in terms of displacement for an S-wave can also be obtained by applying the curl operator to (10.73).

10.5.2 Wave Propagation and Displacement Field Model with Viscosity

With the development of the governing equations for the waves in porous half-space media saturated with Newtonian viscous fluid, a model describes the displacements of the fluid with viscosity and solid of a porous medium; therefore, the displacement field can be established. In developing for such a model, all the assumptions and conditions made in the previous section are valid. The displacements and the waves propagating in the porous medium are also assumed to be generated by vertical circular-cylindrical energy sources. In other words, the waves considered are cylindrical compression waves vertically distributed, and the wave sources are assumed to be uniform along the cylindrical axis in the 3D domain.

To describe the displacements of the solid and fluid with isotropic cylindrical coordinates, employ the operators:

$$\begin{cases} \nabla = \frac{\partial}{\partial r} + \frac{1}{r} \\ \nabla^2 = \frac{\partial^2}{\partial r^2} + \frac{1}{r}\frac{\partial}{\partial r} \end{cases} \tag{10.74a, b}$$

Substituting (10.74) into (10.73), the equations for dilatational waves in a specific x-y plane perpendicular to the vertical cylindrical energy source can be expressed with cylindrical coordinates in the following form:

$$\begin{cases} \left(\frac{d^2}{dr^2} + \frac{1}{r}\frac{d}{dr}\right)(\sigma_{11}\mathbf{u} + \sigma_{12}\mathbf{U}) = \frac{1}{V_c^2}\frac{\partial^2}{\partial t^2}(\gamma_{11}\mathbf{u} + \gamma_{12}\mathbf{U}) + \frac{b}{H}\frac{\partial}{\partial t}(\mathbf{u} - \mathbf{U}) \\ \left(\frac{d^2}{dr^2} + \frac{1}{r}\frac{d}{dr}\right)(\sigma_{12}\mathbf{u} + \sigma_{22}\mathbf{U}) = \frac{1}{V_c^2}\frac{\partial^2}{\partial t^2}(\gamma_{12}\mathbf{u} + \gamma_{22}\mathbf{U}) - \frac{b}{H}\frac{\partial}{\partial t}(\mathbf{u} - \mathbf{U}) \end{cases} \tag{10.75a, b}$$

Notice that the stresses in the equation are in the form of $\sigma_{ij} = \sigma_{ij}^s + s_{ij}$.

For the sake of convenience of the derivation process, the following parameters are used as the same as that defined in the previous section:

$$V_c^2 = H/\rho \tag{10.76}$$

$$\begin{cases} \sigma_{11} = \frac{P}{H}, \quad \sigma_{12} = \frac{Q}{H}, \quad \sigma_{22} = \frac{R}{H} \\ \gamma_{11} = \frac{\rho_{11}}{\rho}, \quad \gamma_{12} = \frac{\rho_{12}}{\rho}, \quad \gamma_{22} = \frac{\rho_{22}}{\rho} \end{cases} \tag{10.77}$$

in which

$$H = P + R + 2Q, \qquad \rho = \rho_{11} + \rho_{22} + 2\rho_{12} \tag{10.78}$$

According to the Sommerfeld Radiation Condition (Pao and Mow 1971), similar as that of Sect. 10.2, the cylindrical waves propagating from a circular-cylindrical source can be assumed:

$$\begin{cases} \mathbf{u} = C_1 H_0^{(1)}(lr) \ \exp(-i\omega t) \\ \mathbf{U} = C_2 H_0^{(1)}(lr) \ \exp(-i\omega t) \end{cases} \tag{10.79a, b}$$

It should be noted that the wave expression is now in the form of displacements of the fluid and solid, in comparison with the volume strains given by Biot (1956a).

By substituting (10.79) into (10.75), the following equations can be obtained:

$$\begin{cases} -l^2\left(\sigma_{11}C_1 + \sigma_{12}C_2\right) = -\frac{1}{V_c^2}\omega^2\left(\gamma_{11}C_1 + \gamma_{12}C_2\right) - \frac{i\omega b}{H}\frac{\partial}{\partial t}\left(C_1 - C_2\right) \\ -l^2\left(\sigma_{12}C_1 + \sigma_{22}C_2\right) = -\frac{1}{V_c^2}\omega^2\left(\gamma_{12}C_1 + \gamma_{22}C_2\right) + \frac{i\omega b}{H}\frac{\partial}{\partial t}\left(C_1 - C_2\right) \end{cases} \tag{10.80a, b}$$

The general equation for the velocities of these waves can be expressed as

$$V = \omega / l, \tag{10.81}$$

and making use of the parameter:

$$\xi = V_c^2 / V^2 \tag{10.82}$$

Equation (10.64) can be rewritten as

$$\begin{cases} \xi\left(\sigma_{11}C_1 + \sigma_{12}C_2\right) = \gamma_{11}C_1 + \gamma_{12}C_2 + \frac{ibV_c^2}{H\omega}\frac{\partial}{\partial t}\left(C_1 - C_2\right) \\ \xi\left(\sigma_{12}C_1 + \sigma_{22}C_2\right) = \gamma_{12}C_1 + \gamma_{22}C_2 - \frac{ibV_c^2}{H\omega}\frac{\partial}{\partial t}\left(C_1 - C_2\right) \end{cases} \tag{10.83a, b}$$

Substitution of (10.79) into (10.66) and elimination of the constants C_1 and C_2 yield the relation

$$\begin{aligned} \left(PR - Q^2\right)\frac{l^4}{\omega^4} - \left(P\rho_{11} + P\rho_{22} - 2Q\rho_{12}\right)\frac{l^2}{\omega^2} + \rho_{11}\rho_{22} \\ - \rho_{12}^2 + \frac{ib}{\omega}\left[\left(P + R + 2Q\right)\frac{l^2}{\omega^2} - \rho\right] = 0 \end{aligned} \tag{10.84}$$

With the variables already introduced, substitution of (10.79) into (10.83) and elimination of the constants C_1 and C_2 yield a nondimensional equation with one single variable ζ,such that

$$\begin{aligned} \left(\sigma_{11}\sigma_{12} - \sigma_{12}^2\right)\zeta^2 - \left(\sigma_{11}\gamma_{12} + \sigma_{22}\gamma_{11} - 2\sigma_{12}\gamma_{12}\right)\zeta \\ + \left(\gamma_{11}\gamma_{22} - \gamma_{12}^2\right) + \frac{ib}{\omega\rho}\left(\zeta - 1\right) = 0 \end{aligned} \tag{10.85}$$

where

$$\zeta = \frac{l^2}{\omega^2} V_c^2 \tag{10.86}$$

In this case l and ζ are complex variables. $V_c = H/\rho$ is the reference velocity. Denote ζ_I and ζ_{II} as the roots of (10.85), which correspond to the velocities of the purely elastic waves as given by (10.66). Assume ζ_I as the root corresponding to the first compression wave, while ζ_{II} corresponds to the second wave. As such, ζ_I and ζ_{II} are expressible in the following complex form:

$$\begin{aligned}(\zeta_I)^{\frac{1}{2}} &= R_I + iT_I \\ (\zeta_{II})^{\frac{1}{2}} &= R_{II} + iT_{II}\end{aligned} \tag{10.87}$$

The phase velocities of the compression waves can be given by equations

$$\begin{aligned}v_I/V_c &= 1/|R_I| \\ v_{II}/V_c &= 1/|R_{II}|\end{aligned} \tag{10.88}$$

Two complex roots can be obtained by solving the quadratic equations of (10.85), corresponding to the velocities. The image parts of (10.87) reflect the attenuation, while the real parts designate the propagation velocities of the waves. It may need to note that the velocities are the phase speeds and not the speed of the particle vibration. The ratio of the image part to the real part is an important factor since it describes the degree of damping of the wave.

As shown in Fig. 10.7, the multiple circular-cylindrical sources are considered to continuously generate harmonic cylindrical waves. Furthermore, in numerically simulating the waves, only the steady state of the wave motions is considered.

Although fluid viscosity is considered in this section, the format of expressions for the displacements of fluid and solid of porous media is identical to that given in (10.79) and those derived in (10.46) in Sect. 10.3, though the wave velocities are in different forms. Therefore, the model to be used for quantifying the displacements and displacement field under the excitations of multiple wave sources will be the same as that developed in Sect. 10.4, such that

$$\begin{cases}\mathbf{u} = \sum_{j=1}^{n} \mathbf{u}_j \\ \mathbf{U} = \sum_{j=1}^{n} \mathbf{U}_j\end{cases} \tag{10.89a, b}$$

as the expression shown in (10.64).

It should be noticed, however, fluid viscosity can be taken into account only if the previous equations described for the model in this section are implemented, for evaluating the displacements and displacement filed. In implementing the equations of the model, material properties including the viscosities for both solid and fluid of the porous medium considered and the characteristics of the waves and wave sources must first be determined.

Table 10.3 Properties of the porous medium with viscous liquid

ϕ	μ	v_s	μ_s	K_f
0.246	5cp	0.29	10.0 GPa	2.4 GPa
ρ_s	ρ_f	C_1/C_2	β	k
2,700 kg/m^3	1,000 kg/m^3	1.308	3.0×10^{-3} Pa s	2.0×10^{-11} M2

10.5.3 Effects of Viscosity on Wave Dispersion in Porous Half-Space Under Multiple Energy Sources

To demonstrate the effects of viscosity on the wave propagation and displacement field, numerical simulations can be performed with employment of the wave model and displacement equations developed in the above section. One may also gain the methodology of applying the established model through numerical simulations.

Consider a numerical simulation for the wave propagations and displacement field in a porous medium containing a single-phase fluid with viscosity. All the basic parameters to be used in the simulation, such as A, P, Q, and R, are taken as the same as that used in the former model without viscosity. Once the physical parameters are given, the coefficient values of the waves can be determined by the established wave model. As an example, the following table gives the parameter values determined and to be used in the numerical computation (Table 10.3).

As per the descriptions in the previous sections and Biot's theory (Biot 1956a, b), the wave velocity can be calculated by standard elastic theory from the composite density and bulkmodulus of the solid and liquid. The composite properties needed for the calculations can be determined from the porosity and the elastic properties of the fluid and the solid framework. In determining for different frequencies of wave propagations, the frequency, permeability of the solid, and viscosity of the fluid of the porous medium considered need to be determined as per the descriptions in Sect. 10.5.1.

The phase velocity of the wave propagating in a porous half-space medium can be firstly considered. Figure 10.23 shows the wave propagation velocity versus the frequency of the wave, corresponding to various viscosities of the fluid saturated in the pores of the porous medium. Figure 10.24 demonstrates the variations of the wave propagation velocity corresponding to different values of permeability of the porous medium with varying frequencies of the wave. The reference velocity from the wave source in these two figures is 3,277 m/s. In all the cases, the velocity of the wave rises with an increase in frequency. When viscosity or permeability is small, the phase velocity increases, for both the figures, very quickly until the curve reaches a plateau, and then the phase velocity increases its value gradually as the frequency increases. As the viscosity or permeability becomes larger, the plateau becomes less noticeable even though the velocities are increasing with the increasing frequencies, as can be seen from Figs. 10.19 and 10.20.

A comparison is made between the results considering fluid with very small viscosity and the results considering fluid with larger viscosity. The effects of

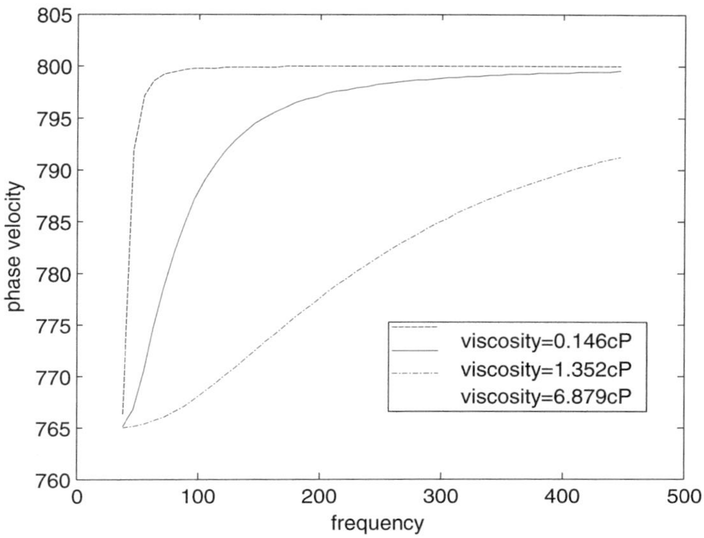

Fig. 10.23 The relationship between phase velocity and wave frequency corresponding to different viscosities

viscosity to the phase velocity are significant. As can be seen from Fig. 10.23, the larger is the viscosity of the fluid, the greater will be the velocity corresponding to the same frequency. In other words, viscosity of the fluid reduces the phase velocity of the wave propagating in porous media. Also, as can be seen from Fig. 10.23, the rate of increasing phase velocity with respect to the increasing frequency decreases as viscosity of the fluid increases.

The similar phenomenon can be observed from Fig. 10.24 where various values of permeability are considered; the higher is the permeability of the porous half-space medium, the larger will be the phase velocity. An asymptotic characteristic can also be observed from Fig. 10.24, especially when the permeability of the porous medium considered is small. The increasing rate of the phase velocity also drops as the permeability increases.

Figure 10.25 shows a case of the maximum nondimensional relative displacement amplitudes along the line connecting the two circular-cylindrical sources, which generate the waves. Again, three cases of single left or right wave source and the combination of two wave sources are considered. Same as in previous section, the nondimensional relative displacement is defined by $(U-u)/u0$. It should be noted that $U/u0$ and $u/u0$ are the nondimensional displacements of the fluid and solid, respectively. Although the amplitudes of the two sources are identical for the case of two wave sources, in actual numerical simulation, the amplitudes of the sources can be set differently if so desired.

The two wave sources for the numerical simulations shown in Fig. 10.21 are located at $x=0$, $y=0$ and $x=1{,}500$, $y=0$, respectively. In general, the wave

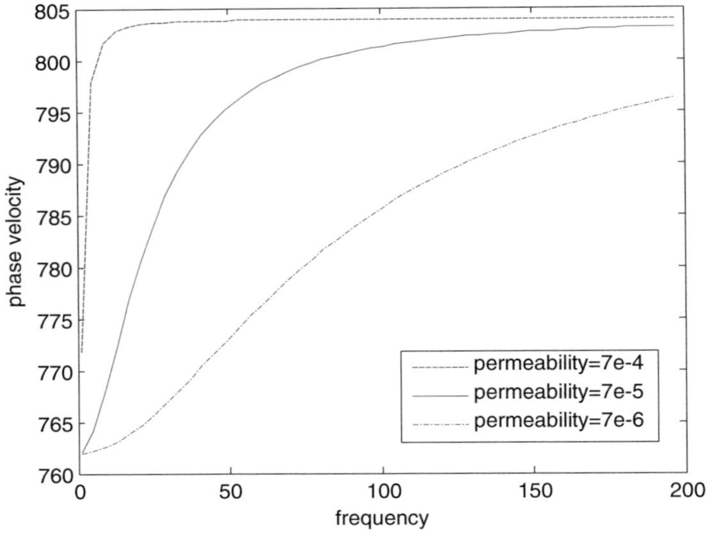

Fig. 10.24 The relationship between phase velocity and wave frequency corresponding to different permeability

amplitudes and the relative displacements of the porous medium decrease with increasing distance from the circular-cylindrical energy sources. This can be seen from Fig. 10.25 for the cases that only one energy source is considered. It should be noticed, however, the amplitude of the combined wave in a steady state is not a simple summation of the amplitudes of two wave sources.

In Fig. 10.25 the amplitude of maximum relative displacement due to the combined wave is smaller than that of the single source in certain areas, while for some other areas, the amplitude is larger than that of the single source. Also, the maximum relative displacement of a single wave source is monotonically decreasing from the location of the source, and the curve of combined waves is in a waveform. It is important to note that the maximum relative displacement of combined waves is generally larger than the maximum relative displacement of either one of the maximum relative displacement generated by the left or right wave source, except in a small region at the middle of the line connecting the two sources. For the mobilization of fluid in the pores of a porous medium, larger relative displacements between the fluid and solid pore walls are preferred. In particular, it will be much beneficial in the EOR practice with vibration and wave stimulations if the relative displacement can be made larger in a controllable manner at a localized position. Two or multiple energy or wave sources therefore provide great advantage over single energy source in promoting liquid mobilization.

It should be noted that the curves shown are the maximum relative displacement. This is why the curves of the maximum relative displacements are monotonic smooth lines for the case of a left or right single wave source. However, the

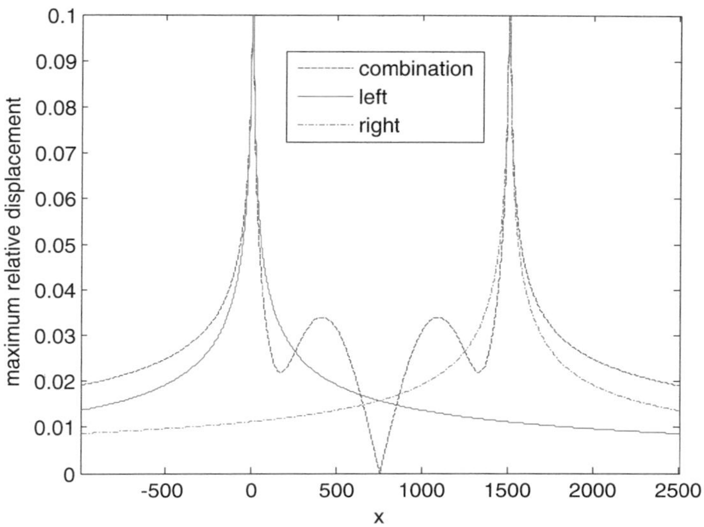

Fig. 10.25 The maximum relative displacements along the connecting line of the two wave sources. The frequency and wave velocity are identical in the case of combination of two wave sources

combined amplitude of two joint waves at a given moment is actually the combination of the two relative displacements caused by the combined waves from the two wave sources. This generates the wave curve for the case of combined two waves. One may also find from the figure that the amplitude of the wave can be zero at the middle of the line connecting the two wave sources and the curves are symmetric as shown in Fig. 10.25. It is also seen that the zero displacement of the combined waves is right located at the position where the two curves of single wave source intersect, along the x-axis. The reason for this symmetrical shape is because all the reference parameters of the two source waves are identical. Specifically, the frequencies of the two sources are identical (5 Hz), and the wave velocities of the two sources are also the same (3,277 m/s).

Obviously, the shape of the maximum relative displacement curve depends on the characteristics of the two wave sources. For demonstrating the effects of different wave sources to the relative displacement of a porous medium, Fig. 10.26 is plotted by changing the wave velocity of right source to 4,577 m/s while keeping the other one as the same as that of Fig. 10.25.

As can be seen from Fig. 10.26, the shapes of the maximum relative displacement curves are changed from that of identical wave sources as shown in Fig. 10.21; the intersection of two single waves is pushed into the right side of the domain due to the high velocity of the right source. The maximum relative displacement curve of combined waves becomes unsymmetrical, and the bottom point of zero relative displacement moves to the left side. Again, the zero displacement of the combined

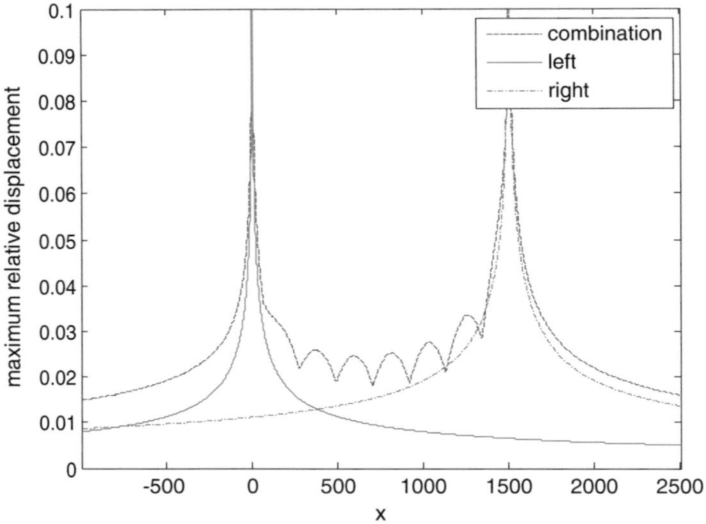

Fig. 10.26 The maximum relative displacements along the connecting line of the two wave sources. The wave velocities are different in the case of combination of two wave sources

waves in this case is located at the position where the two curves of single wave source intersect, along the x-axis. It is evident that changing the properties of the energy sources may significantly change the characteristics of the maximum relative displacements due to the combined waves.

The effects of different wave sources to the relative displacement between fluid and solid of a porous medium are also shown in Fig. 10.27, where the velocities of the two wave sources are kept unchanged as 3,277 m/s as that of Fig. 10.25, while the frequency is set with different values of 5 and 15 Hz for the right and left sources, respectively.

In Fig. 10.27, the intersection of two single waves is pushed into the right side of the domain due to the high frequency of the left source. The maximum relative displacement curves are not symmetric for all the three cases. In this case, almost no maximum relative displacement of combined waves is lower than either one of the maximum relative displacements generated by a single wave source. The variation of the curve of combined waves is still sinusoidal cut; the variation is much smaller than that of Figs. 10.25 and 10.26. It should be noticed, however, the sinusoidal curve in Fig. 10.27 is the combination of the maximum relative displacements from the two wave sources. Its properties such as amplitude, frequency, and wave velocity are in general not the same as that of the wave sources. Also, there is no zero maximum relative displacement for the curve of combined waves, and the lowest maximum relative displacement of combined waves is not at the location of the intersection of the two curves of single wave source.

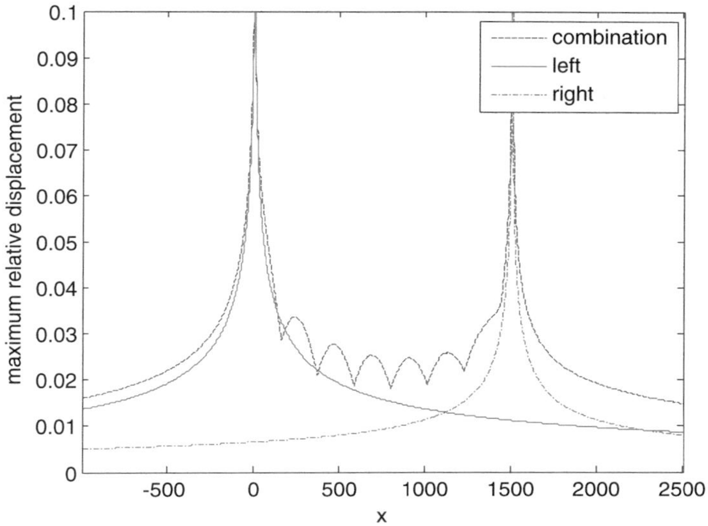

Fig. 10.27 The maximum relative displacements along the connecting line of the two wave sources with different frequencies and velocities

Indeed, the variation of the maximum relative displacement of combined waves can be made even smaller, and the maximum relative displacement of combined wave sources can be made greater than any of the maximum relative displacement of single wave source outside of the wave sources, only through simply controlling the properties of the wave sources. Such a case is shown in Fig. 10.28 in which both the frequencies and velocities of the two simultaneous wave sources are different. For the curves, the frequency of the left source is 15 Hz, whereas the right source is 5 Hz, and the wave velocity for the left source is 2,277 and 4,577 m/s for the right source. It may also be observed from Fig. 10.28 that the intersection point of the two curves of the single wave sources is located almost at the middle point along the connection line of the two sources although the frequencies and the velocities of the sources are not the same; this is due to the fact that high frequency of the left source pushes the intersection to the right and in the meantime the high velocity of the right source pushes the intersection to the left. However, the lowest point of the combined wave curve is apart from the intersection point. Also, the curve of combined waves is nearly symmetrical though neither the frequency nor the velocity of the two sources is the same.

The discussions above are all on the maximum relative displacements along the connection line between the two wave sources, and the maximum relative displacements are from single and two wave sources.

Let us now consider the vibrations of the particles outside of the line connecting the two wave sources in the porous medium, with focus on the characteristics of the maximum relative displacements of combined waves with respect to the distance

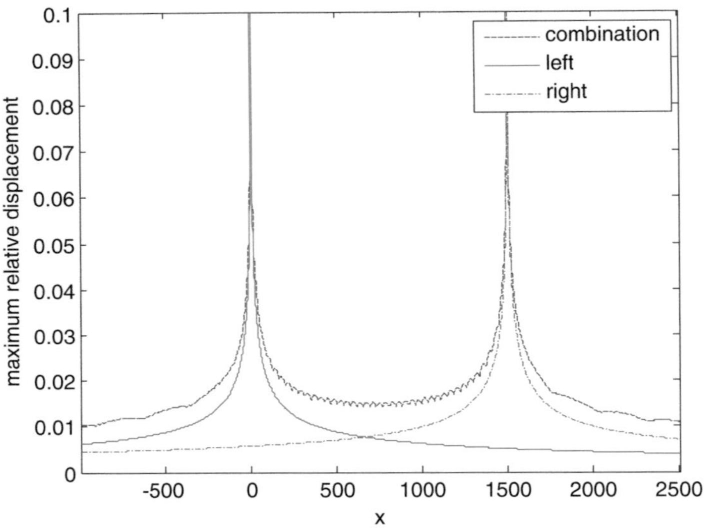

Fig. 10.28 The maximum relative displacements along the connecting line of the two wave sources of different frequency and velocity

from a particle considered to both the wave sources. Figure 10.29 illustrates the maximum relative displacement of a point considered as a porous unit along the M-P line (see point P in Fig. 10.14) at a given moment of $t = 35$ s, where M is the middle point of the line connecting the two wave sources which are located at $x = 0,\ y = 0$ and $x = 1500,\ y = 0$, respectively. For this figure, the wave velocity of the left source is 3,277 m/s, and the velocity of the right source is 4,577 m/s, whereas the frequencies of the two sources are 5 Hz for the left source and 10 Hz for the right source. The horizontal axis indicates the distance between the points P and M. Notice that the point P is along the line perpendicular to the line connecting the two sources, and its instant distance from either one of the sources is the same along the line M-P.

As can be seen from Fig. 10.29, the relative displacements of point P are in a sinusoidal form for all the three cases of single and combined wave sources. The amplitudes of the relative displacements are decreasing with the increase of the distances from the sources, for all the three cases. Also, the relative displacement of combined sources is generally larger when compared with that of the single sources except in some areas; the relative displacement of combined waves is smaller than that of one or both of the single wave sources. This indicates that two wave source arrangements may generate a localized larger relative displacement when compared with that of a single wave source.

Shifting the frequency values for the two wave sources by taking 10 Hz for the left source and 5 Hz for the right source, while maintaining all the other parameter as the same, a similar diagram is plotted in Fig. 10.30. Again, the amplitude of the

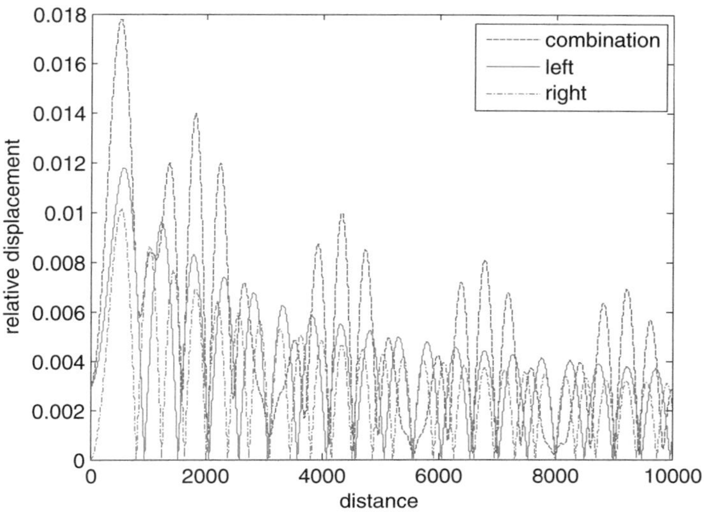

Fig. 10.29 The relative displacements along the M-P line at a given time

relative displacements is decreasing with the increase of the distances as expected, for all the three cases. The relative displacement of combined waves is greater than that of single wave cases all the way along the M-P line. However, in this case, the relative displacement of combined waves is much smooth. This is due to the arrangement of the wave velocities at the two wave sources. Once again, the two-wave-source case shows great advantages over the single-wave-source cases. It also implies that the desired wave and relative displacement can be reached by adjusting either or both the wave frequency and velocity of the wave sources.

As discussed previously, it is the maximum relative displacement that is more significant in mobilizing the fluid in the pores of a porous medium. In terms of maximum relative displacements, Fig. 10.31 shows a case in which the frequency for both the left and right wave sources is identical while the other parameters are maintained as unchanged from that used for Figs. 10.29 and 10.30. The time interval considered for the maximum relative displacements is between 0 and 50 s.

It can be seen from Fig. 10.31 that the maximum relative displacements produced by both the single and double sources decrease rapidly as the distance between points M and P increases in the whole time range considered. In this case, the maximum relative displacements decrease all the time without periodic fluctuation, whereas that of the combined waves shows sinusoidal variation along the M-P line. This results in the smaller maximum relative displacements of the combined waves in some areas. Figure 10.32 shows a similar case as that of Fig. 10.31 except that the frequencies at the two wave sources are different, namely, 10 Hz for the right and 5 Hz for the left sources. In Fig. 10.32, all the curves of the maximum relative displacement are decreased with the increase of the distance. However,

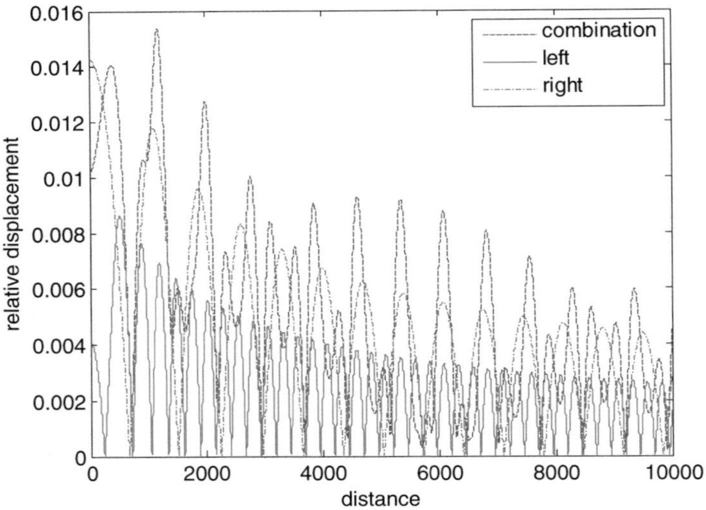

Fig. 10.30 The relative displacements along the M-P line at a given time (shifted frequencies)

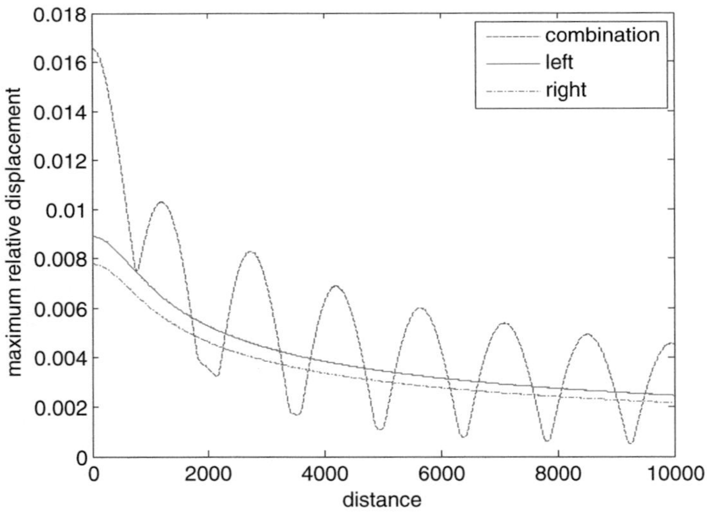

Fig. 10.31 Comparison between the single and multiple wave sources with identical frequencies (15 Hz) in terms of maximum relative displacement

the maximum relative displacement of the combined waves is all the way larger than that of the single wave sources along the M-P line, and the variation of the maximum relative displacement amplitude is much smaller, when compared with that of Fig. 10.31. This implies that a desired or preferred wave and displacement

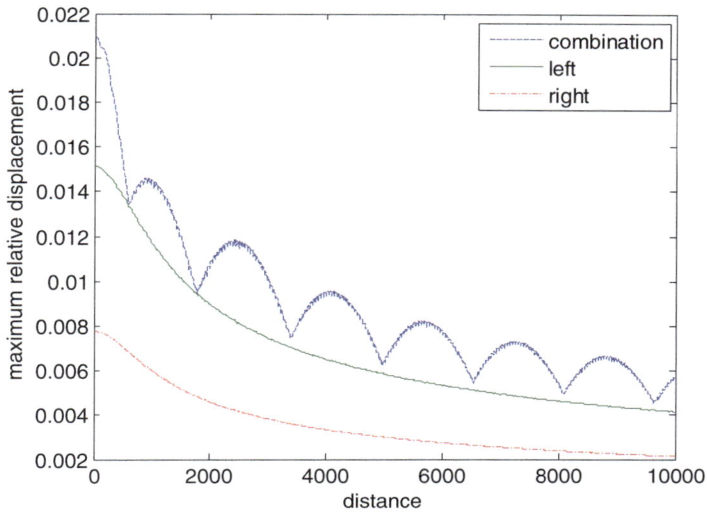

Fig. 10.32 Comparison between the single and multiple wave sources with different frequencies (10 Hz for right and 5 Hz for left sources) in terms of maximum relative displacement

situation can be reached for a porous medium if the properties of the wave sources can be properly adjusted.

Theoretically, the difference between the two maximum relative displacements of single and two or multiple wave sources will vanish at the position infinite from point M. This is not difficult to understand as the distance between the two wave sources is finite and very small when compared with infinity. In practice of vibration and seismic wave stimulations, however, the field considered to be affected by the wave sources is limited in size. A larger maximum relative displacement of two or multiple wave sources is practically significant, which is caused by that the different frequencies move the curves upward and apart from each other. More importantly, the maximum relative displacements in the field can be controlled with two or multiple wave sources. However, this is not the case if only a single wave source is considered.

Let us now consider the effects of the distance between the two wave sources on wave motion and displacements of a porous half-space medium. Consider the maximum relative displacement of a fixed point at $x = 300\ m,\ y = 100m$ with respect to the coordinate system set previously. The porous unit at this point can be excited by the waves from either or both of the left and right circular-cylindrical wave sources. The point (porous unit) considered is still at a fixed point (300, 100) in space which implies a constant distance between the point and the left wave source, whereas the distance of the point to the right wave source can be varied. As such the effect of the distance between the two wave sources on the motion of the porous unit can be examined.

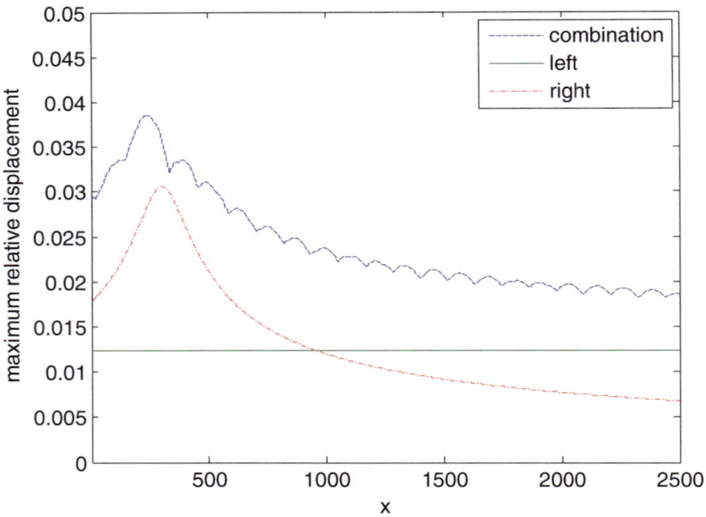

Fig. 10.33 Maximum relative displacements versus location of the right source under different frequencies of the wave sources

Figure 10.33 shows the maximum relative displacements of the porous unit. The horizontal axis represents the distance between the two wave sources, of which the wave velocity of the left wave source is 3,277 m/s and the right has the velocity of 4,577 m/s. The frequency of the left wave source is 5 Hz, whereas the right wave source's frequency is 10 Hz. As can be seen from the figure, the maximum relative displacement due to the single wave source at left is a constant as the distance between the point and the source is unchanged. The maximum relative displacement is getting smaller as the distance between the two wave sources therefore the distance between the point and the right source increases, as the effect of the right wave source reduces with increase of the distance. The combined wave sources always create larger maximum relative displacement when compared with that of a single source, though the maximum relative displacement reduces as the distance increases and they vary periodically with an increase in distance between the right source and the point considered.

Keeping an identical frequency for both the left and right wave sources with a larger value of 15 Hz while all the other parameters remaining as the same as that of Figs. 10.33 and 10.34 shows a large periodic variation of the maximum relative displacement for the combined wave sources, and the displacement is smaller than that of a single wave source in some areas, similar as that of Fig. 10.27. As can be seen from Figs. 10.31, 10.32, 10.33, and 10.34, different and smaller frequencies of the wave sources may generate larger maximum relative displacements of less variability, which is favorable for porous medium fluid mobilization in practice.

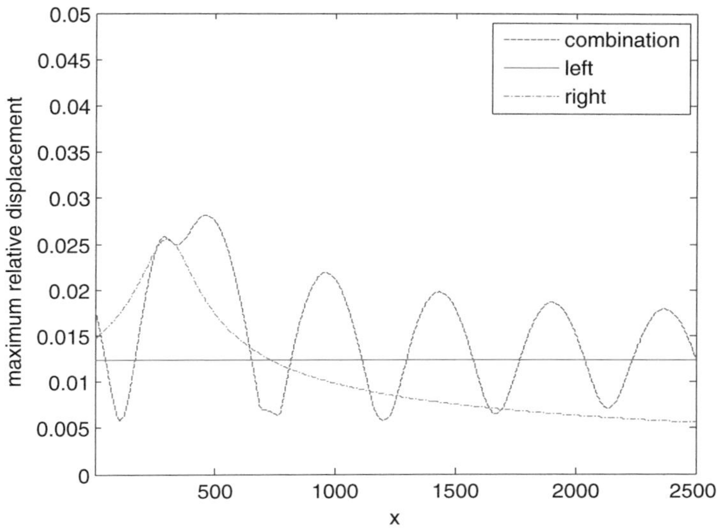

Fig. 10.34 Maximum relative displacements versus location of the right wave source of identical frequency as of the left source

In Fig. 10.35, the left wave frequency is fixed as 1 Hz, while the right wave source varies from 1 to 200 Hz. The effect of the single left wave source on the maximum relative displacement is a constant as the frequency of the source is fixed, and the effect of the single right wave source decreases rapidly from 1 Hz and then stabilized as the frequency increases. When two wave sources are applied, the corresponding maximum relative displacement also decreases quickly and then stabilized as it is affected by the right wave source. However, the maximum relative displacement of combined wave sources is greater than either of the single sources.

Figure 10.35 also shows that the maximum relative displacement of a single wave source will be low when its frequency is high, and the maximum relative displacement of combined wave sources will be relatively low when one of the frequencies of the two sources is high. This implies that the maximum relative displacement will be even lower if both the wave sources are with higher frequencies. Figure 10.36 shows a case where the left wave source is fixed with a higher frequency value of 200 Hz whereas the right wave source is varying from 1 to 200 Hz. As can be seen from Fig. 10.36, the maximum relative displacement of the fixed wave source is much lower than that of Fig. 10.35, where the fixed frequency is 1 Hz. The maximum relative displacement of the combined waves stabilizes at lower values from the initiated value around 0.025, which is approximately the lowest value of the corresponding maximum relative displacement in Fig. 10.35. Figures 10.35 and 10.36 indicate that the frequencies of the wave sources should be lower if higher maximum relative displacements in a porous medium are desired.

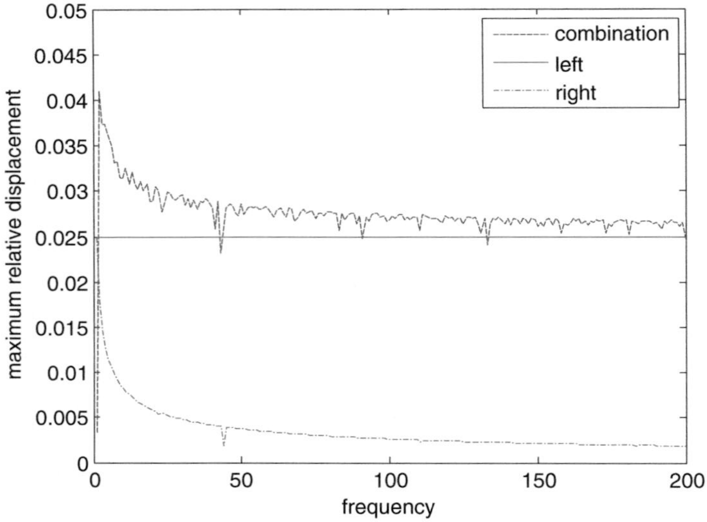

Fig. 10.35 Maximum relative displacements versus frequency of the right wave source (frequency of the left source is 1 Hz)

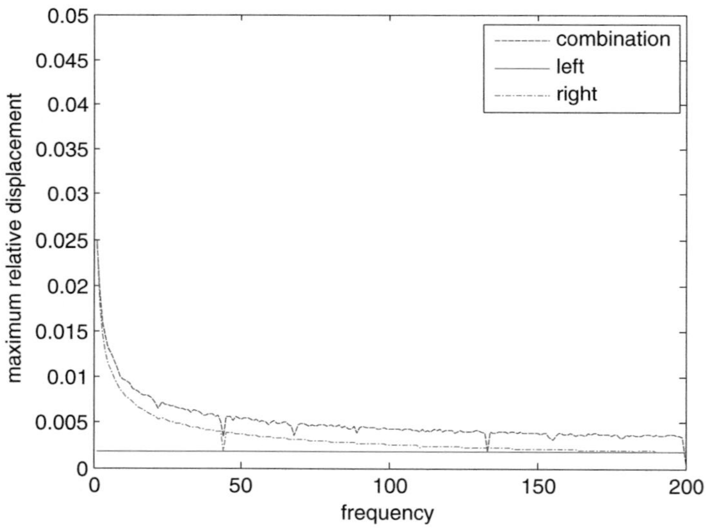

Fig. 10.36 Maximum relative displacements versus frequency of the right wave source (frequency of the left source is 200 Hz)

The relative displacement response is also detected in a large frequency range for the two given resources. Figure 10.37 shows the response within a time range as 0 to 50 s. For the maximum relative displacement, the large value also happens at

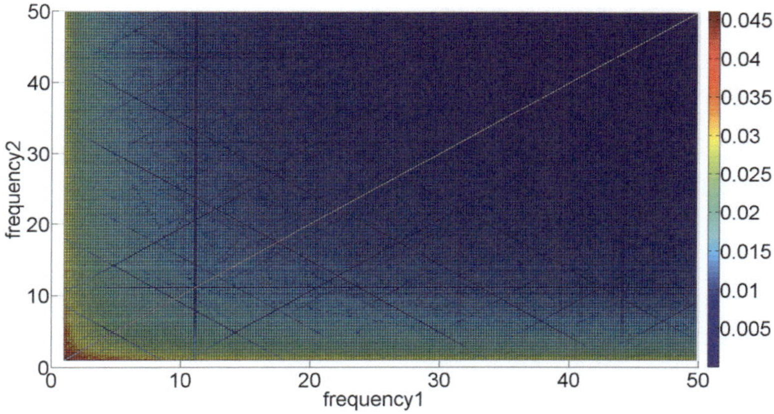

Fig. 10.37 Maximum relative displacements detected under different frequencies at the two wave sources

the boundary areas. This is to say the ideal case happens when one wave source is with very small value, and the frequency of the other source should not be the same as the former one (as the diagonal blank line indicates). Compare to Fig. 10.21 for the wave model without viscosity; this time the bottom lines become more regular appealed as shown on those blue straight lines.

Figure 10.38 shows the instant response at $t = 35$ s. For the maximum relative displacement, the large value also happens at the boundary areas. Compare to Fig. 10.22 for the wave model without viscosity; it exhibits more diverse behavior as observed.

10.6 Wave Field of a Porous Half-Space Medium Saturated with Two Immiscible Fluids Under the Excitations of Multiple Wave Sources

In this section, the wave field of a porous half-space medium saturated by two immiscible fluids is investigated with excitations from several compressible waves in the low frequency range. The cylindrical coordinate system is used to describe multiple circular-cylindrical energy sources so the propagating waves can be expressed with utilization of the Hankel function and the model of superposition of multiple waves is feasible to set up. The combined effect from the circular-cylindrical sources on the wave field is investigated and compared with a single-phase fluid. The relationships between the relative displacements, the porosity of the porous medium, and the saturations of the fluids are also addressed. Numerical simulations are provided to demonstrate the implementation of the model developed in this chapter. The results contribute to the comprehension of fluid and solid interaction under the excitations of waves, such as seismic, electromagnetic, and other artificial vibrations.

10.6.1 Volume Averaging Method

The macroscopic equation can be developed from the microscopic equations (Slattery 1972) with the application of the volume averaging method. For the averaging volume V, in a multiphase porous medium, the volume average of a field quantity of phase ζ, B_ζ, which can be a scalar, vector, or second-order tensor, is

$$\langle B_\zeta \rangle = \frac{1}{V}\int_{V_\zeta} B_\zeta dV \tag{10.90}$$

in which V_ζ is the volume saturated by phase ζ, representing the fluid or solid.

The theorems of the volume averaging method for a gradient, divergence, and time derivative are (Slattery 1972)

$$\langle \nabla B_\zeta \rangle = \nabla \langle B_\zeta \rangle + \frac{1}{V}\int_{S_w} \mathbf{n}_{w\zeta} B_\zeta ds \tag{10.91}$$

$$\langle \nabla \cdot B_\zeta \rangle = \nabla \cdot \langle B_\zeta \rangle + \frac{1}{V_\zeta}\int_{S_w} \mathbf{n}_{w\zeta} \cdot B_\zeta ds \tag{10.92}$$

$$\left\langle \frac{\partial B_\zeta}{\partial t} \right\rangle = \frac{\partial \langle B_\zeta \rangle}{\partial t} - \frac{1}{V}\int_{S_w} \mathbf{n}_{w\zeta} \cdot B_\zeta dA \tag{10.93}$$

where $\mathbf{n}_{w\zeta}$ is the outward normal of the interface S_w.

10.6.2 Governing Equation Development

Tuncay and Corapcioglu (1996) gave the governing equations for compressible waves propagating in porous media saturated by two immiscible fluids:

$$\begin{aligned}
\langle \rho_s \rangle \frac{\partial^2 \varepsilon_s}{\partial t^2} &= a_{11}\nabla^2\varepsilon_s + a_{12}\nabla^2\varepsilon_1 + a_{13}\nabla^2\varepsilon_2 \\
&\quad + C_1\left(\frac{\partial \varepsilon_1}{\partial t} - \frac{\partial \varepsilon_s}{\partial t}\right) + C_2\left(\frac{\partial \varepsilon_2}{\partial t} - \frac{\partial \varepsilon_s}{\partial t}\right) \\
\langle \rho_1 \rangle \frac{\partial^2 \varepsilon_1}{\partial t^2} &= a_{21}\nabla^2\varepsilon_s + a_{22}\nabla^2\varepsilon_1 + a_{23}\nabla^2\varepsilon_2 - C_1\left(\frac{\partial \varepsilon_1}{\partial t} - \frac{\partial \varepsilon_s}{\partial t}\right) \\
\langle \rho_2 \rangle \frac{\partial^2 \varepsilon_2}{\partial t^2} &= a_{31}\nabla^2\varepsilon_s + a_{32}\nabla^2\varepsilon_1 + a_{33}\nabla^2\varepsilon_2 - C_2\left(\frac{\partial \varepsilon_2}{\partial t} - \frac{\partial \varepsilon_s}{\partial t}\right)
\end{aligned} \tag{10.94}$$

where $\varepsilon_\zeta = \nabla \cdot \bar{\mathbf{u}}_\zeta$, $\langle \rho_s \rangle$, $\langle \rho_1 \rangle$, and $\langle \rho_2 \rangle$ are the volume averaging densities of the solid and fluids, which consist of phase one and phase two fluids, respectively. In the present research, it is assumed the phase one fluid is oil and phase two is water and the material parameters will be given in the following table. In (10.94), C_1 and C_2 allowing the consideration of relative saturation and relative permeability of the fluids are coefficients and can be expressed as the following (Tuncay and Corapcioglu 1996):

$$\begin{aligned} C_1 &= \tfrac{(1-\alpha_s)^2 S_1^2 \mu_1}{K k_{r1}} \\ C_2 &= \tfrac{(1-\alpha_s)^2 (1-S_1)^2 \mu_2}{K k_{r2}} \end{aligned} \tag{10.95}$$

in which α_s is the volume fraction of the solid phase, S_1 is the saturation of the fluid one oil, μ_1 and μ_2 are the viscosities, k_{r1} and k_{r2} are the relative permeability of fluids, and K is the intrinsic permeability of the medium.

The coefficients a_{ij}, in the above equations, are defined as (Tuncay and Corapcioglu 1996)

$$\begin{aligned} a_{11} &= K_s\left[A_1\alpha_s\left(K_1A_2S_1 + K_1K_2 + K_2A_2\left(1-S_1\right)\right) + \right. \\ &\left. K_sK_{fr}\left(1-\alpha_s\right)\left(K_1\left(1-S_1\right) + S_1K_2 + A_2\right)\right]/A_3 + 4G_{fr}/3 \\ a_{12} &= K_1K_sA_1S_1\left(1-\alpha_s\right)\left(A_2+K_2\right)/A_3 \\ a_{13} &= K_2K_sA_1\left(1-S_1\right)\left(1-\alpha_s\right)\left(A_2+K_1\right)/A_3 \end{aligned} \tag{10.96}$$

$$\begin{aligned} a_{22} &= \{K_1S_1^2\left(1-\alpha_s\right)[\,K_s^2\left(1-\alpha_s\right)\left(K_2 + A_2/S_1\right) \\ &\quad + K_2A_2A_1\left(1-S_1\right)/S_1\,]\}/A_3 \\ a_{23} &= -K_1K_2S_1\left(1-S_1\right)\left(1-\alpha_s\right)\left[-K_s^2(1-\alpha_s)+A_1A_2\right] \\ a_{33} &= \left\{K_2(1-S_1)^2\left(1-\alpha_s\right)[\,K_s^2\left(1-\alpha_s\right)\left(K_1 + A_2/\left(1-S_1\right)\right)\right. \\ &\quad + K_1A_1A_2S_1/\left(1-S_1\right)]\}/A_3 \\ a_{21} &= a_{12}, \qquad a_{31} = a_{13}, \qquad a_{32} = a_{23} \end{aligned} \tag{10.97}$$

$$\begin{aligned} A_1 &= \alpha_sK_s - K_{fr} \\ A_2 &= \tfrac{dP_{cap}}{dS_1}S_1\left(1-S_1\right) \\ A_3 &= A_1\left(K_1A_2S_1 + K_1K_2 + K_2A_2\left(1-S_1\right)\right) \\ &\quad + K_s^2\left(1-\alpha_s\right)\left(K_1(1-S_1)+A_2+K_2S_1\right) \end{aligned} \tag{10.98}$$

In (10.98), P_{cap} is the capillary pressure, which is equal to the pressure difference between the two fluids, and this quantity is assumed to be a function of saturation of the non-wetting fluid.

Though the commonly used governing equations for compressible waves are expressed in terms of volume strains, they can be transformed into the form in terms of displacements of the solid and fluid by employing the Helmholtz decomposition of the displacement vector as shown below:

$$\mathbf{u} = \text{grad}\,(\varphi) + \text{curl}\,(\boldsymbol{\psi}) \tag{10.99}$$

where φ and $\boldsymbol{\psi}$ are the scalar and vector potentials of the displacement vector and the condition $\nabla \cdot \boldsymbol{\psi} = 0$. For a P-wave, also known as a compressible wave, the displacement corresponds to the scalar potentials without rotation. This implies $\nabla \times \mathbf{u} = 0$. Applying the divergence operation to (10.94), the equations for dilatational waves in terms of displacements can be expressed in the following form:

$$\begin{aligned}
\langle \rho_s \rangle \frac{\partial^2 u_s}{\partial t^2} &= a_{11}\nabla^2 u_s + a_{12}\nabla^2 u_1 + a_{13}\nabla^2 u_2 \\
&\quad + C_1\left(\frac{\partial u_1}{\partial t} - \frac{\partial u_s}{\partial t}\right) + C_2\left(\frac{\partial u_2}{\partial t} - \frac{\partial u_s}{\partial t}\right) \\
\langle \rho_1 \rangle \frac{\partial^2 u_1}{\partial t^2} &= a_{21}\nabla^2 u_s + a_{22}\nabla^2 u_1 + a_{23}\nabla^2 u_2 - C_1\left(\frac{\partial u_1}{\partial t} - \frac{\partial u_s}{\partial t}\right) \\
\langle \rho_2 \rangle \frac{\partial^2 u_2}{\partial t^2} &= a_{31}\nabla^2 u_s + a_{32}\nabla^2 u_1 + a_{33}\nabla^2 u_2 - C_2\left(\frac{\partial u_2}{\partial t} - \frac{\partial u_s}{\partial t}\right)
\end{aligned} \tag{10.100}$$

The compressional plane harmonic waves propagating along the radial direction can be expressed as

$$\overline{u}_\zeta = H_\zeta H_0^{(1)}\,(\xi r)\ \exp\,(-i\omega t) \tag{10.101}$$

in which H_ζ is the wave amplitude, $H_0^{(1)}(\cdot)$ is the third kind of Bessel function of zeroth order, r is the radial coordinate, ω is the frequency of the wave, i is the imaginary number, t is the time, and $\xi = \xi_r + \xi_i$ is a complex wave number. For a dissipative wave, the imaginary part of ξ is larger than zero and is usually called the attenuation coefficient, while the real part of ξ is used to define the phase velocity of the wave by $c = \omega/\xi_r$. A series of matrix equations can be obtained by submitting (10.101) into governing (10.100):

$$\left[-\omega^2 \begin{bmatrix} \langle \rho_s \rangle & & \\ & \langle \rho_1 \rangle & \\ & & \langle \rho_2 \rangle \end{bmatrix} + \xi^2 \begin{bmatrix} a_{11} & a_{12} & a_{13} \\ a_{21} & a_{22} & a_{23} \\ a_{31} & a_{32} & a_{33} \end{bmatrix} + i\omega \begin{bmatrix} -C_1 - C_2 & C_1 & C_2 \\ C_1 & -C_1 & \\ C_2 & & -C_2 \end{bmatrix} \right] \begin{Bmatrix} H_s \\ H_1 \\ H_2 \end{Bmatrix} = \begin{Bmatrix} 0 \\ 0 \\ 0 \end{Bmatrix} \tag{10.102}$$

If there are nonzero solutions existing for H_ζ, the determinant of the coefficient matrix in the above equation must be equal to zero. This leads to a dispersion equation. With the employment of a new variable $X = \omega^2/\xi^2$, the dispersion equation can then be written as

$$Z_1 X^3 + Z_2 X^2 + Z_3 X + Z_4 = 0 \tag{10.103}$$

The coefficients in the equation above are expressed as (Tuncay and Corapcioglu 1996)

$$\begin{aligned} Z_1 &= \frac{C_1C_2(\langle\rho_s\rangle+\langle\rho_1\rangle+\langle\rho_2\rangle)-\langle\rho_s\rangle\langle\rho_1\rangle\langle\rho_2\rangle\omega^2}{\omega^2} \\ &\quad - i\,\frac{\langle\rho_1\rangle\langle\rho_2\rangle(C_1+C_2)+\langle\rho_s\rangle(C_1\langle\rho_2\rangle+C_2\langle\rho_1\rangle)}{\omega} \end{aligned} \tag{10.104}$$

$$\begin{aligned} Z_2 &= \frac{\left(a_{11}\langle\rho_1\rangle\langle\rho_2\rangle+a_{22}\langle\rho_s\rangle\langle\rho_2\rangle+a_{33}(\langle\rho_s\rangle\langle\rho_1\rangle)\omega^2\right.}{\omega^2} \\ &\quad - \frac{C_1C_2\sum_{j=1}^{3}\sum_{k=1}^{3}a_{jk}}{\omega^2} + i\,\frac{a_{22}(C_2(\langle\rho_s\rangle+\langle\rho_2\rangle)+C_1\langle\rho_2\rangle)}{\omega} \\ &\quad + i\,\frac{a_{11}(C_2\langle\rho_1\rangle+C_1\langle\rho_2\rangle)+2a_{12}C_1\langle\rho_2\rangle+2a_{13}C_2\langle\rho_1\rangle}{\omega} \\ &\quad + i\,\frac{a_{33}(C_2\langle\rho_1\rangle+C_1(\langle\rho_s\rangle+\langle\rho_1\rangle))}{\omega} \end{aligned} \tag{10.105}$$

$$\begin{aligned} Z_3 &= -a_{11}\left(a_{22}\langle\rho_2\rangle + a_{33}\langle\rho_1\rangle\right) + a_{12}^2\langle\rho_2\rangle + a_{13}^2\langle\rho_1\rangle \\ &\quad - \langle\rho_s\rangle\left(a_{22}+a_{33}-a_{23}^2\right) - i\,\frac{C_1\left(a_{33}(a_{11}+2a_{12}+a_{22})-(a_{13}+a_{23})^2\right)}{\omega} \\ &\quad - i\,\frac{C_2\left(a_{22}(a_{11}+2a_{13}+a_{33})-(a_{12}+a_{23})^2\right)}{\omega} \end{aligned} \tag{10.106}$$

$$Z_4 = a_{11}\left(a_{22}a_{33}-a_{23}^2\right) - a_{12}^2a_{33} + a_{13}\left(2a_{12}a_{23}-a_{13}a_{22}\right) \tag{10.107}$$

Three complex roots can be obtained from (10.103) with the given coefficients. Therefore, there are six available complex roots for wave number ξ. However, only three roots among these six solutions are reasonable to be the wave numbers for the compressional wave in the porous media. Nevertheless, these three roots must have imaginary parts larger than zero.

The ratios between H_1 and H_s, and H_2 and H_s can be obtained by substituting (10.101) into (10.100) and expressed as follows:

$$\frac{H_1}{H_s} = \frac{\{(a_{13}a_{21}-a_{23}a_{11})\,\xi^4+i\omega\xi^2\,(C_1a_{13}+C_2a_{21}+a_{23}\,(C_1+C_2-i\omega\,\langle\rho_s\rangle))-C_1C_2\omega^2\}}{\{(a_{12}a_{23}-a_{13}a_{22})\,\xi^4+i\omega\xi^2\,(C_1a_{23}-C_2a_{22}+a_{13}\,(C_1-i\omega\,\langle\rho_1\rangle))+\omega^2C_2\,(i\omega\,\langle\rho_1\rangle-C_1)\}}$$

$$\frac{H_2}{H_s}=\frac{\{(a_{12}a_{13}-a_{11}a_{32})\,\xi^4+i\omega\xi^2\,(C_1a_{13}+C_2a_{21}+a_{23}\,(C_1+C_2-i\omega\,\langle\rho_s\rangle))-C_1C_2\omega^2\}}{\{(a_{31}a_{32}-a_{12}a_{33})\,\xi^4+i\omega\xi^2\,(C_2a_{32}-C_1a_{33}+a_{12}\,(C_2-i\omega\,\langle\rho_2\rangle))+\omega^2C_1\,(i\omega\,\langle\rho_1\rangle-C_2)\}} \tag{10.108}$$

The displacements of the solid grain and the fluid phases can be described as

$$\begin{aligned} DisS &= H_sH_0^{(1)}(\xi r)\ \exp(-i\omega t)\\ Dis1 &= H_1H_0^{(1)}(\xi r)\ \exp(-i\omega t)\\ Dis2 &= H_2H_0^{(1)}(\xi r)\ \exp(-i\omega t)\end{aligned} \tag{10.109}$$

10.6.3 Multisource Model

As shown in Fig. 10.7, if the global coordinate system is located at one source center (the bottom left source in Fig. 10.7), the coordinates of other source locations, which, as well as the bottom left source, are on the plane at the depth $z=z_{depth}$ in the cylindrical coordinate system, can be expressed by a complex number $d_j=r_{j0}(\cos\theta_{j0}+i\sin\theta_{j0})$, where r_{j0} is the distance between the jth source and the origin of the global coordinate system and θ_{j0} is the azimuthal angle. All the waves considered in this section are supposed to be continuous and harmonic cylindrical waves generated by circular-cylindrical sources. Furthermore, only steady state is considered. The waves can be expressed in the following form:

$$\begin{cases} DisS = \mathrm{Re}\left[H_sH_0^{(1)}(\xi r)\exp\left(-i\omega t\right)\right](\cos\theta+i\sin\theta)\\ Dis1 = \mathrm{Re}\left[H_1H_0^{(1)}(\xi r)\exp\left(-i\omega t\right)\right](\cos\theta+i\sin\theta)\\ Dis2 = \mathrm{Re}\left[H_2H_0^{(1)}(\xi r)\exp\left(-i\omega t\right)\right](\cos\theta+i\sin\theta)\end{cases} \tag{10.110}$$

in which the term $(\cos\theta+i\sin\theta)$ is introduced to represent the direction of the displacement. This term can also be written as $[z/|z|]$. z will be given in the following part.

Thus, the waves propagating from each of the circular-cylindrical sources can be expressed by the following formulas:

$$\begin{cases} DisS_1 = \mathrm{Re}\left[D_1H_sH_0^{(1)}(\xi_1r_1)\exp\left(-i\omega_1t\right)\right]\left[\frac{z_1}{|z_1|}\right]\\ Dis1_1 = \mathrm{Re}\left[D_1H_1H_0^{(1)}(\xi_1r_1)\exp\left(-i\omega_1t\right)\right]\left[\frac{z_1}{|z_1|}\right]\\ Dis2_1 = \mathrm{Re}\left[D_1H_2H_0^{(1)}(\xi_1r_1)\exp\left(-i\omega_1t\right)\right]\left[\frac{z_1}{|z_1|}\right]\end{cases} \tag{10.111}$$

$$\cdots$$

$$\begin{cases} DisS_n = \mathrm{Re}\left[D_n H_s H_0^{(1)}\,(\xi_n r_n)\exp\left(-i\omega_n t\right)\right]\left[\frac{z_n}{|z_n|}\right] \\ Dis1_n = \mathrm{Re}\left[D_n H_1 H_0^{(1)}\,(\xi_n r_n)\exp\left(-i\omega_n t\right)\right]\left[\frac{z_n}{|z_n|}\right] \\ Dis2_n = \mathrm{Re}\left[D_n H_2 H_0^{(1)}\,(\xi_n r_n)\exp\left(-i\omega_n t\right)\right]\left[\frac{z_n}{|z_n|}\right] \end{cases} \tag{10.112}$$

D_j (j = 1, 2, ..., n) are the nondimensional relative displacement amplitudes of the jth source. $DisS_j$, $Dis1_j$, and $Dis2_j$ ($j = 1, 2, \ldots,$) are the displacement vectors of the solid, fluid one, and fluid two excited by the jth source. In (10.112), $r_j = |z_j|$ is the distance from a point P to the jth wave source. The term $[z_j/|z_j|]$ is introduced to describe the direction of the relative displacements.

The expression for each wave is written in a common coordinate system using the moving-coordinate method to conveniently investigate the superposed action of multiple waves.

Express wave j in the xyz-coordinate system as shown in Fig. 10.7 and let $z_j = z - d_j$. The above equation can be written as

$$\begin{cases} DisR1_j = \mathrm{Re}\left[D_j\,(H_1 - H_s)\,H_0^{(1)}\left(\xi_j\,|z - d_j|\right)\exp(-i\omega_j t)\right]\left[\frac{z-d_j}{|z-d_j|}\right] \\ DisR2_j = \mathrm{Re}\left[D_j\,(H_2 - H_s)\,H_0^{(1)}\left(\xi_j\,|z - d_j|\right)\exp(-i\omega_j t)\right]\left[\frac{z-d_j}{|z-d_j|}\right] \end{cases} \tag{10.113}$$

where d_j are the coordinates of the origin in the coordinate system located at the jth wave source. As such, d_j can be considered as the coordinate in a common coordinate system.

The displacements of any given point P in the domain can be described in a common coordinate system with the developed equations. The xyz-coordinates are considered as the common coordinates (also named global coordinates). This implies $d_1 = 0$. The combined relative displacements due to several waves can now be presented by

$$\begin{cases} DisR1 = \sum_{j=1}^{n} DisR1_j = \mathrm{Re}\left[D_1\,(H_1 - H_s)\,H_0^{(1)}\left(\xi_1\,|z|\right)\exp(-i\omega_1 t)\right]\left[\frac{z}{|z|}\right] \\ \qquad + \cdots + \mathrm{Re}\left[D_n\,(H_1 - H_s)\,H_0^{(1)}\left(\xi_n\,|z - d_n|\right)\exp(-i\omega_n t)\right]\left[\frac{z-d_n}{|z-d_n|}\right] \\ DisR2 = \sum_{j=1}^{n} DisR2_j = \mathrm{Re}\left[D_1\,(H_2 - H_s)\,H_0^{(1)}\left(\xi_1\,|z|\right)\exp(-i\omega_1 t)\right]\left[\frac{z}{|z|}\right] \\ \qquad + \cdots + \mathrm{Re}\left[D_n\,(H_2 - H_s)\,H_0^{(1)}\left(\xi_n\,|z - d_n|\right)\exp(-i\omega_n t)\right]\left[\frac{z-d_n}{|z-d_n|}\right] \end{cases} \tag{10.114}$$

Table 10.4 Material parameters used in the calculation

Parameters	Unit	Sym.	Value
Bulk modulus of solid	GPa	*Kfr*	1.02
Bulk modulus of solid grains	GPa	*Ks*	35.00
Shear modulus of solid matrix	GPa	*Gfr*	1.44
Density of solid grain	kg/m^3	ρs	2650.0
Intrinsic permeability	M2	*K*	2.0×10^{-11}
Volume fraction of solid phase		αs	0.90
Density of water	kg/m^3	$\rho 2$	997.0
Bulk modulus of water	GPa	$K2$	2.25
Viscosity of water	Pa s	$\mu 2$	1.0×10^{-3}
Density of oil	kg/m^3	$\rho 1$	762.0
Bulk modulus of oil	GPa	$K1$	0.57
Viscosity of oil	Pa s	$\mu 1$	3.0×10^{-3}

The displacement field generated from multiple circular-cylindrical sources can be quantified by using the model provided above. The characteristics of the wave field can be analyzed quantitatively once the parameters of the porous half-space medium, the sources, and the locations of the sources are specified (Table 10.4).

10.6.4 Numerical Analyses

The amplitudes of both waves for the solid grain are assumed to be H_0. The nondimensional relative displacements are defined as $(H_1 - H_s)/H_0$ and $(H_2 - H_s)/H_0$ for fluid phases one and two, respectively.

Figure 10.39 shows the maximum value of the nondimensional relative displacements between the fluid phases and solid grain. The symmetry can be seen along the line connecting the two circular-cylindrical sources when the frequencies and velocities of the two sources have the same value. One can also find the maximum displacement of the middle point is zero. This means this point does not vibrate at all. This is because the two waves, which have the same properties, propagate in opposite directions and kill each other. Thus, the combined effect is zero vibration. The difference between Figs. 10.39 and 10.40 is due to the difference in frequencies of the two source waves. Compared with the displacement values in Fig. 10.39, the relative displacements reduce with a decrease in frequency. The relative displacements will be increased if the corresponding relative saturation S1 is reduced. A comparison between Figs. 10.41 and 10.42 shows how the relative displacements will increase with the decreasing saturation of nonwettable fluid phase one (Fig. 10.43).

Figure 10.44 shows the maximum value for the nondimensional relative displacement of the points on the connecting line between the two wave sources. The frequencies of the two wave sources are $\omega_1 = 5$Hz and $\omega_2 = 15$Hz, respectively. We find that the value close to the right source is a little bit larger than the value

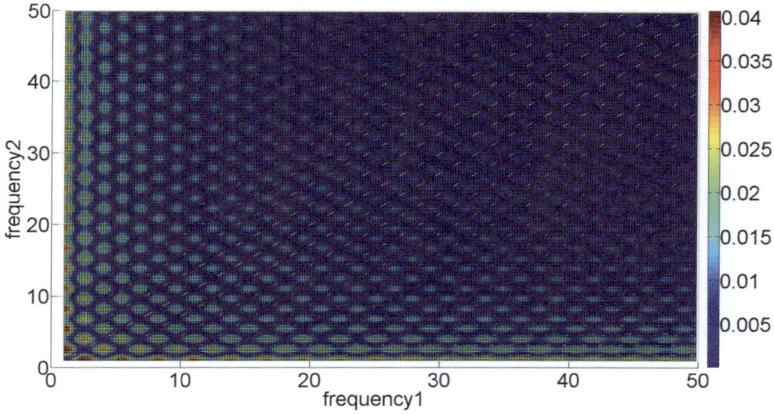

Fig. 10.38 Relative displacements detected corresponding to different frequencies at the two wave sources

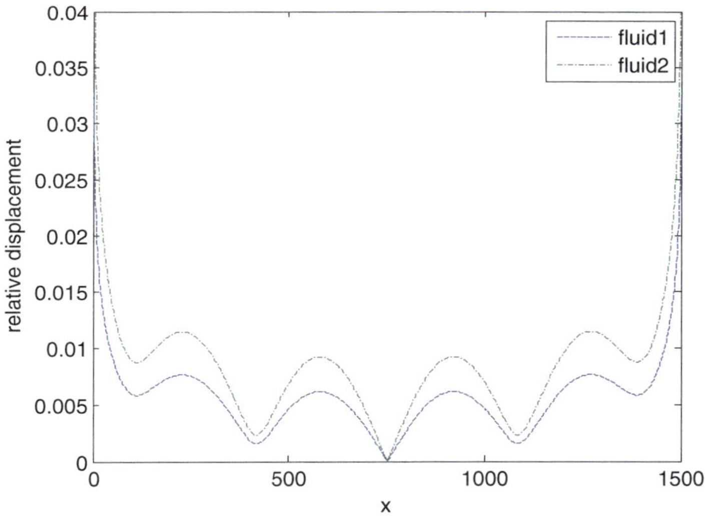

Fig. 10.39 The maximum relative displacements along the connecting line ($S1 = 0.43$, $\omega_1 = \omega_2 = 5\text{Hz}$)

close to the left source with the same distance from the sources. The source with lower frequency has a less effect on the relative displacement. The same case can be observed in Fig. 10.41 when saturation reduced to 0.24 and the frequencies of the two wave sources are $\omega_1 = 15\text{Hz}$ and $\omega_2 = 5\text{Hz}$.

Figures 10.45 and 10.46 show the maximum nondimensional relative displacement versus the position of the right source to 2,500 m at a location (300, 10) in the domain investigated. The effect of the right source will vanish with an increase in distance from the right source to the point under consideration.

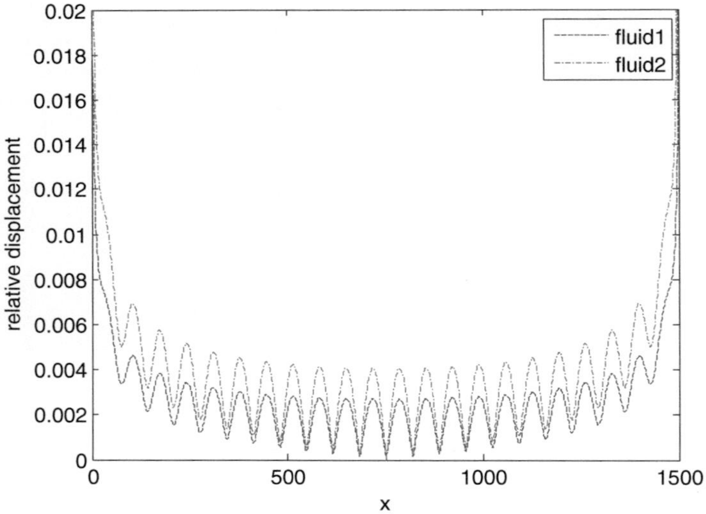

Fig. 10.40 The maximum relative displacements along the connecting line (S1 = 0.43, $\omega_1 = \omega_2 = 25$Hz)

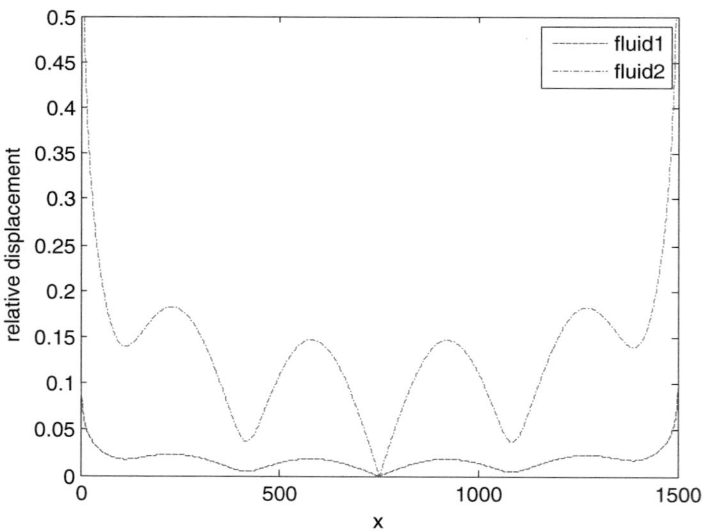

Fig. 10.41 The maximum relative displacements along the connecting line (S1 = 0.24, $\omega_1 = \omega_2 = 5$Hz)

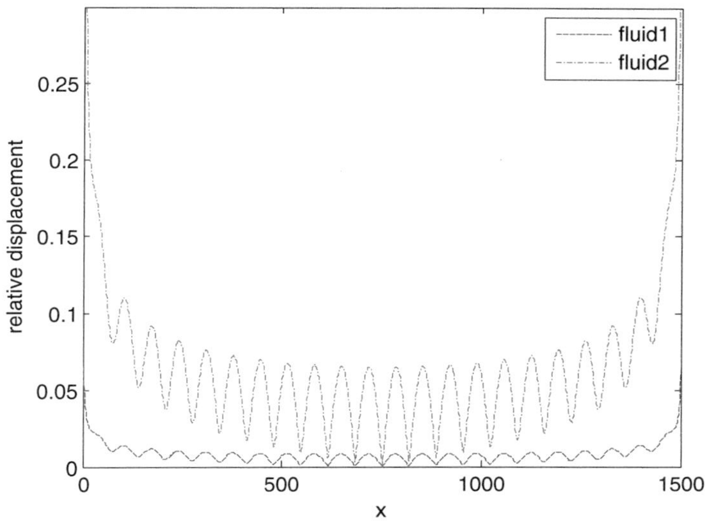

Fig. 10.42 The maximum relative displacements along the connecting line (S1 = 0.24, $\omega_1 = \omega_2 = 25\text{Hz}$)

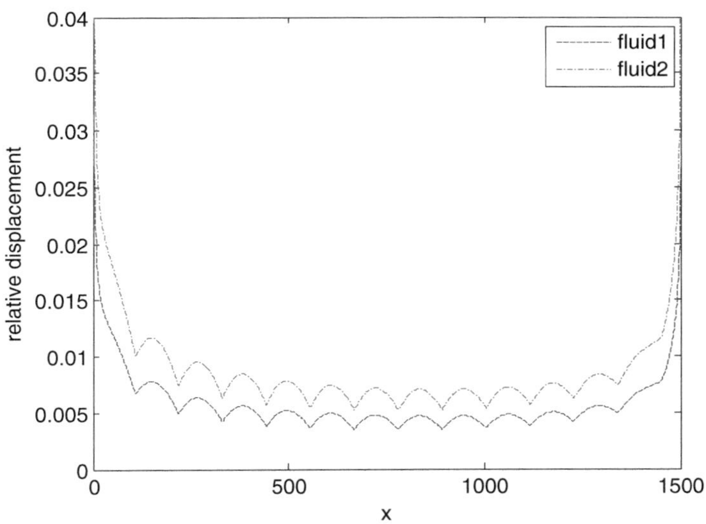

Fig. 10.43 The maximum relative displacements along the connecting line (S1 = 0.43, $\omega_1 = 5\text{Hz}, \omega_2 = 15\text{Hz}$)

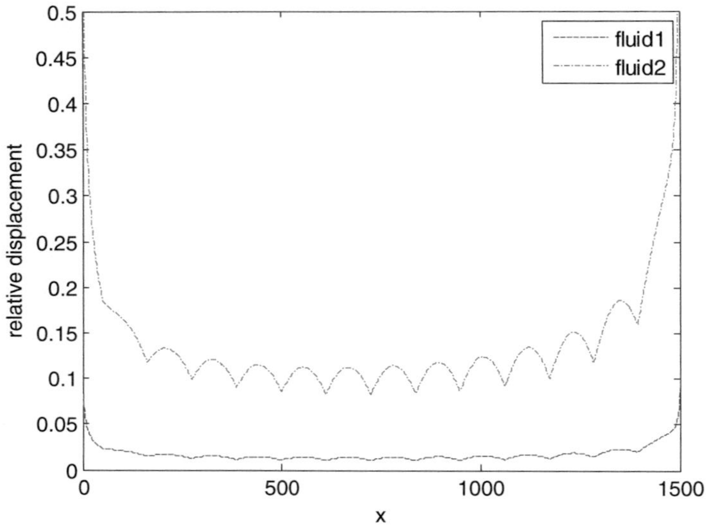

Fig. 10.44 The maximum relative displacements along the connecting line ($S1 = 0.24$, $\omega_1 = 15\text{Hz}, \omega_2 = 5\text{Hz}$)

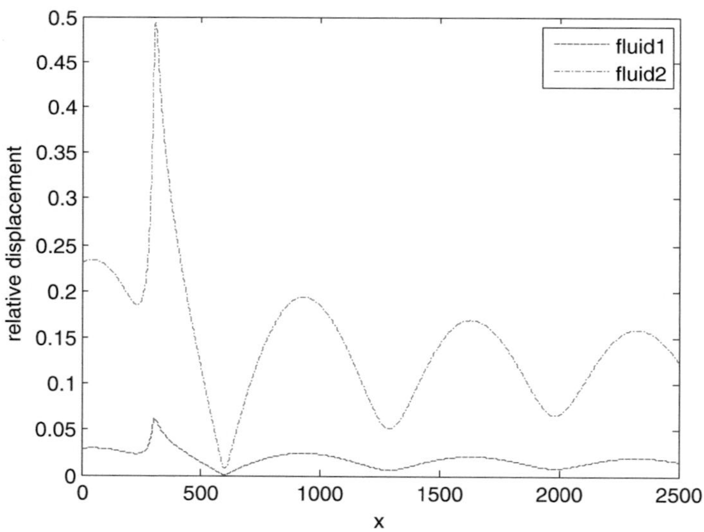

Fig. 10.45 The maximum relative displacements with the changing of the location of the right source ($S1 = 0.24$, $\omega_1 = 10\text{Hz}, \omega_2 = 10\text{Hz}$)

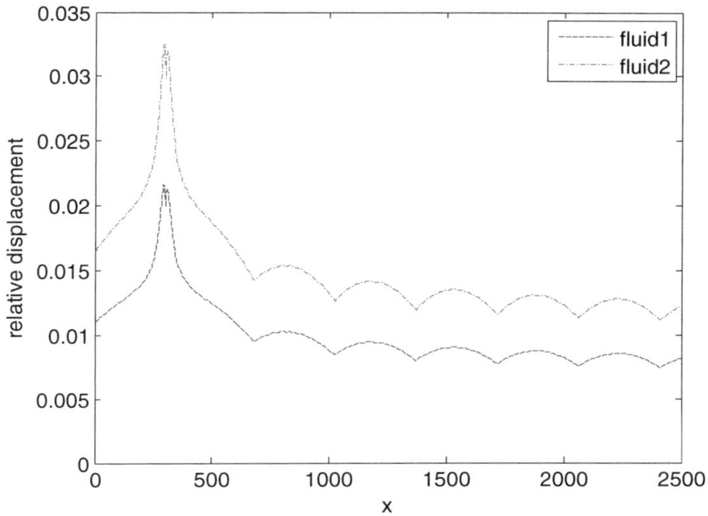

Fig. 10.46 The maximum relative displacements with the changing of the location of the right source ($S1 = 0.43$, $\omega_1 = 5\text{Hz}$, $\omega_2 = 10\text{Hz}$)

Considering all the graphics, the relative displacement of phase two, the wetting phase, is always larger than that of phase one. The relative displacement can be zero all the time for some points in the domain as a result of the superposition of the effect from the two waves. Furthermore, the relative displacements will decrease with increasing saturation of non-wettable fluid phase one (Fig. 10.47).

In Figs. 10.48 and 10.49, they show how the maximum nondimensional relative displacements of fluid one change with the varying location of the right source with different relative saturations. The porosities and frequencies are indicated in the figure. Obviously, the amplitude of the relative placement is larger with the identical frequencies and the curve is more regular periodic.

Figures 10.50, 10.51, 10.52, and 10.53 show how the maximum relative displacements of the fluids change along the connecting line between the two circular-cylindrical sources with different porosities. The porosities and frequencies are indicated in the figures.

1. The higher the relative saturation of fluid one (the non-wetting phase), the smaller the value of the relative displacement.
2. The waves with higher frequencies have a larger effect on the relative displacement compared with the wave source with lower frequencies.

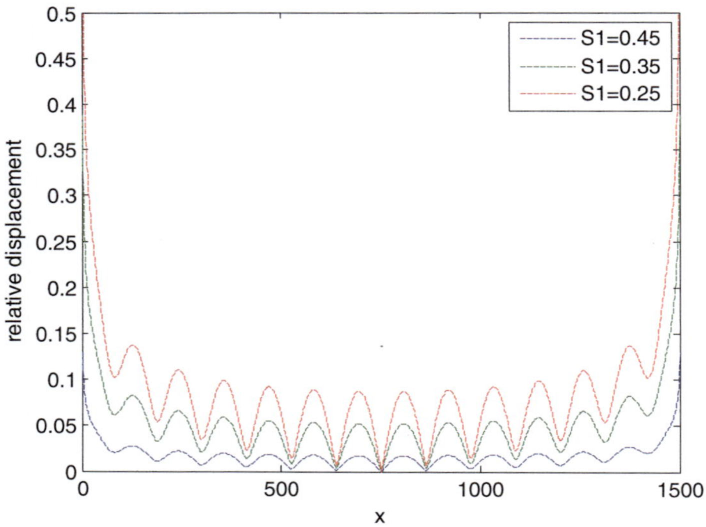

Fig. 10.47 The maximum relative displacements of fluid one along the connecting line with different relative saturation ($\omega_1 = \omega_2 = 15\text{Hz}$)

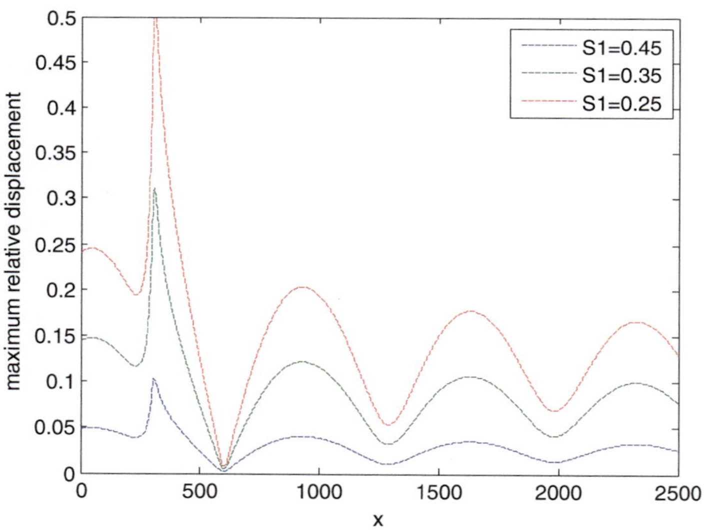

Fig. 10.48 The maximum relative displacements of fluid one versus the changing of the location of the right source with different relative saturation at same frequency ($\omega_1 = \omega_2 = 5\text{Hz}$)

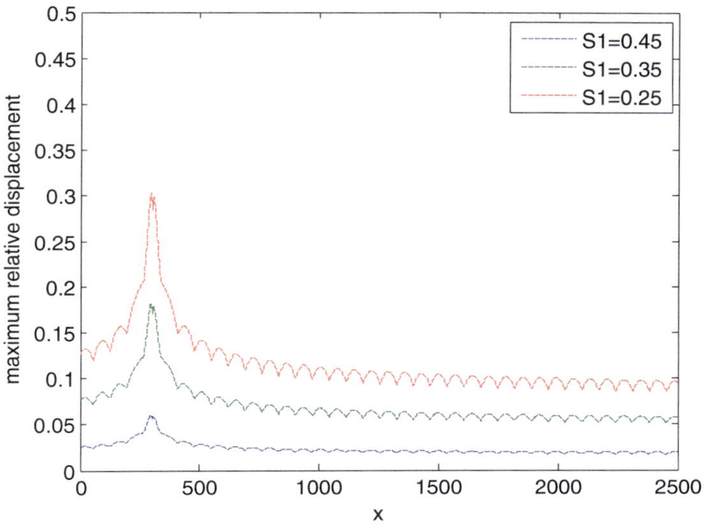

Fig. 10.49 The maximum relative displacements of fluid one versus the changing of the location of the right source with different relative saturation at different frequencies ($\omega_1 = 25\text{Hz}, \omega_2 = 50\text{Hz}$)

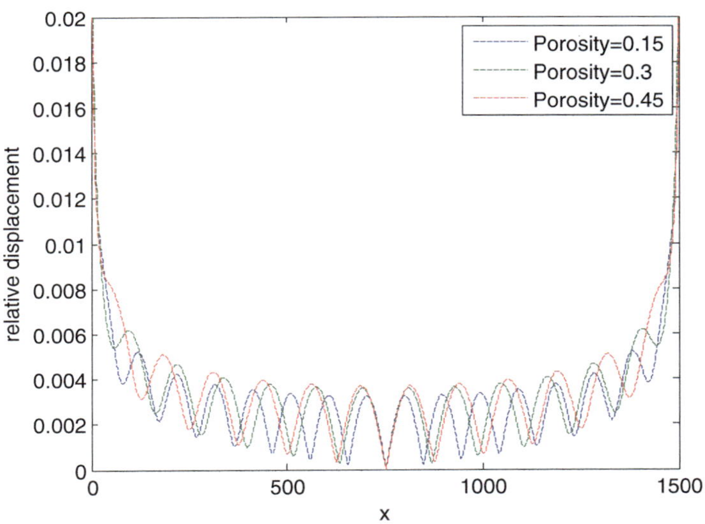

Fig. 10.50 The maximum relative displacements of fluid one along the connecting line with different porosity ($\omega_1 = \omega_2 = 15\text{Hz}, \text{S1} = 0.43$)

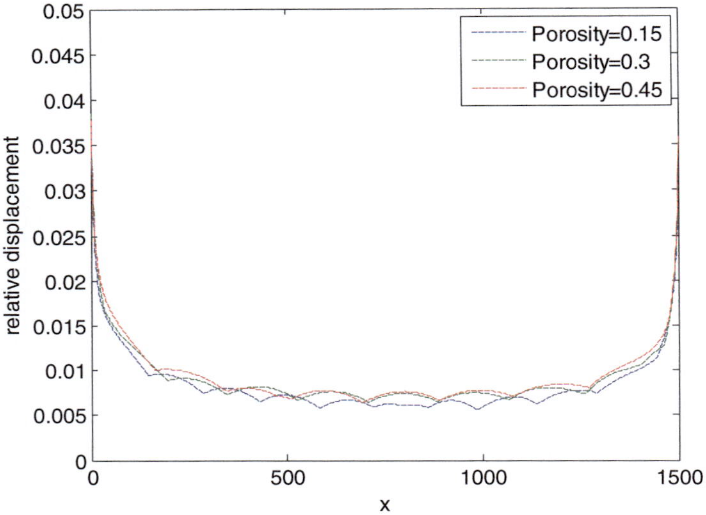

Fig. 10.51 The maximum relative displacements of fluid one along the connecting line with different porosity ($\omega_1 = 2.5\text{Hz}, \omega_2 = 5\text{Hz}, \text{S1} = 0.43$)

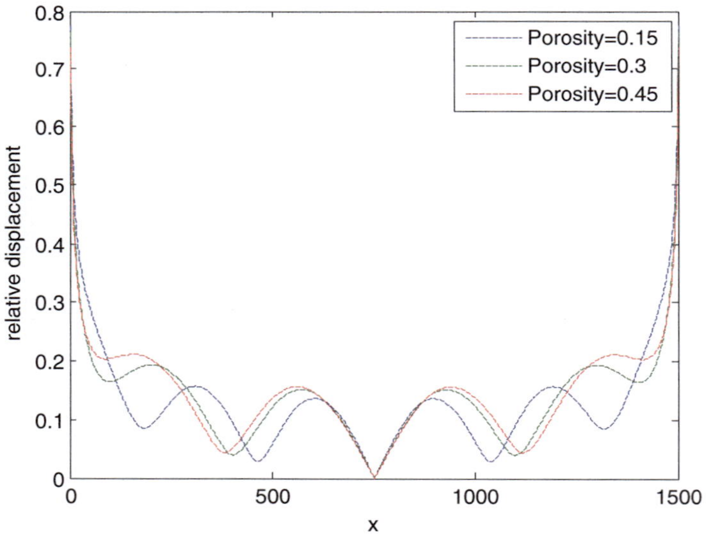

Fig. 10.52 The maximum relative displacements of fluid two along the connecting line with different porosity ($\omega_1 = \omega_2 = 5\text{Hz}, \text{S1} = 0.24$)

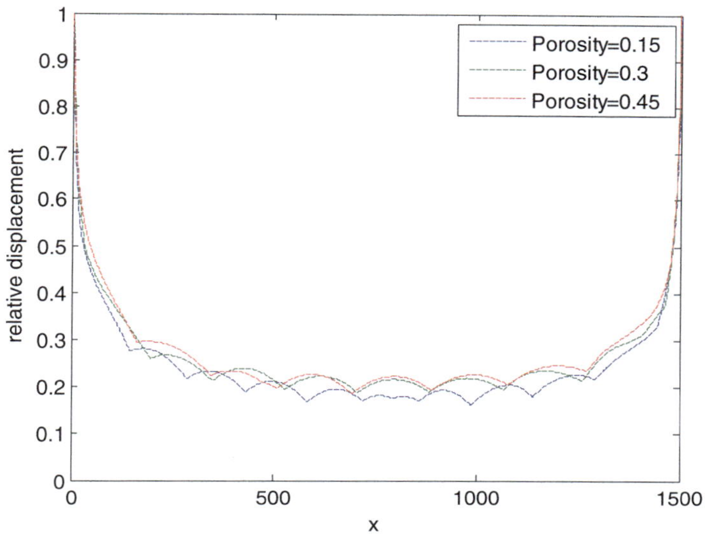

Fig. 10.53 The maximum relative displacements of fluid two along the connecting line with different porosity ($\omega_1 = 2.5\text{Hz}, \omega_2 = 5\text{Hz}, \text{S}1 = 0.24$)

Acknowledgments I am grateful to acknowledge many students and faculty for their help with this chapter of the book and their support and encouragement during the development of this book. Specific thanks to Lu Han, Guoqing Wang, Xiaojie Wang, and Lin Sun. Many contents of the chapter are the research results of our research team under my supervision. I would also like to thank the University of Regina for its encouragement in performing the research involved in the book.

Appendix A: Double Complex Fourier Transform

The important properties of the Fourier transformation of functions of several variables which are of most frequent in the early chapters of this book are presented in this Appendix.

A.1 Fourier Transform of Function

If f(x,y,z) is a function of the three independent variables x, y, and z, then the Fourier transform of this function with respect to x is

$$f_1(p, y, z) = \frac{1}{\sqrt{2\pi}} \int_{-\infty}^{\infty} f(x, y, z) \exp(ipx) dx \tag{A.1}$$

The Fourier transform of the above equation with respect to z can also be presented as

$$\overline{f}(p, y, q) = \frac{1}{\sqrt{2\pi}} \int_{-\infty}^{\infty} f_1(p, y, z) \exp(iqz) dz \tag{A.2}$$

Therefore, the last equation in terms of f(x,y,z) can be written by the combination of the last two transformations:

$$\overline{f}(p, y, q) = \frac{1}{2\pi} \int_{-\infty}^{\infty}\int_{-\infty}^{\infty} f(x, y, z) \exp(ipx) \exp(iqz) dx\, dz \tag{A.3}$$

The last equation is the definition of a double Fourier transform of (x,y,z), and p and q are the transform variables.

H.R. Hamidzadeh et al., *Wave Propagation in Solid and Porous Half-Space Media*,
DOI 10.1007/978-1-4614-9269-6,

A.1.1 Fourier Transform of Derivatives of Functions

By definition the double Fourier transform of $\frac{\partial^r}{\partial x^r} f(x, y, z)$ can be given by the following expression:

$$\overline{f}^r_x(p, y, q) = \frac{1}{2\pi}\int_{-\infty}^{\infty}\int_{-\infty}^{\infty} \frac{\partial^r}{\partial x^r} f(x, y, z) \exp(ipx)\exp(iqz)dx\,dz \tag{A.4}$$

Separating exponential components in the above equations and using by part integration, it yields

$$\overline{f}^r_x(p, y, q) = \int_{-\infty}^{\infty}\left\{\begin{array}{l}\left[\frac{1}{\sqrt{2\pi}}\frac{\partial^{r-1}}{\partial x^{r-1}} f(x, y, z)\exp(ipx)\right]_{-\infty}^{+\infty} \\ -\int_{-\infty}^{\infty}\frac{1}{\sqrt{2\pi}}\frac{\partial^{r-1}}{\partial x^{r-1}} f(x, y, z)(ip)\exp(ipx)\,dx\end{array}\right\} x\exp(iqz)\,dz \tag{A.5}$$

If the limit of $\frac{\partial^{r-1}}{\partial x^{r-1}} f(x, y, z)$ tends to zero as $|x|$ tends to infinity, by repetition of this rule, one can conclude the following expression:

$$\overline{f}^r_x(p, y, q) = (-ip)^r \int_{-\infty}^{\infty}\int_{-\infty}^{\infty} f(x, y, z)\exp(ipx)\exp(iqz)dx\,dz \tag{A.6}$$

or

$$\overline{f}^r_x(p, y, q) = (-ip)^r\,\overline{f}(p, y, q) \tag{A.7}$$

This means that the Fourier transform of the function $\frac{\partial^r}{\partial x^r} f(x, y, z)$ is $(-ip)^r$ times the Fourier transform of the function $f(x,y,z)$, provided that the first (r-1) derivatives of $f(x,y,z)$ vanish as $|x|$ tends to infinity.

A.1.2 Inverse of Fourier Transform

The relation between a function $f(x,y,z)$ and its Fourier transform

$$f(x, y, z) = \frac{1}{2\pi}\int_{-\infty}^{\infty}\int_{-\infty}^{\infty}\overline{f}(p, y, q)\exp(-ipx)\exp(-iqz)\,dp\,dq \tag{A.8}$$

This equation provides the inverse transform of the double Fourier transform of $f(x,y,z)$.

A.1.3 Fourier Transform of the Dirac Delta Function

In order to obtain the Fourier transform of the Dirac delta function, the following definition is used:

$$\int_{-\infty}^{+\infty} \delta\left(t - t_0\right) \phi\ (t) = \phi\ (t_0) \tag{A.9}$$

where $\phi\ (t)$ is an arbitrary function, continuous at a given point to. Delta functions for two variables can be separated as $\Delta(x,z) = \delta(x)\ \delta(z)$. Therefore, the double complex Fourier transform of $\Delta(x,z)$ can be presented by the following equation:

$$\overline{\Delta}\,(p,q) = \frac{1}{2\pi} \int_{-\infty}^{\infty}\int_{-\infty}^{\infty} \Delta\,(x,z) \exp(ipx) \exp(iqz) dx\ dz \tag{A.10a}$$

or

$$\overline{\Delta}\,(p,q) = \frac{1}{2\pi} \int_{-\infty}^{\infty} \delta(x) \exp(ipx)\ dx \int_{-\infty}^{\infty} \delta(z) \exp(iqz)\ dz \tag{A.10b}$$

According to equation (A.9), one can write $\overline{\Delta}\,(p,q) = \frac{1}{2\pi}$.

In order to examine the validity of the inversion formula for this function, the following equality should be proved:

$$\Delta\,(x,z) = \frac{1}{2\pi} \int_{-\infty}^{\infty}\int_{-\infty}^{\infty} \frac{1}{2\pi} \exp\left(-ipx\right) \exp\left(-iqz\right) dp\ dq \tag{A.11}$$

The right-hand side of the above equality is

$$I = \left(\frac{1}{2\pi}\right)^2 \int_{-\infty}^{+\infty} \exp\left(-ipx\right) dp \int_{-\infty}^{+\infty} \exp\left(-iqz\right) dq \tag{A.12a}$$

or

$$I = \left(\frac{1}{2\pi}\right)^2 \int_{-\infty}^{+\infty} \cos px \, dp \int_{-\infty}^{+\infty} \cos qz \, dq \tag{A.12b}$$

Papoulis (1962), on page 281 (by using Riemann-Lebesgue lemma), showed that

$$\int_{-\infty}^{+\infty} \cos \omega t \, d\omega = 2\pi \, \delta(t) \tag{A.13}$$

Therefore, using the Papoulis result, the function Ican be represented by

$$I = \delta(x)\delta(z) \tag{A.14}$$

The above equation indicates that the function I is the separated form of the delta function for two variables. This means that the inversion formula for the delta function is valid.

Appendix B: Evaluation of Certain Infinite Integrals

Integral representation of displacement in equations 3.15 and 4.12 for both forms of disturbances cannot be evaluated by direct methods of integration due to the singular point and branch points of its integrand. To deal with these integrals, four different integrals which are given below should be evaluated:

$$I_1 = \int_{-\infty}^{\infty} \Psi_1 (\xi)\, d\xi = \int_{-\infty}^{\infty} \frac{2\xi^2 - k^2 - 2\sqrt{(\xi^2 - h^2)(\xi^2 - k^2)}}{F(\xi)} \xi \exp(i\xi x)\, dx \tag{B.1}$$

$$I_2 = \int_{-\infty}^{\infty} \Psi_2 (\xi)\, d\xi = \int_{-\infty}^{\infty} \frac{k^2\sqrt{\xi^2 - h^2}}{F(\xi)} \exp(i\xi x)\, dx \tag{B.2}$$

$$I_3 = \int_{-\infty}^{\infty} \Psi_3 (\xi)\, d\xi = \int_{-\infty}^{\infty} \frac{k^2\sqrt{\xi^2 - k^2}}{F(\xi)} \exp(i\xi x)\, dx \tag{B.3}$$

$$I_4 = \int_{-\infty}^{\infty} \Psi_4 (\xi)\, d\xi = \int_{-\infty}^{\infty} \frac{1}{\sqrt{\xi^2 - k^2}} \exp(i\xi x)\, dx \tag{B.4}$$

The first three integrals have three singularities at the branch points ($\pm$ k, o), ($\pm$ h, o) and at the Rayleigh poles ($\pm$ k, o) which are the principle roots of the Rayleigh equation. A suitable method for integrating the first three integrals utilizes contour integration. The last integrals can be separated into their parts, and each part can be evaluated. In the following sections, contour of integration and method of integration for the last integral will be discussed.

A suitable contour in the plane of the complex variable $\xi = \sigma + i\eta$. If this contour does not include either poles ($\pm$ k, o) or branch points of the function to

H.R. Hamidzadeh et al., *Wave Propagation in Solid and Porous Half-Space Media*,
DOI 10.1007/978-1-4614-9269-6,

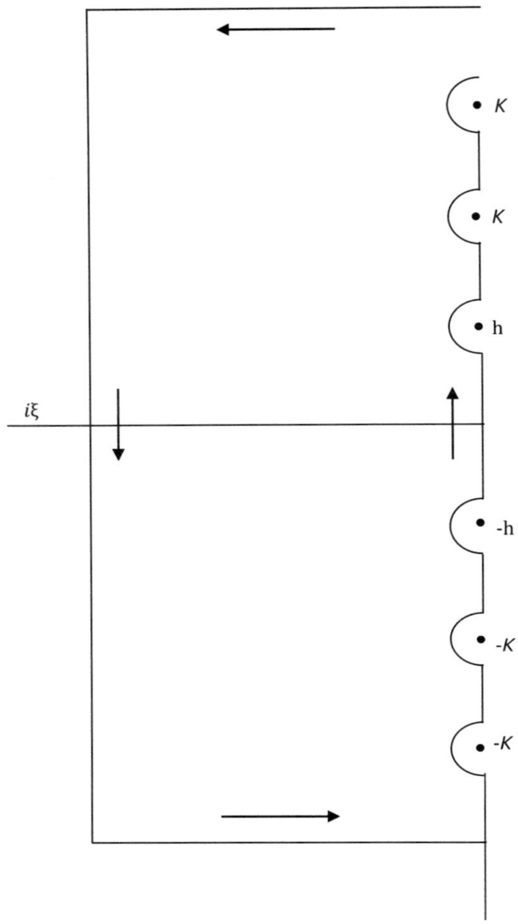

Fig. B.1 Contour of Integration

be integrated, the result will be zero. A convenient contour for this purpose is a rectangle, one side of which consists of the axis of σ except for small semicircular indentations surrounding the singular points specified, while the remaining sides are at an infinite distance on the side $\eta > \text{o}$. It can easily be seen that the parts of the integral due to this infinitely distant side will vanish. If we adopt for the radicals, $\sqrt{\xi^2 - k^2}$ and $\sqrt{\xi^2 - h^2}$ are the points of the axis of σ, the consistent system of values indicated in Fig. B.1.

Evaluations of the four integrals by means of this specified contour are as follows:

1. If the integral $\Psi_1(\xi)$ along the real axis of contour is separated into three intervals of $(-\infty, -k), (-k, -h)$, and $(-h, +\infty)$,

$$\int_{-\infty}^{\infty} \Psi_1(\xi)\, d\xi = \int_{-\infty}^{-k} \xi \frac{\left(2\xi^2 - k^2 - 2\gamma_1\gamma_2\right)}{F(\xi)} \exp(i\xi x)\, d\xi \\ - i\pi \frac{-K\left(2K^2 - k^2 - 2\gamma_{1R}\gamma_{2R}\right)}{F'(-K)} \exp(-iKx) \tag{B.5}$$

$$\int_{-k}^{-h} \Psi_1(\xi)\, d\xi = \int_{-k}^{-h} \xi \frac{\left(2\xi^2 - k^2 - 2\gamma_1\gamma_2\right)}{F(\xi)} \exp(i\xi x)\, d\xi \tag{B.6}$$

$$\int_{-h}^{+\infty} \Psi_1(\xi)\, d\xi = \mathbb{C}P \int_{-h}^{+\infty} \xi \frac{\left(2\xi^2 - k^2 - 2\gamma_1\gamma_2\right)}{F(\xi)} \exp(i\xi x)\, d\xi \\ - i\pi \frac{K\left(2K^2 - k^2 - 2\gamma_{1R}\gamma_{2R}\right)}{F'(K)} \exp(iKx) \tag{B.7}$$

The sign $\mathbb{C}P\int$ is used to indicate the Cauchy principle value of integration.

Summation of the left-hand sides of these integrals should be zero. Then, the Cauchy principal value of integral I_1 becomes

$$I_1 = \mathbb{C}P \int_{-\infty}^{+\infty} \xi \frac{2\xi^2 - k^2 - 2\gamma_1\gamma_2}{F(\xi)} exp(i\xi x)\, d\xi = -2i\pi H_1 \cos Kx \\ - 4k^2 \int_{h}^{k} \xi \frac{\left(2\xi^2 - k^2\right)\gamma_1\gamma_2}{F(\xi)\, f(\xi)} \exp(i\xi x)\, d\xi \tag{B.8}$$

where

$$H_1 = -\frac{K\left(2K^2 - k^2 - 2\gamma_{1R}\gamma_{2R}\right)}{F'(K)} \tag{B.9}$$

And γ_{1R} and γ_{2R} are values of γ_1 and γ_1 at $\xi = \pm K$. It should also be noted that the Rayleigh function is an even function.

2. Integration of $\Psi_2(\xi)$ along the real axis of the contour can be made by the same separation of integration which was used for the first integral; thus

$$\int_{-\infty}^{-k} \Psi_2(\xi)\, d\xi = \mathbb{C}P \int_{-\infty}^{-k} \frac{-K^2\gamma_1}{F(\xi)} \exp(i\xi x) - i\pi \frac{-K^2\gamma_{1R}}{F'^{(-K)}} \exp(-iKx) \tag{B.10}$$

$$\int_{-k}^{-h} \Psi_2(\xi)\, d\xi = \int_{-k}^{-h} \frac{k^2\gamma_1}{f(\xi)} \exp(i\xi x)\, d\xi \tag{B.11}$$

$$\int_{-h}^{+\infty} \Psi_2(\xi)\,d\xi = \mathbb{C}P\int_{-h}^{+\infty} \frac{k^2\gamma_1}{F(\xi)}d\xi - i\pi\frac{k^2\gamma_{1R}}{F'(K)}\exp(i\xi x) \tag{B.12}$$

Adding these integrals and equating to zero and bearing in mind that $F'(-K)$ is equal to $-F'(K)$, then the integral I_2 will be given in the following form:

$$\begin{aligned} I_2 = \mathbb{C}P\int_{-h}^{+\infty} \frac{k^2\gamma_1}{F(\xi)}\exp(i\xi x)\,d\xi &= -2i\pi H_2\cos Kx + \\ +2\int_{k}^{\infty} \frac{k^2\gamma_1}{F(\xi)}\exp(-i\xi x)\,d\xi &+ 2k^2\int_{h}^{k} \frac{\left(2\xi^2-k^2\right)^2\gamma_1}{F(\xi)\ f(\xi)}\exp(-i\xi x)\,d\xi \end{aligned} \tag{B.13}$$

$$H_2 = \frac{k^2\gamma_1}{F'(K)} \tag{B.14}$$

3. The integral I_3 over the real axis between the three intervals is

$$\circ\int_{-\infty}^{-k} \Psi_3(\xi)\,d\xi = \int_{-\infty}^{-k} \frac{-k^2\gamma_2}{F(\xi)}\exp(i\xi x)\,d\xi - i\pi\frac{k^2\gamma_{2R}}{F'(-K)}\exp(-iKx) \tag{B.15}$$

$$\int_{-k}^{-h} \Psi_3(\xi)\,d\xi = \int_{-k}^{-h} \frac{-k^2\gamma_2}{f(\xi)}\exp(i\xi x)\,d\xi \tag{B.16}$$

$$\circ\int_{-h}^{\infty} \Psi_3(\xi)\,d\xi = \mathbb{C}P\int_{-h}^{\infty} + \frac{k^2\gamma_2}{F(\xi)}\exp(i\xi x)\,d\xi - i\pi\frac{k^2\gamma_{2r}}{F'(K)}\exp(iKx)\ \circ \tag{B.17}$$

Equating the sum of the left-hand side to zero, it results to

$$\begin{aligned} I_3 = \ \mathbb{C}P\int_{-\infty}^{+\infty} \frac{k^2\gamma_2}{F(\xi)}\exp(i\xi x)\,d\xi &= -2i\pi H_3\cos Kx \\ +\mathbb{C}P\int_{-\infty}^{-k} \frac{2k^2\gamma_2}{F(\xi)}\exp(i\xi x)\,d\xi + &\int_{-h}^{k}\left(\frac{1}{F(\xi)}+\frac{1}{f(\xi)}\right)k^2\gamma_2\exp(i\xi x)\,d\xi \end{aligned}$$

or

$$I_3 = -2i\pi H_3 \cos Kx + 2\mathbb{C}P\int_k^{\infty} \frac{k^2\gamma_2}{F(\xi)} \exp(-i\xi x)\, d\xi + \\ + 2k^2 \int_h^k \frac{\left(2\xi^2 - k^2\right)^2 \gamma_2}{F(\xi)\ f(\xi)} \exp(-i\xi x)\, d\xi \tag{B.18}$$

where

$$H_3 = \frac{-k^2\gamma_{2R}}{F'(K)} \tag{B.19}$$

4. Finally the last integral can be split into

$$I_4 = \int_{-\infty}^{+\infty} \frac{1}{\sqrt{\xi^2 - k^2}} \exp(i\xi x)\, d\xi = \int_{-\infty}^{+\infty} \frac{\cos(\xi x)\, d\xi}{\sqrt{\xi^2 - k^2}} + i\int_{-\infty}^{+\infty} \frac{\sin(\xi x)\, d\xi}{\sqrt{\xi^2 - k^2}} \tag{B.20}$$

Since the integrand in the last right-hand side is an odd function, I_4 is equal to the first integral in the right-hand side integral:

$$I_4 = \int_{-\infty}^{+\infty} \frac{\cos(\xi x)\, d\xi}{\sqrt{\xi^2 + k^2}} = 2\int_0^{\infty} \frac{\cos(\xi x)\, d\xi}{\sqrt{\xi^2 + k^2}} \tag{B.21}$$

By changing the variable ξ to $k\eta$ and the splitting the interval into (0,1) and $(1, \infty)$, the result will be

$$I_4 = -2i\int_0^1 \frac{\cos(k\eta x)\, d\eta}{\sqrt{1 - \eta^2}} + 2\int_1^{\infty} \frac{\cos(k\eta x)\, d\eta}{\sqrt{\eta^2 + 1}} \tag{B.22}$$

Two integrals in the right-hand side are given by I.S. Gradshteyn and I.M. Ryzhik (1965) as

$$\int_0^1 \frac{\cos(k\eta x)\, d\eta}{\sqrt{1 - \eta^2}} = \frac{\pi}{2} J_o(kx) \tag{B.23}$$

$$\int_1^{\infty} \frac{\cos(k\eta x)\, d\eta}{\sqrt{\eta^2 - 1}} = -\frac{\pi}{2} Y_o(kx) \tag{B.24}$$

Therefore, the final result for I_4 is

$$I_4 = -\pi i J_o(kx) - \pi Y_o(kx) = -\pi i H_o^2(kx) \tag{B.25}$$

Appendix C: Numerical Evaluation of Certain Integrals

In this appendix, the different numerical techniques which have been used in the more involved integration are discussed. These techniques are as follows.

C.1 The Numerical Evaluation of Cauchy Principal Values of the Integral

This section is concerned with the numerical evaluation of integrals of the form

$$\int_{a}^{b} f(x)dx \ (a, b \ \textit{finite}) \tag{C.1}$$

where $f(x)$ is a function of the real variable x, analytic in the region containing the interval $[a, b]$ except at a single point $x = \frac{1}{2}(a + b)$ in the middle of the interval (a,b). For this purpose, Longman's (1958) technique is employed. In this technique, the integral can be reduced to the form

$$\int_{-1}^{1} f(t)dt \tag{C.2}$$

By means of the standard transformation

$$x = \frac{1}{2}[(b - a)\,t + (b + a)] \tag{C.3}$$

where the singular point is at t = 0.

After the range of integration is made symmetrical about origin, the function f(t) can be split up into its even and odd components

H.R. Hamidzadeh et al., *Wave Propagation in Solid and Porous Half-Space Media*, DOI 10.1007/978-1-4614-9269-6, © Springer Science+Business Media New York 2014

$$f(t) = \frac{1}{2}[f(t) + f(-t)] + \frac{1}{2}[f(t) - f(-t)] \tag{C.4}$$

Then, the final form of the integral is

$$\int_0^1 [f(t) + f(-t)]\,dt \tag{C.5}$$

Longman showed that in this integral either there is no singularity at t = 0 or, if there is, it will not make the integral diverge. In order to evaluate the integral (C.5), a table of Gaussian quadrature formula by Stroud and Secrest (1966), with ten places of decimals, is used.

C.2 Integral of the Form $\int_0^b (b - x)^{\alpha}(x - a)^{\beta} f(x)dx$

The kind of integrals where ($\alpha = 0, \frac{1}{2}$ and $\beta = \frac{1}{2}$) and $f(x)$ is an analytic function cannot be evaluated by the direct Gaussian quadrature formula. In fact it has been shown by Krylov (1962) that the remainder error in the usual method of integration is proportional nth derivative of the integrand, and since these derivatives, in this case, are infinite at the boundaries of [a,b], the remainder error is also infinite. Krylov considered this problem and he developed the Gaussian quadrature formula to evaluate these integrals, by going through the orthogonal system of the Jacobi polynomials. In this section the evaluation method for two different cases of $\left(\alpha = \beta = \frac{1}{2}\right)$ and $\left(\alpha = 0, \beta = \frac{1}{2}\right)$ is considered.

1.

$$\alpha = \beta = \frac{1}{2}$$

In this case the integral is

$$I_1 = \int_a^b (b - x)^{1/2}(x - a)^{1/2}\ f(x)dx \tag{C.6}$$

Transforming the integral by the linear transformation

$$x = \frac{1}{2}[(b + a) + t\,(b - a)] \tag{C.7}$$

then I_1 can be written in the general form

$$I_1 = \int_{-1}^1 (1 - t)\,(1 + t)\,g(t)dt \tag{C.8}$$

A table of the modified Gaussian quadrature formula for this kind of integral is given for n = 2 to 20, by Stroud and Secrest (1966), page 161–163.

2.

$$\alpha = 0, \beta = \frac{1}{2}$$

The general for in this case is

$$I_2 = \int_a^b (x-a)^{1/2} f(x)dx \tag{C.9}$$

Transforming the segment into the appropriate segment for the quadrature formula by linear transformer of

$$x = (b-a)t + a \tag{C.10}$$

then general form of integral I_2 is

$$I_2 = \int_0^1 \sqrt{t}\, g(t)dt \tag{C.11}$$

A table of quadrate formula for this case is given for n = 1, 8 by Krylov (1962), page 119.

Appendix D: Trigonometric Formulae

Definitions in terms of exponentials.

$$\cos z = \frac{e^{iz} + e^{-iz}}{2} \tag{D.1}$$

$$\sin z = \frac{e^{iz} - e^{-iz}}{2i} \tag{D.2}$$

$$\tan z = \frac{e^{iz} - e^{-iz}}{i\,(e^{iz} + e^{-iz})} \tag{D.3}$$

$$e^{iz} = \cos z + i\,\sin z \tag{D.4}$$

$$e^{-iz} = \cos z - i\,\sin z \tag{D.5}$$

Angle sum and difference.

$$\sin(\alpha \pm \beta) = \sin\alpha\cos\beta \pm \cos\alpha\sin\beta \tag{D.6}$$

$$\cos(\alpha \pm \beta) = \cos\alpha\cos\beta \mp \sin\alpha\sin\beta \tag{D.7}$$

$$\tan(\alpha \pm \beta) = \frac{\tan\alpha \pm \tan\beta}{1 \mp \tan\alpha\tan\beta} \tag{D.8}$$

$$\cot(\alpha \pm \beta) = \frac{\cot\alpha\cot\beta \mp 1}{\cot\beta \pm \cot\alpha} \tag{D.9}$$

Symmetry.

$$\sin(-\alpha) = -\sin\alpha \tag{D.10}$$

$$\cos(-\alpha) = \cos\alpha \tag{D.11}$$

$$\tan(-\alpha) = -\tan\alpha \tag{D.12}$$

H.R. Hamidzadeh et al., *Wave Propagation in Solid and Porous Half-Space Media*,
DOI 10.1007/978-1-4614-9269-6, © Springer Science+Business Media New York 2014

Multiple angle.

$$\sin(2\alpha) = 2\sin\alpha\cos\alpha = \frac{2\tan\alpha}{1+\tan^2\alpha} \tag{D.13}$$

$$\cos(2\alpha) = 2\cos^2\alpha - 1 = 1 - 2\sin^2\alpha = \cos^2\alpha - \sin^2\alpha \tag{D.14}$$

$$\tan(2\alpha) = \frac{2\tan\alpha}{1-\tan^2\alpha} \tag{D.15}$$

$$\cot(2\alpha) = \frac{\cot^2\alpha - 1}{2\cot\alpha} \tag{D.16}$$

$$\sin(3\alpha) = -4\sin^3\alpha + 3\sin\alpha \tag{D.17}$$

$$\cos(3\alpha) = 4\cos^3\alpha - 3\cos\alpha \tag{D.18}$$

$$\tan(3\alpha) = \frac{-\tan^3\alpha + 3\tan\alpha}{-3\tan^2\alpha + 1} \tag{D.19}$$

$$\sin(4\alpha) = -8\sin^3\alpha\cos\alpha + 4\sin\alpha\cos\alpha \tag{D.20}$$

$$\cos(4\alpha) = 8\cos^4\alpha - 8\cos^2\alpha + 1 \tag{D.21}$$

$$\tan(4\alpha) = \frac{-4\tan^3\alpha + 4\tan\alpha}{\tan}^{4}\alpha - 6\tan^2\alpha + 1 \tag{D.22}$$

$$\sin(5\alpha) = 16\sin^5\alpha - 20\sin^3\alpha + 5\sin\alpha \tag{D.23}$$

$$\cos(5\alpha) = 16\cos^5\alpha - 20\cos^3\alpha + 5\cos\alpha \tag{D.24}$$

$$\sin(n\alpha) = 2\sin((n-1)\alpha)\cos\alpha - \sin((n-2)\alpha) \tag{D.25}$$

$$\cos(n\alpha) = 2\cos((n-1)\alpha)\cos\alpha - \cos((n-2)\alpha) \tag{D.26}$$

$$\tan(n\alpha) = \frac{\tan((n-1)\alpha) + \tan\alpha}{1 - \tan((n-1)\alpha)\tan\alpha} \tag{D.27}$$

Half angle.

$$\cos\left(\frac{\alpha}{2}\right) = \pm\sqrt{\frac{1+\cos\alpha}{2}} \tag{D.28}$$

$$\sin\left(\frac{\alpha}{2}\right) = \pm\sqrt{\frac{1-\cos\alpha}{2}} \tag{D.29}$$

$$\tan\left(\frac{\alpha}{2}\right) = \frac{1-\cos\alpha}{\sin\alpha} = \frac{\sin\alpha}{1+\cos\alpha} = \pm\sqrt{\frac{1-\cos\alpha}{1+\cos\alpha}} \tag{D.30}$$

$$\sin\alpha = \frac{2\tan\frac{\alpha}{2}}{1+\tan^2\frac{\alpha}{2}} \tag{D.31}$$

$$\cos\alpha = \frac{1-\tan^2\frac{\alpha}{2}}{1+\tan^2\frac{\alpha}{2}} \tag{D.32}$$

Powers of functions.

$$\sin\alpha\cos\alpha = \frac{1}{2}\sin(2\alpha) \tag{D.33}$$

$$\sin^2\alpha = \frac{1}{2}(1-\cos(2\alpha)) \tag{D.34}$$

$$\cos^2\alpha = \frac{1}{2}(1+\cos(2\alpha)) \tag{D.35}$$

$$\tan^2\alpha = \frac{1-\cos(2\alpha)}{1+\cos(2\alpha)} \tag{D.36}$$

$$\sin^3\alpha = \frac{1}{4}(3\sin(\alpha)-\sin(3\alpha)) \tag{D.37}$$

$$\sin^2\alpha\cos\alpha = \frac{1}{4}(\cos\alpha-3\cos(3\alpha)) \tag{D.38}$$

$$\sin\alpha\cos^2\alpha = \frac{1}{4}(\sin\alpha+\sin(3\alpha)) \tag{D.39}$$

$$\cos^3\alpha = \frac{1}{4}(\cos(3\alpha)+3\cos\alpha) \tag{D.40}$$

$$\sin^4\alpha = \frac{1}{8}(3-4\cos(2\alpha)+\cos(4\alpha)) \tag{D.41}$$

$$\sin^3\alpha\cos\alpha = \frac{1}{8}(2\sin(2\alpha)-\sin(4\alpha)) \tag{D.42}$$

$$\sin^2\alpha\cos^2\alpha = \frac{1}{8}(1-\cos(4\alpha)) \tag{D.43}$$

$$\sin\alpha\cos^3\alpha = \frac{1}{8}(2\sin(2\alpha)+\sin(4\alpha)) \tag{D.44}$$

$$\cos^4\alpha = \frac{1}{8}(3+4\cos(2\alpha)+\cos(4\alpha)) \tag{D.45}$$

$$\sin^5\alpha = \frac{1}{16}(10\sin\alpha-5\sin(3\alpha)+\sin(5\alpha)) \tag{D.46}$$

$$\sin^4\alpha\cos\alpha = \frac{1}{16}(2\cos\alpha-3\cos(3\alpha)+\cos(5\alpha)) \tag{D.47}$$

$$\sin^3\alpha\cos^2\alpha = \frac{1}{16}(2\sin\alpha+\sin(3\alpha)-\sin(5\alpha)) \tag{D.48}$$

$$\sin^2\alpha\cos^3\alpha = \frac{1}{16}(2\cos\alpha-3\cos(3\alpha)-5\cos(5\alpha)) \tag{D.49}$$

$$\sin\alpha\cos^4\alpha = \frac{1}{16}\left(2\sin\alpha + 3\sin(3\alpha) + \sin(5\alpha)\right) \tag{D.50}$$

$$\cos^5\alpha = \frac{1}{16}\left(10\cos\alpha + 5\cos(3\alpha) + \cos(5\alpha)\right) \tag{D.51}$$

Products of sin and cos.

$$\cos\alpha\cos\beta = \frac{1}{2}\cos(\alpha-\beta) + \frac{1}{2}\cos(\alpha+\beta) \tag{D.52}$$

$$\sin\alpha\sin\beta = \frac{1}{2}\cos(\alpha-\beta) - \frac{1}{2}\cos(\alpha+\beta) \tag{D.53}$$

$$\sin\alpha\cos\beta = \frac{1}{2}\sin(\alpha-\beta) + \frac{1}{2}\sin(\alpha+\beta) \tag{D.54}$$

$$\cos\alpha\sin\beta = \frac{1}{2}\sin(\alpha+\beta) - \frac{1}{2}\sin(\alpha-\beta) \tag{D.55}$$

$$\sin(\alpha+\beta)\sin(\alpha-\beta) = \cos^2\beta - \cos^2\alpha = \sin^2\alpha - \sin^2\beta \tag{D.56}$$

$$\cos(\alpha+\beta)\cos(\alpha-\beta) = \cos^2\beta + \sin^2\alpha \tag{D.57}$$

Sum of functions.

$$\sin\alpha \pm \sin\beta = 2\sin\frac{\alpha\pm\beta}{2}\cos\frac{\alpha\pm\beta}{2} \tag{D.58}$$

$$\cos\alpha + \cos\beta = 2\cos\frac{\alpha+\beta}{2}\cos\frac{\alpha-\beta}{2} \tag{D.59}$$

$$\cos\alpha - \cos\beta = -2\sin\frac{\alpha+\beta}{2}\sin\frac{\alpha-\beta}{2} \tag{D.60}$$

$$\tan\alpha \pm \tan\beta = \frac{\sin(\alpha\pm\beta)}{\cos\alpha\cos\beta} \tag{D.61}$$

$$\cot\alpha \pm \cot\beta = \frac{\sin(\beta\pm\alpha)}{\sin\alpha\sin\beta} \tag{D.62}$$

$$\frac{\sin\alpha + \sin\beta}{\sin\alpha - \sin\beta} = \frac{\tan\frac{\alpha+\beta}{2}}{\tan\frac{\alpha-+\beta}{2}} \tag{D.63}$$

$$\frac{\sin\alpha + \sin\beta}{\cos\alpha - \cos\beta} = \cot\frac{-\alpha+\beta}{2} \tag{D.64}$$

$$\frac{\sin\alpha + \sin\beta}{\cos\alpha + \cos\beta} = \tan\frac{\alpha+\beta}{2} \tag{D.65}$$

$$\frac{\sin\alpha - \sin\beta}{\cos\alpha + \cos\beta} = \tan\frac{\alpha-\beta}{2} \tag{D.66}$$

Trigonometric relations.

$$\sin^2\alpha - \sin^2\beta = \sin(\alpha + \beta)\sin(\alpha - \beta) \tag{D.67}$$

$$\cos^2\alpha - \cos^2\beta = -\sin(\alpha + \beta)\sin(\alpha - \beta) \tag{D.68}$$

References

Aggarwal HR, Ablow CM (1967) Solution to a class of three-dimensional pulse propagation problems in an elastic half-space. Int J Eng Sci 5:663–679

Ahmed N, Sunada DK (1969) Non-linear flow in porous media. J Hydraul Div 95:1847

Anastasopoulos I, Kontoroupi T (2011) Simplified approximate method for analysis of rocking systems accounting for soil inelasticity and foundation uplifting. Soil Dyn Earthq Eng 56: 28–43

Andersen L (2011) Assessment of lumped-parameter models for rigid footings. Comput Struct 88(23):1333–1347

Amrouche C, Bernardi C, Dauge M, Girault V (1998) Vector potentials in three dimensional non-smooth domains. Math Method Appl Sci 21:823–864

Andrews GE, Askey R, Roy R (2001) Special functions. Cambridge University Press, Cambridge

Apsel RJ (1979) Dynamic Green's functions for layered media and applications to boundary value problems. Ph.D. Dissertation, University of California, San Diego, USA

Arfken G, Weber H, Weber HJ (2005) Mathematical methods for physicists. Academic, San Diego

Arnold RN, Bycroft GN, Warburton GB (1955) Forced vibrations of a body in an infinite elastic solid. J Appl Mech 22:391–400

Arora A, Tomar SK (2010) Seismic reflection from an interface between an elastic solid and a fractured porous medium with partial saturation. Trans Porous Med 85:375–396

Arora A, Tomar SK (2007) Elastic waves along a cylindrical borehole in a poroelastic. J Earth Syst Sci 116:225–234

Ashlock JC (2011) Experimental multi-modal foundation vibrations and comparison with benchmark half-space solutions. Proceedings of geo-frontiers: advances in geotechnical engineering, 3118–3127

Awojobi AO, Grootenhuis P (1965) Vibration of rigid bodies on semi-infinite elastic media. Proc Roy Soc 287: Series A 27–63

Awojobi AO, Tabiowo PH (1976) Vertical vibration of rigid bodies with rectangular bases on elastic media. Int J Earthq Eng Struct Dyn 4:439–454

Awojobi AO (1964) Vibrations of rigid bodies on elastic media. Ph.D. Thesis, University of London

Awojobi AO (1966) Harmonic rocking of a rigid rectangular body on a semi-infinite elastic medium. J Appl Mech ASME 33:547–552

Awojobi AO (1969) Torsional vibration of a rigid circular body on an infinite elastic stratum. Int J Solids Struct 5:369–378

Awojobi AO (1972a) Vertical vibration of a rigid circular body and harmonic rocking of a rigid rectangular body on an elastic stratum. Int J Solids Struct 8:759–774

Awojobi AO (1972b) Vertical vibration of a rigid circular foundation on Gibson soil. Geotechnique 22(2):333–343

H.R. Hamidzadeh et al., *Wave Propagation in Solid and Porous Half-Space Media*, DOI 10.1007/978-1-4614-9269-6, © Springer Science+Business Media New York 2014

Awojobi AO, Sobayo OA (1977) Ground vibration due to seismic detonation of a buried source. J Earthquake Eng and Struct Dyn 5(2): 131–143

Azarhoosh Z, Amiri G (2010) Elastic response of soil-structure systems subjected to near-fault rupture directivity pulses. Soil Dynamics and Earthquake Engineering 153–151

Baidya DK, Mandal A (2006) Dynamic response of footing resting on a layered soil system. West Indian J Eng 28(2):65–79

Banerjee PK, Mamoon SM (1990) A fundamental solution due to a periodic point force in the interior of an elastic half-space. Int J Earthquake Eng Struct Dyn 19:91–105

Bardet JP, Sayed H (1993) Velocity and attenuation of compressional waves in nearly saturated soils. Soil Dyn Earthquake Eng 12(7):391–401

Barkan DD (1962) Dynamics of bases and foundations. McGraw-Hill, New York

Bateman H, Erdelyi A (1954) Table of integral transforms, vol. 2. McGraw Hill, New York

Baviere M (2007) Basic Concepts in Enhanced Oil Recovery Processes. Elsevier Applied Science, London

Bear J (1972) Dynamics of fluids in porous media. American Elsevier Publishing Company, INC, New York

Bear J, Zaslavsky D, Irmay S (1968) Physical principles of water percolation and seepage. UNESCC, Paris

Beeston HE, McEvilly TV (1977) Shear wave velocities from down hole measurements. Int J Earthq Eng Struct Dyn 5:181–190

Beredugo JO, Novak M (1972) Coupled horizontal and rocking vibration of embedded footings. Canadian Geotechnical J 9(4):477–497

Beresnev IA, Johnson PA (1994) Elastic-wave stimulation of oil production: a review of methods and results. Geophysics 59:1000–1017

Bettess P, Zienkiewicz OC (1977) Diffraction and refraction of surface waves using finite and infinite elements. Int J Numer Methods Eng 11:1271–1290

Biot MA, Willis DG (1957) The elastic coefficients of the theory of consolidation. J Appl Mech 24:594–601

Biot MA (1956a) Theory of propagation of elastic waves in a fluid-saturated porous solid, part II: higher frequency range. J Acoust Soc Am 28:179–191

Biot MA (1956b) Theory of propagation of elastic waves in a fluid-saturated porous solid, part I: low frequency range. J Acoust Soc Am 28:168–178

Broad WJ (2010) Tracing oil reserves to their tiny origins. The New York Times

Bycroft GN (1956) Forced vibrations of circular plate on a semi-infinite elastic space and on an elastic stratum. Phil Trans Roy Soc Series A 248(948):327–368

Bycroft GN (1959) Machine foundation vibration. Proc Inst Mech Eng 173(18):30

Bycroft GN (1977) Soil-structure interaction at higher frequency factors. Int J Earthq Eng Struct Dyn 5:235–248

Carstens H (2011) An excellent reservoir. GEO Expro 8:12

Carter JP, Booker JR (1987) Analysis of pumping a compressible pore fluid from a saturated elastic half space. Comput Geotech 4:21–42

Chae YS (1967) The material constants of soils as determined from dynamic testing. Proceeding, international symposium on wave propagation and dynamic properties of earth materials, albuquerque, University of New Mexico 759–771

Chao CC (1960) Dynamical response of an elastic half-space to tangential surface loadings. J Appl Mech ASME 27:559–567

Chapel F, Tsakalidis C (1985) Computation of the Green's functions of elastodynamics for a layered half-space through a Hankel transform—applications to foundation vibration and seismology. Proceeding numerical methods in geomechanics, Nagoya, Japan, 1311–1318

Chau KT (1996) Fluid point source and point forces in linear elastic diffusive half-spaces. Mech Mater 23:241–253

Chen GJ (2005) Steady-State solutions of multilayered and cross-anisotropic poroelastic half-Space due to a point sink. Int J Geomech 5:45–57

Chin CY (2008) Soil-structure interaction: from rules of thumb to reality. Proceedings of the 18th New Zealand geotechnical society symposium on soil- structure interaction, Auckland, 34(1)
Chopra AK (2007) Dynamics of structures, 3rd edn. Pearson, Upper Saddle River, NJ
Choudhury D, Subba Rao KS (2005) Seismic bearing capacity of shallow strip foundations. Geotech Geol Eng 23(4):403–441
Chow YK (1987) Vertical vibration of three-dimensional rigid foundations on layered media. Int J Earthq Eng Struct Dyn 15:585–594
Chuhan Z, Chongbin Z (1987) Coupling method of finite and infinite elements for strip foundation wave problems. Int J Earthq Eng Struct Dyn 15:839–851
Ciarletta M (2003) Reflection of plane waves by the free boundary of a porous elastic half space. Appl Math Comput 176:364–378
Ciarletta M, Sumbatyan MA (2003) Reflection of plain waves by the free boundary porous elastic half-space. J Sound Vib 259(2):253–264
Clemmet JF (1974) Dynamic response of structures on elastic media. Ph.D. Thesis, Nottingham University
Cook RL, Nicholson JW, Sheppard MC, Westlake W (1989) First real time measurements of downhole vibrations, forces, and pressures used to monitor directional drilling operations. SPE/IADC drilling conference, 28 February-3 March, New Orleans, LA
Cunny RW, Fry ZB (1973) Vibratory in situ and laboratory soil moduli compared. J SMFD ASCE 99:1055–1076
Dai L, Wang GQ (2009) Analytical approach on nonlinear vibration and wave response of saturated porous media subjected to multiple excitations. DETC2009-87022, proceedings of ASME 2009 international design engineering technical conferences (IDETC), San Diego
Dai L, Lou Z (2008) An experimental and numerical study of tire/pavement noise on porous and nonporous pavements. J Environ Informatics 11:62–73
Davey AB, Payne AR (1964) Rubber in engineering practice. Maclaren & Sons, London
Davies TG, Banerjee PK (1983) Elastodynamic Green's function for half-space. Report GT/1983/1, Department of Civil Engineering, State University of New York at Buffalo, USA
Dawance G, Guillot M (1963) Vibration des massifs de foundations de machines. Ann Inst Tech Batiment et Travaux Publics 116(185):512–531
Day SM, Frazier GA (1979) Seismic response of hemispherical foundation. J Eng Mech Div ASCE 105:29–41
De Craft-Johnson JWS (1967) The damping capacity of compacted kaolinite under low stresses. Proceedings of international symposium on wave propagation and dynamic properties of earth materials, University of New Mexico, Albuquerque, NM, pp 771–780
Dobry R, Gazetas G (1986) Dynamic response of arbitrary shaped foundation. J Geotech Eng ASCE 112:109–135
Dobry R, Gazetas G, Strohoe KH (1986) Dynamic response of arbitrary shaped foundation II. J Geotech Eng ASCE 112:136–154
Dominguez J, Roesset JM (1978) Dynamic stiffness of rectangular foundations. Report R78-20, Department of Civil Engineering, MIT, Cambridge, MA
Dominguez J (1978) Dynamic stiffness of rectangular foundation. Report R78-20, Department of Civil Engineering, MIT
Duns CS, Butterfield R (1967) The dynamic analysis of soil-structure system using the finite element method. Proceeding, international symposium on wave propagation and dynamic properties of earth materials, Albuquerque, University of New Mexico, 615–631
Eastwood W (1953) Vibrations in foundations. Struct Eng 82:82–98
Elorduy J, Nieto JA, Szekely EM (1967) Dynamic response of bases of arbitrary shape subjected to periodic vertical loading. Proceedings of international symposium on wave propagation and dynamic properties of earth materials, Albuquerque, University of New Mexico, 105–123
Erickson EL, Miller DE, Waters KH (1968) Shear wave recording using continuous signal methods. Geophysics 33(2):240–254
Estorff O, Kausel E (1989) Coupling of boundary and finite elements for soil–structure interaction problems. Earthq Eng Struct Dyn 18:1065–1075

Fry ZB (1963) Development and evaluation of soil bearing capacity, foundation of structures. WES, Tech Rep No. 3—632

Ghorai AP, Samal SK, Mahanti NC (2009) Love waves in a fluid-saturated porous layer under a rigid boundary and lying over an elastic half-space under gravity. Appl Math Model 34: 1873–1883

Gibson RE (1967) Some results concerning displacements and stresses in a Non-. homogeneous elastic half-space. Geotechnique 17(1):58–67

Girard J (1968) Vibrations des massifs sur supports elastiques. Ann Inst Tech Batiment et Trayaux Publics 23–24:407–425

Gladwell GML (1968a) Forced tangential and rotatory vibration of a rigid circular disc on a semi-infinite solid. Int J Eng Sci 6:591–607

Gladwell GML (1968b) The calculation of mechanical impedances relating to an indenter vibrating on the surface of a semi-infinite elastic body. J Sound Vib 8:215–228

Gladwell GML (1969) The forced torsional vibration of an elastic stratum. Int J Eng Sci 7: 1011–1024

Glushkov YV, Glushkova NV, Kirillova YV (1992) The dynamic contact problem for a circular punch adhering to an elastic layer. J Appl Math Mech 56(5):675–679

Grootenhuis P (1970) The dynamics of foundation blocks. Proceedings of the international conference on dynamics and waves, Institution of Civil Engineers, pp 95–105

Grootenhuis P, Awojobi AO (1965) The in-situ measurement of the dynamic properties of soils. Proceeding symposium vibration i civil engineer, Institution of Civil Engineers 181–187.

Hall JR Jr (1967) Coupled rocking and sliding oscillations of rigid circular footings. Proceeding international symposium on wave propagation and dynamic properties of earth materials, Albuquerque, University of New Mexico, 139–149

Hall JR Jr, Kissenpfenning JF, Rizzo PC (1975) Continuum and finite element analyses for soil-structure interaction analysis of deeply embedded foundations. Third International Conference on Structural Mechanics in Reactor Technology, Vol. 4, Part K, Paper K 2/4

Halperin EI, Frolova AV (1963) Study of seismic waves by combination of vertical and horizontal profiling. Bulletin of the Academy of Sciences of USSR Geophysics Series (Columbia Technical Translation), 9:798–807

Hamidzadeh HR (1978) Dynamics of rigid foundations on the surface of an elastic half-space. Ph.D. Dissertation, University of London, England

Hamidzadeh HR, Chandler DE (1991) Elastic waves on semi-infinite solid due to a harmonic vertical surface loading. Proc Canadian Cong Appl Mech 1:370–371

Hamidzadeh HR, Grootenhuis G (1981) The dynamics of a rigid foundation on the surface of an elastic half-space. Int J Earthq Eng Struct Dyn 9:501–515

Hamidzadeh HR, Minor GR (1993) Horizontal and rocking vibration of foundation on an elastic half-space. Proc Canadian Cong Appl Mech 2:525–526

Hamidzadeh HR (1986) Surface vibration of an elastic half-space. Proc SECTAM XIII 2:637–642

Hamidzadeh HR (1987) Dynamics of foundation on a simulated elastic half-space. Proc Int Symp Geotech Eng Soft Soils 1:339–345

Hamidzadeh HR (2010) On analytical methods for vibrations of soils and foundations. Chapter 26 of Dynamical Systems: Discontinuous, Stochasticity and Time-delay, Edited by Luo, A. C. J., Springer, 318–340

Han L, Dai L (2011b) 3D shear wave propagations of multiple energy sources in fluid-saturated elastic porous media. Proceedings of third international conference on dynamics, Vibration and Control (DVC 2011), Calgary

Han L, Dai L (2011a) Spherical wave propagations of multiple energy sources in fluid-saturated elastic porous media. IMECE2011-62110, proceedings of 2010 ASME international mechanical engineering congress & exposition (IMECE), Denver

Hardin BO, Black WL (1968) Vibration modulus of normally consolidated clay. J SMFD ASCE p4(SM2): 353–369

Hardin BO, Drnevich VP (1972a) Shear modulus and damping in soils. I. Measurement and parameter effects. J SMFD ASCE 98:603–624

Hardin BO, Drnevich VP (1972b) Shear modulus and damping in soils. II. Design equations and curves. J SMFD ASCE 98:667–692
Harding JW, Sneddon IN (1945) The elastic stresses produced by the indentation of the plane surface of a semi-infinite solid by a rigid punch. Proc Camb Phil Soc 41:16–26
Harkrider DG (1964) Surface waves in multi-layered elastic media, I, Rayleigh and Love waves from buried sources in multilayered elastic half-space. Bull Sismol Soc Am 54:627–679
Heider Y, Markert B, Ehlers W (2010) Dynamic wave propagation in porous media semi-infinite domains. Proc Appl Math Mech 10:499–500
Heller LW, Weiss RA (1967) Ground motion transmission from surface sources. Proceeding international symposium on wave propagation and dynamic properties of earth materials, Albuquerque, University of New Mexico, 71–84
Holzlohner U (1980) Vibrations of the elastic half-space due to vertical surface loads. Int J Earthq Eng Struct Dyn 8:405–441
Housner GW, Castellani A (1969) Discussion of "Comparison of footing vibration tests with theory. By F.E. Richart, Jr. and R.V. Whitman, J. SMFD, ASCE, 95:360–364
Hsieh TK (1962) Foundation vibration. Proc Inst Civ Eng 22:211–226
Hsu CJ, Schoenberg M (1990) Experiments of elastic wave propagation through stacks of thin plates. J Acoust Soc Am 88:S46
Hull EH (1937) The use of rubber in vibration isolation. J Appl Mech Trans ASME 59:109–114
Iljitchov VA (1967) Towards the soil transmission of vibrations from one foundation to another. Proceeding international symposium on wave propagation and dynamic properties of earth materials, Albuquerque, University of New Mexico, 641–654
Israil ASM, Ahmad S (1989) Dynamic vertical compliance of strip foundations in layered soils. Int J Earthq Eng Struct Dyn 18:933–950
Jin B, Liu H (2000) Transient response of an elastic circular plate on a poroelastic half space. Mech Res Commun 27:149–156
Jeong C, Seylabi E, Taciroglu E (2013) A time-domain substructuring method for dynamic soil structure interaction analysis of arbitrarily shaped foundation systems on heterogeneous media. Computing in Civil Engineering 346–353
Jin B, Liu H (2001) Dynamic response of a poroelastic half space to tangential surface loading. Mech Res Commun 28:63–70
Johnson LR (1974) Green's function for Lamb's problem. Geophys J Roy Astr Soc 37:99–131
Jolly RN (1956) Investigation of shear waves. Geophysics 21(4):905–938
Jones R (1958) In-situ measurement of the dynamic properties of soil by vibration method. Geotechnique 8(1):1–21
Jones R (1959) Interpretation of surface vibrations measurements. Proceeding symposium vibration testing of road and runways, Koninklijke/shell-Laboratorium, Amsterdam
Jordan PM, Puri P (2003) Stokes' first problem for a Rivlin–Ericksen fluid of second grade in a porous half-space. Int J Nonlin Mech 38:1010–1025
Kalpna CR (1997) On the 2D plane strain problem for a harmonic stress applied to an impervious elastic layer resting on a porous elastic half space. Phys Earth Planet In 103:151–164
Kalpna X, Chander R (2000) Green's function based stress diffusion solutions in the porous elastic half space for time varying finite reservoir loads. Phys Earth Planet In 120:93–101
Kanai K, Yoshizawa S (1961) On the period and the damping of vibration in actual buildings. BERI 39:477
Karabalis DI, Beskos DE (1984) Dynamic response of 3-D rigid surface foundations by time domain boundary element. Int J Earthq Eng Struct Dyn 12:73–93
Karasudhi P, Keer LM, Lee SL (1968) Vibratory motion of a body on an elastic half-space. J Appl Mech 35:697–705
Kausel E, Roesset JM (1975) Dynamic stiffness of circular foundations. J Eng Mech Div ASCE 111:771–785
Kausel E, Tassoulas JL (1981) Transmitting boundaries: a closed form comparison. Bull Seism Soc Am 71:143–159

Kausel E (1981) An explicit solution for the Green functions for dynamic loads in layered media. MIT Research Report R 81-13, Cambridge, Massachusetts, USA
Ke L, Wang Y, Zhang Z (2005) Propagation of love waves in an inhomogeneous fluid saturated porous layered half-space with properties varying exponentially. J Eng Mech 131:1322–3128
Kelder O, Smeulders D (1997) Observation of the Biot slow wave in water saturated Nivelsteiner sandstone. Geophysics 62(6):1794–1796
Kobayashi S, Nishimura N (1980) Green's tensors for elastic half-space—an application of boundary integral equation method. Memories-Fac. Eng. Kyoto University, XLII 228–241
Kobori T, Suzuki T (1970) Foundation vibrations on a viscoelastic multilayered medium. Proceeding third Japan earthquake engineering symposium Tokyo, 493–499
Kobori T, Minai R, Suzuki T (1966b) Dynamic ground compliance of rectangular foundation on an elastic stratum. Proceeding second Japan national symposium earthquake engineering 261–266
Kobori T, Minai R, Suzuki T (1971) The dynamical ground compliance of a rectangular foundation on a viscoelastic stratum. Bulletin of the Disaster Prevention Research Institute, Kyoto University, 20, 289–329
Kobori T, Minai R, Suzuki T, Kusakabe K (1966a) Dynamical ground compliance of rectangular foundations. Proceeding 16th Japan national congress applied mechanical engineering 301–315
Kobori T, Minai R, Suzuki T, Kusakabe K (1968) Dynamic ground compliance of rectangular foundation on a semi-infinite viscoelastic medium. Annual Report, Disaster Prevention Research Institute of Kyoto University, No. 11A, 349–367
Kovacs WD, Seed HB, Chan CK (1971) Dynamic moduli and damping ratios for a soft clay. J SMFD ASCE 97:59–75
Krizek RO, Gupta DC, Parmelee RA (1972) Coupled sliding and rocking of embedded foundations. J SMFD ASCE 98:1347–1358
Krylov VL (1962) Approximate calculation of integrals. Macmillan, New York
Lamb H (1904) On the propagation of tremors over the surface of an elastic solid. Phil Trans Roy Soc 203 A:1–42
Lawrence FV Jr (1965) Ultrasonic shear wave velocities in sand and clay. Rep. R65-05., WES, Department of Civil Engineering Massachusetts Institute of Technology Cambridge, Mass
Lee TH, Wesley DA (1973) Soil-structure interaction of nuclear reactor structure considering through-soil coupling between adjacent structures. Nucl Eng Design 24(3):374–387
Lee VW, Chen S, Hsu IR (1999) Antiplane diffraction from canyon above subsurface unlined tunnel. ASCE Eng Mech Div 125:668–674
Le Houédec D (2001) Modelling and analysis of ground vibration problems: a review, Civil and Structural Engineering Computing, 475–485, Saxe-Coburg Publications
Liang J, Fu J, Todorovska MI, Trifunac MD (2013a) Effects of site dynamic characteristics on soil–structure interaction (II): Incident P and SV waves. Soil Dyn Earthq Eng 51:58–76
Liang J, Fu J, Todorovska MI, Trifunac MD (2013b) Effects of the site dynamic characteristics on soil–structure interaction (I): Incident SH waves. Soil Dyn Earthq Eng 44:27–37
Li W, Zhao C (2005) Scattering of plane SV waves by cylindrical canyons in saturated porous medium. Soil Dyn Earthq Eng 25:981–995
Li W, Zhao C, Shi P (2005) Scattering of plane P waves by circular-arc alluvial valleys with saturated soil deposits. Soil Dyn Earthq Eng 25:997–1014
Lin G, Han Z, Li J (2013) An efficient approach for dynamic impedance of surface footing on layered half-space. Soil Dyn Earthq Eng 49:39–51
Lin C, Lee VW, Trufunac MD (2001) On the reflection of waves in a poroelastic half-space saturated with non-viscous fluid, CE 01-04. University of Southern California, Los Angeles, CA
Lin C, Lee VW, Trufunac MD (2005) The reflection of plane waves in a poroelastic half-space saturated with inviscid fluid. Soli Dyn Earthq Eng 25:205–223
Longman IM (1956) Note on a method for computing infinite integrals of oscillatory functions. Proc Cambridge Phil Soc 52:764–768
Longman IM (1958) On the numerical evaluation of Cauchy principal values of integrals. MTAC 12:205–207

Lou M, Wang H, Chen X, Zhai Y (2011) Structure–soil–structure interaction: literature review. Soil Dyn Earthq Eng 31:1724–1730
Lu J, Xu B, Wang J (2009) A numerical model for the isolation of moving-load induced vibrations by pile rows embedded in layered porous media. Int J Solids Struct 46:3771–3781
Luco JE, Apsel RJ (1983) On the Green's function for a layered half space parts I & II. Bull Seism Soc Am 73:909–929, 931–951
Luco JE, Westmann RA (1971) Dynamic response of circular footings. JEMD, ASCE 97:1381–1395
Luco JE, Westmann RA (1972) Dynamic response of a rigid footing bonded to an elastic half-space. J Appl Mech ASME 39:527–534
Luco JE (1976) Vibrations of a rigid disc on a layered visco-elastic medium. Nucl Eng Design 36:325–340
Luco JE, Trifunac MD, Wong HL (1987) On the apparent changes in dynamic behavior of a nine-story reinforced concrete building. Bull Seism Soc Am 77(6):1961–1983
Luco JE, Trifunac MD, Wong HL (1988) Isolation of soil-structure interaction effects by full-scale forced vibration tests. Int J Earthq Eng Struct Dyn 16:1–21
Luco J, Ozcelik O, Conte J (2010) Acceleration tracking performance of the UCSD-NEES shake table. J Struct Eng 1365:481–490
Lysmer J, Kuhlemeyer RL (1971) Closure to finite dynamic model for infinite media. J EMD ASCE 97:129–131
Lysmer J, Richart FE Jr (1966) "Dynamic response of footings to vertical loading. J SMFD Proc ASCE 92:65–91
Lysmer J (1965) Vertical motion of rigid footings. Department of Civil Engineering University of Michigan, Report to WES Contract Report, No. 3.115
Ma X, Wang Z, Cai Y, Xu C (2013) Mixed boundary-value analysis of rocking vibrations of an elastic strip foundation on elastic soil with saturated substrata. J Appl Math 2013:10
MacCalden PB, Matthiesen RB (1973) Coupled response of two foundations. Fifth World Conference on Earthquake Engineering, Rome, 1913–1922
Mandelbrot BB (1982) The fractal geometry of nature. Freeman, New York
Maravas A, Mylonakis G, Karabalis D (2008) Dynamic soil-structure interaction for SDOF structures on footings and piles. Geotech Earthq Eng Soil Dyn 4:1–10
Matsukowa E, Hunter AM (1956) The variation of sound velocity with stress in sand. Proc Phys Soc 69:847–848
Maxwell AA, Fry ZB (1967) A procedure for determining elastic moduli of in-situ soils by dynamic technique. Proceeding international symposium on wave propagation and dynamic properties of earth materials, albuquerque, University of New Mexico, 913–920
McDonal J, Angona FA, Mills RL, Sengbush RL, van Nostrand RG, White JE (1958) Attenuation of shear and compression waves in pierre shale. Geophysics 23(3):421–439
Meral FC, Royston TJ (2009) Surface response of a fractional order viscoelastic halfspace to surface and subsurface sources. J Acoust Soc Am 126(6):3278–3285
Miller GF, Pursey H (1955) On the partition of energy between elastic waves in a semi-infinite solid. Proc Roy Soc Lond Series A 233:55–69
Miller GF, Pursey H (1954) The field and radiation impedance of mechanical radiators on the free surface of a semi-infinite isotropic solid. Proc Roy Soc Lond Series A 223:521–541
Morand HJ-P, Ohayon R (1995) Fluid structure interaction: applied numerical methods. John Wiley & Sons, Chichester
Nakano H (1930) Some problems concerning the propagation of the disturbances in and on semi-infinite elastic solid. Geophys Mag Tokyo 2:189–348
Nautiyal CM (1972) Seismic wave propagation in a layer over a half space. Massachusetts Institute of Technology
Navarro C (1992) Vertical radiation damping for a circular footing resting on a simple layered half-space. Soil Dyn Earthq Eng 11:249–255

Ojetola D, Hamidzadeh H (2012) Dynamic response of a rigid foundation subjected to a distance blast. Proceedings of 2012 ASME international mechanical engineering congress & exposition (IMECE), Houston, TX, IMECE 2012-86282, pp 83–88

Oner M, Janbu N (1975) Dynamic soil-structure interaction in offshore storage tank. Proceeding international conference on soil mechanics and foundation engineering. March, Istanbul

Pak RYS, Soudkhah M, Ashlock JC (2012) Dynamic behavior of a square foundation in planar motion on a sand stratum. Soil Dyn Earthq Eng 42:151–160

Pak RYS, Ashlock JC, Kurahashi S, Abedzadeh F (2008) Parametric Gmax sounding of granular soils by vibration methods. Geotechnique 58(7):571–580

Pak RYS, Ashlock JC, Kurahashi S, Soudkhah M (2001) Physical characteristics of dynamic vertical–horizontal-rocking response of surface foundations on cohesionless soils. Geotechnique 61(8):687–697

Pak RYS, Ashlock JC (2011) A fundamental dual-zone continuum theory for dynamic soil–structure interaction. Earthq Eng Struct Dyn 40(9):1011–1025

Pao YH, Mow CC (1971) Diffraction of elastic waves and dynamics stress concentrations. Crane Russak & Company Inc., New York

Papadopulus M (1963) The use of singular integrals in wave propagation problems with application to the point source in a semi-infinite elastic medium. Proc Roy Soc Lond A 276:204–237

Pekeris CL (1955a) The seismic buried pulse. Proc Natl Acad Sci U S A 41:629–639

Pekeris CL (1955b) The seismic surface pulse. Proc Natl Acad Sci U S A 41:469–480

Peplow AT, Jones CJC, Petyt M (1999) Surface vibration propagation over a layered elastic half-space with an inclusion. Appl Acoust 56(4):283–296

Pitilakis D, Moderessi-Farahmand-Razavi A, Clouteau D (2013) Equivalent-linear dynamic impedance functions of surface foundations. J Geotech Geoenviron Eng 1397:1130–1139

Plona TJ, Johnson DL (1984) Acoustic Properties of Porous Systems 1. Phenomenological Description. Paper presented at American Institute of Physics Conference Proceedings, physics and chemistry of porous media, New York

Pradhan PK, Baidya DK, Ghosh DP (2004) Dynamic response of foundations resting on layered soil by cone model. Soil Dyn Earthq Eng 24(6):425–434

Pridea RP, Garambois S (2002) The role of Biot slow waves in electroseismic wave phenomena. J Acoust Soc Am 11(2):697

Puangnak H, Choy B, Mason H, Kutter B, Bray J (2012) Constructive and destructive footing-soil-footing interaction for vertically vibrating footings. GeoCongress 2012:1849–1858

Puswewala UG, Rajapakse RK (1988) Axisymmetric fundamental solutions for a completely saturated porous elastic solid. Int J Eng Sci 26:419–436

Quinlan PM (1953) The elastic theory of soil dynamics. Symp on Dynamic Testing of Soils, ASTM STP, No. 156, 3–34

Ratay RT (1971) Sliding-rocking vibration of body on elastic medium. J SMFD ASCE 97:177–192

Rayhani M, Naggar HE (2008) Physical and numerical modeling of dynamic soil-structure interaction. Geotech Earthq Eng Soil Dyn 4:1–11

Reissner E, Sagoci HF (1944) Forced torsional oscillations of an elastic half-space. J Appl Phys 15:652–662

Reissner E (1936) Stationare, axialsymmetrische durch eine schuttelnde masseerregte schwingungen eines homogenen elastischen habraumes. Ingenieur Archiv., 7, Part 6, 381–397

Richardson JD (1969) Forced vibrations of rigid bodies on a semi-infinite elastic medium. Ph.D. Thesis, University of Nottingham

Richardson JD, Webster JJ, Warburton GB (1971) The response on the surface of an elastic half-space near to a harmonically excited mass. J Sound Vib 14:307–316

Richart FE Jr, Whitman RV (1967) Comparison of footing vibration test with theory. J SMFD ASCE 93:143–168

Richart FE Jr, Woods RD, Hall JR Jr (1970) Vibrations of soils and foundations. Prentice-Hall, Englewood Cliffs, NJ

Riggs ED (1955) Seismic wave types in borehole. Geophysics 20(1):53–67

Rivlin RS (1948) Some applications of elasticity theory to rubber engineering. Proceedings of the second technical conference, London
Robertson IA (1966) Forced vertical vibration of a rigid circular disc on a semi-infinite elastic solid. Proc Camb Phil Soc 62 A:547–553
Robertson IA (1967) On a proposed determination of the shear modules of an isotropic, elastic half-space by the forced torsional oscillations of a circular disc. Appl Sci Res 17:305–312
Roesset JM, Ettouney MM (1977) Transmitting boundaries: a comparison. Int J Numer Anal Methods Geomech 1:151–176
Roesset JM, Whitman RV, Dobry R (1973) Model analysis for structures with foundation interaction. J STD ASCE 99:399–416
Rucker W (1982) Dynamic behavior of rigid foundations of arbitrary shape on a half-space. Int J Earthq Eng Struct Dyn 10:675–680
Safak E (2006) Time-domain representation of frequency-dependent foundation impedance functions. Soil Dyn Earthq Eng 26(1):65–70
Scheidegger AE (1972) The physicals of flow through porous media, 3rd edn. University of Toronto Press, Toronto
Seed HB, Lysmer J, Whitman RV (1975) Soil structure interaction effects on the design of nuclear power plants. Proceeding symposium on structural and geotechnical mechanics, Honoring N.M. Newmark, University of Illinois, 220–241
Serdyukov SV, Kurlenya MV (2007) Seismic stimulation of oil reservoirs. Russ Geol Geophys 48(11):960–967
Shahi R, Noorzad A (2011) Dynamic response of rigid foundations of arbitrary shape using half-space Green's function. Int J Geomech 11:391–398
Shama MD, Kumar R, Gogna ML (1991) Surface wave propagation in a liquid-saturated porous layer overlying a homogeneous transversely isotropic half-space and lying under a uniform layer of liquid. Int J Solids Struct 27:1255–1267
Sharma MD (2004) Surface waves in a general anisotropic poroelastic solid half-space. Geol J Int 159:703–710
Shekhter OY (1948) Consideration of inertial properties of soil in the computations of vertical forced vibrations of massive foundation. NII Symposium 12, Vibratasii, Osnovaniy i Fundementov, Moscow
Shima E, Yanagisawa M, Allam A (1968) Experimental study on generation and propagation of S-waves: IV. S-wave prospecting by means of well shooting. Bulletin of the Earthquake Research Institute of Japan 46(pt 3):517–528
Simkin EM, Surguchev ML (1991) Advanced vibroseismic techniques for water flooded reservoir stimulation, mechanism and field results. Proceedings of sixth European symposium on improved oil recovery, Stavanger, Norway, vol i, Book 1, pp 233–241
Simkina EM, Chilingarb GV, Katz SA (1998) Improved oil recovery by application of vibro-energy to waterflooded sandstones. J Pet Sci Eng 19(3–4):191–200
Singh D, Tomar SK (2006) Wave propagation in micropolar mixture of porous media. Int J Eng Sci 44:1304–1323
Slattery JC (1972) Momentum, energy, and mass transfer in continua. McGraw-Hill, New York
Smeulders DM (1992) On wave propagation in saturated and partially saturated porous media. Ph.D. Thesis, Technische Universiteit Eindhoven, Eindhoven, Netherlands
Sneddon IN (1959) Fourier transform. McGraw Hill, New York
Snyder MD, Shaw DE, Hall JR Jr (1975) Structure-soil-structure interaction of nuclear structures. Third International Conference on "Structural Mechanics in Reactor Technology," Vol. 4, Park K, K2/9
Southwell RV (1944) Introduction to the theory of elasticity, 2nd edn. Oxford University Press, London
Spyrakos CC, Beskos DE (1986a) Dynamic response of rigid strip foundation by time domain boundary element method. Int J Numer Meth Eng 23:1547–1565
Spyrakos CC, Beskos DE (1986b) Dynamic response of flexible strip foundation by boundary and finite elements. Soil Dyn Earthq Eng 5:84–96

Stokoe KH, Woods RD (1972) In-situ shear wave velocity by cross-hole method. J SMFD ASCE 98(SM5):443–460

Stokoe KH, Richart FE Jr (1974) Dynamic response of embedded machine foundations. J GTD, ASCE, Vol. 100, No. GT4, April, Proc. Paper 10499, 427–447

Stroud AH, Secrest D (1966) Gaussian quadrature formulas. Prentice-Hall, Englewood Cliffs, NJ

Sung TY (1953) Vibration in semi-infinite solids due to periodic surface loadings. Symposium on Dynamic Testing of Soils, ASTM, STP No. 156, 35–64

Swaddiwudhipong S, Chow YK, Tan SC, Phoon KF (1991) Dynamic response of surface foundations on layered media. Earthq Eng Struct Dyn 20(11):1065–1081

Swain TY (1962) Recent techniques for determination of in-situ elastic properties and measurement of motion amplification in layered media. Geophysics 27:237–241

Tabiowo PH (1973) Vertical vibration of rigid bodies with rectangular bases on elastic media. Ph.D. Thesis, University of Lagos, Nigeria

Tavarez FA, Plesha ME (2007) Discrete element method for modelling solid and particulate materials. Int J Numer Meth Eng 70:379–404

Taylor P, Hughes J (1965) Dynamic properties of foundation subsoils as determined from laboratory tests. Proceedings of the third world conference on earthquake engineering, 1:196–212

Terzaghi K (1943) Theoretical soil mechanics. John Wiley & Sons Inc., New York

Terzaghi K (1955) Evaluation of coefficients of subgrade reaction. Geotechnique 5:297–326

Theirs GR, Seed HB (1968) Cyclic stress–strain characteristics of clay. J SMFD ASCE 94:555–569

Thomson WT, Kobori T (1963) Dynamical compliance of rectangular foundations on an elastic half-space. J Appl Mech Trans ASME 30:579–584

Tileylioglu S, Nigbor R, Stewart J (2008) Determination of soil-structure interaction effects for a model test structure using parametric system identification procedures. Geotech Earthq Eng Soil Dyn 4:1–10

Titchmarsh EC (1937) Introduction to the theory of Fourier integrals. Oxford, New York

Triantafyllidis TH, Prange B (1988) Rigid circular foundation: dynamic effects of coupling to the half-space. Soil Dyn Earthq Eng 7:40–52

Triantafyllidis TH, Prange B (1989) Dynamic subsoil-coupling between rigid, circular foundations on the halfspace. Soil Dyn Earthq Eng 8:9–21

Triantafyllidis T (1986) Dynamic stiffness of rigid rectangular foundations on the half-space. Int J Earthq Eng Struct Dyn 14:391–411

Trifunac MD (1973) Scattering of plane SH-waves by a semi-cylindrical canyon. Earthq Eng Struct 1:267–281

Truong H (2010) Effects of damping and dynamic soil mass on footing vibration. Soil Dynamics and Earthquake Engineering 178–184

Tuncay K, Corapcioglu MY (1996) Body waves in poroelastic media saturated by two immiscible fluids. J Geophys Res 101:25149–25159

Vashishth AK, Khurana P (2002) Inhomogeneous waves in anisotropic porous layer overlying solid bedrock. J Sound Vib 258:577–594

Veletsos AS, Verbic B (1974) Basic response functions for elastic foundation. J EMD ASCE 100:189–201

Veletsos AS, Wei YT (1971) Lateral and rocking vibration of footings. J SMFD ASCE 97: 1227–1248

Veletsos AS (1975) Dynamics of structure-foundation systems. Proceeding symposium on structural and geotechnical mechanics, Honoring N.M. Newmark, University of Illinois, 333–361

Verruijt A (2010) An introduction to soil dynamics. Springer, New York

Wang G (2008) Wave scattering and propagation in porous media with excitations from multiple wave sources. University of Regina

Wang G, Dai L, Dong M (2007a) Response of fluid saturated elastic porous media to excitations of multiple energy sources. J Multi-Body Dyn 221(K2):319–330

Wang G, Dai L, Dong M (2009) Attenuated wave field in fluid-saturated porous medium with excitations of multiple sources. Trans Porous Med 79(3):359–375

Wang G, Wang Z, Meng F (2007b) Vertical vibrations of elastic foundation resting on saturated half-space. Appl Math Mech 28:1071–1078
Wang H, Liu W, Zhou D, Wang S, Du D (2013) Lumped-parameter model of foundations based on complex Chebyshev polynomial fraction, Soil Dyn Earthq Eng 50:192–203
Wang J, Fang S (2003) State space solution of non-axisymmetric Biot consolidation problem for multilayered porous media. Int J Eng Sci 41:1799–1813
Warburton GB, Richardson JD, Webster JJ (1971) Forced vibrations of two masses on an elastic half-space. J Appl Mech ASME 38:148–156
Warburton GB, Richardson JD, Webster JJ (1972) Harmonic response of masses on an elastic half-space. J EI ASME 94:193–200
Weissmann GF (1966) A mathematical model of a vibrating soil-foundation system. Bell Syst Tech J 45(1):177–228
Westermark RV et al (2001) Enhanced oil recovery with downhole vibration stimulation. SPE production and operations symposium, 24-27 March, Oklahoma City, OK
Whitman RV (1969) The current status of soil dynamics. Appl Mech Rev 22(1):1–8
Whitman RV, Richart FE (1967) Design procedures for dynamic loaded foundations. J SMFD ASCE 93:169–193
Winkler E (1987) Die lehre von elastizitat and festigkeit (on elasticity and fixity), Praguc, 182
Wolf JP, Somaini DR (1986) Approximate dynamic model of embedded foundation in time domain. Int J Earthq Eng Struct Dyn 14:683–703
Wolf JP (1975) Approximate soil-structure interaction with separation of base mat from soil lifting-off. Third International Conference on "Structural Mechanics in Reactor Technology," 4, park K, pp. K 3/6
Wolf JP (1985) "Dynamic soil-structure interaction. Prentice-Hall, Englewood Cliffs, NJ
Wolf JP (1997) Spring-dashpot-mass models for foundation vibrations. Earthq Eng Struct Dyn 26:931–949
Wolf JP (2002) Response of unbounded soil in scaled boundary finite-element method. Earthq Eng Struct Dyn 31:15–32
Wong HL, Luco JE (1986) Dynamic interaction between rigid foundations in a layered half-space. Soil Dyn Earthq Eng 5:149–158
Wong HL, Luco JE (1976) Dynamic response of rigid foundations of arbitrary shape. Int J Earthqu Eng Struct Dyn 4:579–587
Wong HL, Luco JE (1978) Table of impedance functions and input motions for rectangular foundations. Report CE 78-15, Department of Civil Engineering, USC, Los Angeles, CA
Wong HL, Luco JE, Trifunac MD (1977) Contact stresses and ground motion generated by soil-structure interaction. Int J Earthq Eng Struct Dyn 5:67–79
Wong HL, Luco JE (1985) Tables of impedance functions for square foundations on layered media. Soil Dyn Earthq Eng 4:64–81
Xu B, Lu J, Wang J (2008) Dynamic response of a layered water-saturated half space to a moving load. Comput Geotech 35:1–10
Yang YB, Hung HH (2009) Wave propagation for train-induced vibrations—a finite/infinite element approach. World Scientific, Singapore
Yin S, Rothenburg L, Dusseault MB (2006) 3D coupled displacement discontinuity and finite element analysis of reservoir behavior during production in semi-infinite domain. Trans Porous Med 65:425–444
Longman, I. M., (1956) Note on a method for computing infinite integrals of oscillatory functions, Proc. Cambridge Phil. Soc., 52, 764–768

Index

H.R. Hamidzadeh et al., *Wave Propagation in Solid and Porous Half-Space Media*,
DOI 10.1007/978-1-4614-9269-6, © Springer Science+Business Media New York 2014

MIX
Papier aus verantwortungsvollen Quellen
Paper from responsible sources
FSC® C105338
FSC
www.fsc.org

If you have any concerns about our products,
you can contact us on
ProductSafety@springernature.com

In case Publisher is established outside the EU,
the EU authorized representative is:
Springer Nature Customer Service Center GmbH
Europaplatz 3, 69115 Heidelberg, Germany

Printed by Libri Plureos GmbH
in Hamburg, Germany